The Springfield Gas Machine

The Springfield Gas Machine: Illuminating Industry and Leisure, 1860s–1920s

Donald W. Linebaugh

The University of Tennessee Press / Knoxville

First Edition.

Support for this book was provided in part by The Rushlight Club, Inc.; The Fitzpatrick Family and The Red Lion Inn, Stockbridge, MA, redlioninn.com; and by Furthermore: a program of the J.M. Kaplan Fund.

The paper in this book meets the requirements of American National Standards Institute / National Information Standards Organization specification Z39.48-1992 (Permanence of Paper). It contains 30 percent post-consumer waste and is certified by the Forest Stewardship Council.

Library of Congress Cataloging-in-Publication Data

Linebaugh, Donald W.
The Springfield gas machine: illuminating industry and leisure, 1860s–1920s / Donald W. Linebaugh. — 1st ed.
p. cm.
Includes bibliographical references and index.
ISBN-13: 978-1-57233-491-5 (hardcover)
ISBN-10: 1-57233-491-6 (hardcover)
1. Gas-lighting.
2. Gas-machines.
3. Lighting—United States—History.
I. Title.

TH7910.L56 2011
621.32'40973—dc23
2011020836

To Suzanne and Claire, the lights of my life

CONTENTS

TABLES

FIGURES

ACKNOWLEDGMENTS

The idea for this book grew out of a small contract research project with colleagues and fellow archaeologists Nancy O'Malley and Jay Stottman of the University of Kentucky. While conducting an archaeological survey at Ashland, the Henry Clay Estate in Lexington, Kentucky, we discovered a feature that appeared to be part of an early lighting system. Undergraduate student Jennie Duwan provided an excellent summary of lighting systems of the period, particularly portable gas lighting units; she and Nancy also helped excavate a brick access shaft that turned out to hold an intact Springfield Gas Machine generator. Nancy O'Malley also performed exemplary documentary research on the property and lighting system, which yielded intriguing clues to the system's use and configuration. A coauthored article on the system (Linebaugh, O'Malley, and Duwan), published in *Historical Archaeology*, led to further research on the system and its manufacturers. With this publication, historical sites across the country began to contact me to inquire if they might "have a gas machine." After several visits to sites with gas systems, including the Alexander Stephens site in Georgia and Sagamore in the Adirondacks of New York, it became apparent that Gilbert and Barker and their Springfield Gas Machine presented an important case study of technological and business development and of the social applications of new technology.

The many curators, historians, and preservationists who invited me to visit and explore their sites deserve great thanks for their interest in what initially might have seemed a rather obscure topic. I appreciate their support of my efforts to better understand and interpret this unique lighting system. Both former curator Jeff Meyer and the present curator, Eric Brooks, of Ashland provided extensive research help, locating images and documents that proved critical to the project. At Alexander Stephens State Park in Crawfordville, Georgia, park superintendent Judd Smith was an interested host and helpful researcher and facilitator. He opened the site to me for several days and took me to the Robert Toombs site in Washington, Georgia. Likewise, Russ Boudreau at Mount Holyoke College in western Massachusetts "caught the bug" and proved to be both a skilled researcher and extremely helpful and genial colleague. Russ found the Springfield generator vault and subsequently provided careful documentation of the system's removal; he also scoured the records of the college's Maintenance and Grounds Department for maps and other records of the system.

In the company of Russ Boudreau, I also visited the beautiful Hill-Stead Museum in Farmington, Connecticut. Curator Cynthia Cormier, curatorial assistant Melanie Anderson Bourbeau, and archivist Sandy Wheeler were extremely helpful and allowed me full access to the house to examine the evidence for gas light. As we were leaving, they happened to mention a rolled-up document that "might be of interest." It turned out to be a plan for the installation of the Springfield system at Hill-Stead, one of only two known to exist.

Historians and site managers Michael Wilson and Beverly Bridger of Sagamore, an Adirondack great camp in Raquette Lake, New York, were both marvelous hosts on my research trips there. They graciously accommodated my many requests for documents and photographs. Both shared a great deal about the history and social setting of Adirondack great camps, providing stimulating discussion and helpful references. Another intriguing site was Shelburne Farms in Shelburne, Vermont, where Curator of Collections Julie Eldridge Edwards gave me valuable assistance on very short notice. Although she wasn't familiar with the site's Springfield system, she quickly uncovered a wealth of pamphlets and catalogs in Shelburne's extensive catalog collection. As far as I know, Shelburne holds the only extant copies of these Gilbert and Barker publications. They were tremendously important in understanding the system, the company in general, and the Shelburne site in particular. After my visit, Julie continued her exploration and, with the help of intern Devin Colman, found an undated plan of the Springfield system at Shelburne.

Dr. Greg Galer, former curator of the Stonehill Industrial History Center at Stonehill College in North Easton, Massachusetts, has supported this project almost from the start. In particular, he provided introductions to the owners of many Ames family buildings in North Easton and opened the superb collections at the center to me on several occasions. Greg has also talked with me at length about the project, offering many excellent suggestions and pointing me to several very important sources at other institutions. For example, he directed me to the R. G. Dun and Company Collection at Harvard's Baker Library, which turned out to be a goldmine in terms of the business history of Gilbert and Barker.

I have been very lucky to have worked with a number of skilled and dedicated librarians at many institutions. In particular, the reference staff at the Historical Collections Department of Baker Library at the Harvard Business School, Jean Marie Procious, Nicole Hayes, Lisa LaPierre, and Christine Riggle, gave me tremendous help. In addition to their assistance with the Dun and Company Collection, they also helped me navigate their vast trade catalog collection, which yielded extremely valuable information. The staff of the library at the Connecticut Valley

Museum in Springfield, Massachusetts, particularly Cynthia Murphy, were also very helpful in assisting in my research relative to Gilbert and Barker Manufacturing and the broader context of this western Massachusetts community.

In addition to the help from colleagues at these and other institutions, I must also thank the many students who have helped with research, data entry, and draw ing. In particular, I offer my sincere thanks to graduate assistants Laura Mancuso and Pamela Watson from the University of Maryland.

Dr. Patrick Martin, Dr. Greg Galer, and one anonymous reviewer provided incisive comments and useful suggestions that improved the project and final book significantly. I greatly appreciate their insightful and honest critique and advice.

Finally, I owe a tremendous debt to my family, who have supported and encouraged this project for many years. My wife Suzanne has provided both moral and intellectual support, discussing the project at length and reading and commenting on the final manuscript; her expert editing skills have made this a much better book. My daughter Claire, just a baby when the book began, is now a teenager. She provides constant joy and energy to me, and is herself now a budding writer.

INTRODUCTION

For most, if not all, twenty-first-century Americans, trips to the gas station to refuel their cars and trucks have become a regular part of their routines. Relatively cheap gasoline has driven the modern American economy and fueled a level of mobility and independence that Americans have taken for granted for almost one hundred years and appear unwilling to relinquish. So complete is the United States' dependence on oil and gasoline that Daniel Yergin, writing in *The Prize: The Epic Quest for Oil, Money, and Power,* has characterized the twentieth century as "the Age of Hydrocarbon Man."[1] Nevertheless, global economic and political developments of the last fifty years have forced the United States to confront its heavy dependence on petroleum, particularly gasoline.

Despite Americans' reliance on fossil fuels, most know almost nothing of the early history of oil and its derivatives. Many associate the introduction of gasoline with the automobile and early inventors such as Henry Ford. It would no doubt surprise them to learn that gasoline has been produced since about 1863, well before the advent of either the internal combustion engine or the automobile. Most surprisingly, gasoline was a little used and often discarded byproduct of refining crude oil for its first twenty years. Initially used as an industrial cleaner and solvent, it developed fairly quickly into fuel for self-contained air-gas generating machines used to light homes and businesses. The story of this early and astonishing use of gasoline begins with the Springfield Gas Machine Company, the firm that developed the most popular gasoline lighting system.

The discovery of an intact Springfield Gas Machine during investigations at Ashland, the Henry Clay Estate, in Lexington, Kentucky, provided an unexpected opportunity to examine this unique commercial and domestic gas lighting system, investigate significant technological changes related to advances in lighting during the nineteenth and twentieth centuries, and explore some of the social and psychological consequences of these changes. Patented by the Springfield Gas Machine Company of Springfield, Massachusetts (later the Gilbert and Barker Manufacturing Company), about 1867, this gas generating unit initially was sold and installed throughout New England but quickly spread across the United States and even to Europe and South America.

The Springfield was a self-contained, automatic gas machine designed principally for use in homes and businesses located outside the reach of city gas works.

City gas works have operated in America since the 1820s, but their use was limited geographically by their need for extensive systems of distribution pipes. Gas machine systems like the Springfield, essentially do-it-yourself gas plants, operated by forcing air across liquid naphtha, and later gasoline, which then vaporized and flowed through pipes to burners throughout the house and business. In the early 1860s, naphtha and gasoline were little-used byproducts of the oil-refining process that produced kerosene, which had become the universal fuel for lamps. Although several types of self-contained gas machines that used naphtha and gasoline existed prior to the introduction of Gilbert and Barker's system, the Springfield Gas Machine's remarkably simple design and subsequent refinements made it easier to use and eliminated many of the hazards of earlier machines.

To a twenty-first-century sensibility, a system that used gasoline to light interiors sounds incredibly dangerous and counterintuitive. In answer to an early query for help in understanding the system, archaeologist Dan Mouer questioned "that gasoline was involved. Frankly it's hard to imagine why gasoline fumes under pressure would not have blown up the house once ignited. . . . Sounds more like the description of a bomb than anything else!"[2] While many have echoed Mouer's concerns, nineteenth-century insurers actually considered the Springfield system and ones like it as safe as city gas lighting. Moreover, they rated such systems as safer than kerosene and other "portable" lighting devices, which contained flammable, liquid fuels and were easily knocked over with disastrous results. In fact, the Springfield system eventually earned the approval and endorsement of the National Board of Fire Underwriters.

The Springfield's excellent quality of light had significant social consequences. With an estimated thirty thousand machines in operation in the United States by the early twentieth century, the Springfield system was a means by which the gas lighting revolution could have broader distribution geographically and socially. In fact, the system's portability and self-contained design was perfectly suited to the vast rural and suburban landscape of the United States as well as the country's focus on individualism and independence; as such, it is no surprise that portable gas machine technology was an American invention. The system allowed its users to retain control of their light production at a time when corporate and monopolistic control of energy was on the minds of American consumers.

From a purely technological perspective, the use of a waste byproduct created a new, efficient, and inexpensive type of gas light that competed successfully against municipal gas lighting, kerosene, and, for a short while, electricity. The relatively simple, self-contained design of the system led to much greater use of gas light. Gas lighting systems also opened the market for other innovative products, such as gas ranges, broilers, laundry iron heaters, gas stoves, water heaters, and gas radiators.

Gas lighting in general and the Springfield machine in particular were part of an extended technological revolution in the conceptualization and use of light and time. This form of artificial light lengthened the day, reduced the time involved in lighting maintenance, and altered people's routines and activities in both private and public spheres.

While the invention and success of the Springfield Gas Machine alone presents a compelling and important story, other factors distinguish it from the myriad inventions and lighting devices of the period. Its manufacturer, Gilbert and Barker, earned an important place in American business history by developing a strategy to capitalize on the early, pre-automobile market for gasoline. While the system was sold for a modest profit, the parts, service, and particularly the fuel were sold at profitable markups. Gilbert and Barker's ingenuity attracted the attention of the Standard Oil Trust, which bought the company and made it an important component in the trust's (and later Standard Oil of New Jersey's) business strategy. Such plans became famous with the introduction of Gillette safety razors in the 1860s and computer printer companies continue to employ this tactic in the twenty-first century. Unlike many of its competitors, the company distinguished itself by very quickly adjusting to changes in the market. Faced with the increasing availability of electric light, Gilbert and Barker compensated for losses in the lighting market by pioneering fuel oil systems in the late nineteenth century and service station gas pumps in the early twentieth century. In fact, many contemporary American drivers use versions of the gasoline pump developed by the Gilbert and Barker firm, now known as Gilbarco, Inc.

Unlike the straightforward business and economic history of the Springfield system and Gilbert and Barker, the social significance of this lighting system has proven a bit more challenging to document. From humanity's very beginnings, light—or its absence—has had far-reaching implications for how people live their daily lives. Lighting's technological history parallels its role in the social realm. For instance, light has assumed great importance in the rituals and ceremonies of many societies. The importance of light and lighting, particularly as metaphor, is deeply imbedded in our cultural institutions. Light represents goodness, progress, and civilization and defends against darkness, untamed nature, and barbarous civilizations. It serves a symbolic role in many of the world's religions, denoting religious truth and revelation as well as deities' power and ability to redeem. It is not going too far to say that light is power in every sense of the word. Light has controlled our daily activities and social gatherings and is prominently represented in a range of literary traditions, art forms, and social performance over many centuries.

The Springfield Gas Machine system represents one in a series of major developments that revolutionized lighting in America and contributed to changes in technology and society. The introduction of new devices that intensified interior lighting clearly promoted simple functional changes in day-to-day individual and group behavior and ultimately had an effect on social interaction and performance at many levels. As Barbara Carson has observed, "[Social] performance depends upon objects. The possibility of behaving in some ways but not others may be intimately bound up with the straightforward presence or absence of an item, or it may depend upon specific characteristics of design and choice of materials. Often these variations related closely to technological capabilities, regional availability, and cost to consumers."[3] Without going too far down the road leading to technological determinism, it seems that the adoption of gas lighting appliances had just such an effect on nineteenth-century society.

The Springfield Gas Machine made good quality gas light available to homes and businesses beyond the fairly restricted range of city gasworks. Although it principally impacted the upper middle and upper classes in terms of domestic use, the system's installation in large mills and commercial manufacturing establishments brought the revolution of gas light to the entire social spectrum. The commercial and industrial use of the system impacted the working classes in ways both positive, such as enhancing workplace safety, and negative, for instance, the lengthening work day. Working-class men and women in service at country estates and grand homes that utilized the system also would have witnessed a clear difference in lighting. For instance, at Alfred Vanderbilt's Adirondack camp, Sagamore, the workers' houses and service buildings were lit with kerosene lamps, in stark contrast to the buildings of the family and their guests, which sported a range of gas fixtures powered by the Springfield system. It is also likely that Gilbert and Barker's system and ones like it influenced the expansion of suburban development around the country's major cities by enhancing the amenities in those areas. Furthermore, this type of system in particular and gas light in general also seem to have had some impact on residential and industrial building design, particularly in the case of mills and large manufacturing plants.

The story of the Springfield Gas Machine's part in changing the way people lived, worked, and played starts in chapter 1 with a general discussion of lighting. This chapter establishes a framework for exploring the place of gas lighting, particularly the Springfield Gas Machine, in the overall history of lighting. Chapter 1 also focuses on the history of gas lighting in the United States, with a particular emphasis on the invention of gasoline lighting systems. The chapter briefly compares various types of self-contained gas lighting systems, both "hot" and "cold" process units, available in the nineteenth century, and considers the companies that produced systems that competed with the Springfield Gas Machine.

Chapter 2 details the extraordinary story of inventors and entrepreneurs John F. Barker and Charles N. Gilbert, their firm, Gilbert and Barker Manufacturing Company of Springfield, Massachusetts, and its immediate predecessors. The discussion highlights the period from Gilbert and Barker's initial collaboration on the Springfield Gas Machine to the end of the system's production in the early twentieth century. Although their initial venture, the New England Portable Gasworks Company, did not pan out due to design problems with their first machine, Gilbert and Barker reworked their invention and created the Springfield Gas Machine Company in 1867 (reorganized as Gilbert and Barker in 1869). The design of the new machine succeeded because it placed the gas generator and liquid gasoline in an underground vault that was separated from the building to be lit. Within a few years, the company manufactured and installed hundreds of Springfield Gas Machine units across the country. Chapter 2 also explores the company's success as a major supplier of gasoline and its 1884 acquisition by the newly formed Standard Oil Company.

Chapter 3 explores the design of the Springfield Gas Machine system through an examination of patents and company catalogs. While the system's basic design stayed largely the same, the company made many changes and improvements to individual components. The multiple patents on various units attest to the company's continuous experimentation and innovations. In addition, Gilbert and Barker developed a line of special burners and fixtures for use with the Springfield system and eventually marketed a range of household appliances. Likewise, the company created equipment that utilized the Springfield system for commercial soldering and brazing purposes and to produce a gaseous fuel for many other applications such as laundries.

This chapter also discusses the operation of the Springfield Gas Machine and issues related to its economics, maintenance, safety, and quality of light. Gilbert and Barker advertised the machine as simple to operate, requiring minimal maintenance, and economical to run, since "carbureted air" cost less than gas from acetylene systems and coal gas plants. Insurance underwriters, such as the National Board of Fire Underwriters, considered the machine safe because the liquid gas was stored away from the house or workplace. The light from the Springfield system was reported to be quite good, offering a steady white light comparable to coal gas light.

Chapters 2 and 3 rely heavily on literature produced by Gilbert and Barker, which raises the question of appropriateness and accuracy of these sources. In fact, much of the available information on the system comes from catalogs and customer's testimonial letters. While this type of promotional literature requires careful analysis and screening for use in a project of this type, several factors mitigate the inherent bias of company marketing materials. First and foremost, Gilbert and Barker had a solid reputation for honesty, quality, and safety, among outside evaluators. Reports by the credit firm R. G. Dun and Company thoroughly document the status, integrity, and credit worthiness of the firm and its principals. Similarly, various insurance underwriting groups, including the National Board of Fire Underwriters, consistently gave Gilbert and Barker high ratings.

Another factor to consider is the firm's longevity and the number of repeat customers, and glowing testimonials, it enjoyed. In the 1860s, several hundred firms made gas machines, but by the early twentieth century only a handful remained in business, including Gilbert and Barker. Much of the firm's success appears based on its established reputation for a quality product and honest dealing.

A review of customers' testimonial letters indicates that the company regularly published letters that talked about flaws in the machine (usually referencing earlier versions/models). In addition, the limited material available from other sources, particularly letters of and diaries written by customers, seems to support the company's claims and to offer criticisms that differ little from negative comments in the testimonial letters. When used thoughtfully in the context of the firm's history, balanced by available external sources and physical evidence of the system itself, company literature and testimonials offer a reliable picture of an otherwise little known or recorded technology.

Chapter 4 draws on fieldwork at a number of homes that used the Springfield system, including Ashland, the Henry Clay Estate in Lexington, Kentucky; Hill-Stead, the home of Cleveland industrialist Alfred Pope in Farmington, Connecticut; Liberty Hall, the home of Alexander Stephens, vice-president of the Confederacy, in Crawfordville, Georgia; Sagamore, the Adirondack great camp owned by Alfred Vanderbilt in Raquette Lake, New York; several homes of Ames family members in North Easton, Massachusetts; and Shelburne Farms, the Vermont home of businessman Dr. W. Seward Webb in Shelburne, Vermont. Several of these homes still contain largely intact Springfield systems, allowing careful documentation. This evidence, along with available historic documents, plans and drawings, and photographs, permit a detailed view of the installations from a technological standpoint. Using Gilbert and Barker catalogs from 1867 to 1906, this chapter also discusses the geographic distribution of these machines and the types of homes and installations found in various regions of the country.[4] The distribution of Springfield machines in domestic households across the country clearly represents a geography of considerable power, wealth, and influence.

Chapter 4 provides an extended discussion of the social implications of the introduction of gas lighting and the clear relationship between technological development and social usage. The introduction of these new lighting devices and the intensified level of interior lighting prompted changes in daily life and made possible the transfer of selected social behaviors and performance from urban settings to the rural and suburban countryside; this was critical to the increase of lavish country estates and the expansion of middle-class suburbs. Residents of both estates and suburbia sought to replace the problems of the city with the "simplicity" of country living yet ironically hesitated to give up urban luxuries like gas light. This chapter explores the introduction of gas lighting in these homes, including the use of gas to power appliances such as stoves, irons, fireplaces, and water heaters, and examines the social implications of these uses within the context of each household. For example, the Springfield Gas Machines at Sagamore and in similar settings present an intriguing and socially significant use of gas light in the relative wilderness, where the system served as a true "machine in the garden."[5]

Domestic Springfield systems had an impact on the lives of those in service in the country homes of the wealthy, as well as women's roles within middle-class households, by providing extensive lighting and fueling new "time-saving" appliances.[6] This chapter also examines changes in space utilization and the affect of the self-contained gas system on household structure and operation. Similarly, the system contributed to changes in the use of urban and suburban space in late-nineteenth- and early-twentieth-century America, as the availability of amenities like gas lighting increased the lure of areas beyond the city's borders.[7]

In chapter 5, the discussion moves from lighting domestic structures to the illumination of commercial and institutional buildings such as factories, hotels, libraries, and colleges. These installations represent the sale of large, complex Springfield systems and of significant volumes of gasoline and thus formed a central part of the company's marketing strategy focused on upper-middle- and upper-class clientele. The use of the Springfield Gas Machine in many of the East Coast's finest hotels and mineral spas—from the Mount Washington Hotel in New Hampshire's White Mountains to Henry Flagler's famous Royal Poinciana Hotel in Palm Beach, Florida—indicates a deliberate strategy to introduce the wonders of gas light and the Springfield Gas Machine system to wealthy guests. It is no accident that Flagler, one of the founders of Standard Oil, installed a

Springfield Gas Machine in his home at Mamaroneck, Long Island, after Standard Oil had acquired control of Gilbert and Barker. Hotels and inns like the Royal Poinciana enjoyed increasing popularity among the upper class and the upwardly mobile middle class in the nineteenth century. The gas light from Springfield systems made certain types of social performance and interaction possible in these often remote inns and hotels with significant social and economic consequences.[8]

In addition, chapter 5 closely examines Springfield systems at a number of colleges and schools, including Mount Holyoke College; several industrial settings, such as the Springfield Armory; a range of New England textile mills and manufacturing plants; and several public buildings in North Easton, Massachusetts. Springfield systems at several of these institutions and businesses remain largely intact and have been carefully documented; other systems have been studied through historic records and drawings. Both types of evidence provide a window into the technology of large, complex installations. Examples of the Springfield system in commercial and institutional settings range from the late 1860s to the early 1900s. For example, the system at the Springfield Armory, installed circa 1864, was reportedly the first Springfield Gas Machine system to separate the air pump and generator. Later generation systems at Mount Holyoke College and the Royal Poinciana Hotel delivered the "carbureted air" to thirteen individual buildings and twelve hundred rooms, respectively. These installations represent the largest machines built by the company and thus bear witness to the evolution of the system and its components.

As well as detailing the installation and operation of these large, intricate systems, this chapter explores the economic, safety, and social implications of gas lighting for large institutions, factories, and businesses. The introduction of the Springfield self-contained lighting system had an impact on labor and manufacturing by allowing the lighting of large mills and factories, thereby altering working schedules and conditions. The Springfield Gas Machine system also promoted the proliferation of businesses like canning factories, which had to be located near remote fishing towns and farming communities to ensure freshness and economy.

Using Gilbert and Barker's multiple catalogs, this chapter discusses the geographic distribution of these machines and the types of businesses, factories, schools, and government buildings and installations found in various regions of the country.[9] The commercial and institutional installations represent a geographic distribution of capital and power, entrepreneurship and ingenuity. For example, Springfield units at Mount Holyoke College and Middlebury College indicate a desire to provide the children of upper-middle- and upper-class families with the latest and best technology, just as the magnificent tourist hotels and inns that employed the Springfield system catered to money and power by providing the best and latest in accommodations and amenities.

Chapter 6 explores the transition to the next major lighting technology, electric lighting, in the final decades of the nineteenth century and the slow abandonment of thousands of Springfield machines across the county. While gas lighting was a major innovation and improvement in domestic and commercial lighting, its impact was generally limited to the upper middle and upper classes, particularly in terms of domestic use. Electricity, on the other hand, brought lighting and "time saving" appliances to the homes and workplaces of the masses, democratizing the availability of good quality lighting.

This chapter also explores the Springfield system in the context of industrialization and the growth of industrial capitalism, with particular emphasis on Gilbert and Barker's purchase and ownership by Standard Oil in the mid-1880s. Although use of the Springfield system was discontinued in the early twentieth century, Gilbert and Barker successfully built on its expertise with gasoline- and fuel-handling systems to develop new product lines that have continued to make it a vibrant company into the twenty-first century.

As early as 1889, the company began to manufacture new products, such as burners and commercial heating systems, part of the larger strategy of Standard Oil to expand the uses of their fuels. As a major supplier of gasoline, the company developed another new line in 1910 that would become one of their signature products: the gasoline pump. Gilbert and Barker, ever attentive to the marketplace and their customers, went on to become a leading manufacturer of commercial and domestic oil burners and of gasoline pumps for the growing number of gas stations across the country. At the end of the twentieth century, neither the company nor the general public had an inkling of the firm's original products or its impact on the homes and businesses of late Victorian America.

CHAPTER 1

LIGHTING IN AMERICA: FROM RUSH LAMPS TO GASOLIERS

Lighting systems used in the United States from the colonial period to the present evolved from candles and rush lamps to various types of oil lamps (eventually including kerosene), to gas lighting systems, and finally to electricity. This evolution represents a range of important technical and social innovations. As researcher Richard Rhodes has written, "Along its growth trajectory, an innovation interacts with existing techniques, depends on the development of a mediating framework for its effective absorption into the sociotechnical system, and changes its technological, economic, and social characteristics."[1] The Springfield Gas Machine clearly overlapped with existing kerosene lamp technology and with municipal gas systems in urban areas. Importantly, the earlier development of municipal gas lighting provided the mediating framework that allowed the ready acceptance and, in fact, anticipation of the Springfield system in homes and businesses outside the distribution area of urban gas companies. As the technology of portable gas systems improved with the introduction of mixing regulators and new more efficient burners, the social and economic characteristics of the system altered in a way that would allow it to compete effectively, albeit briefly, with the next innovation, electric light.

Gas lighting researcher Mark Grisi has proposed three phases for conceptualizing the trajectory of innovation and acceptance: transition, equilibrium, and dominance. As he explains, in the "transition phase, two distinct lighting applications are advancing due to technological developments. Each application has its advantages and disadvantages, but they are primarily competing for economic proficiency and cultural acceptance." In the next stage, "each of the systems are equal in stature . . . either mechanism may excel based upon the validity of the technology, the comparable economic incentives, and/or acceptance into the cultural mainstream." Finally, in the dominance phase, the "newer-based technology . . . causes the downturn or elimination of the previous technology-based mechanism." In this final phase, Grisi reports, "there can also be absorption of the previous technology by the dominant force that may or may not supplement the newer technology."[2]

A large portion of the nineteenth century was one of transition for gas lighting, in that while gas was becoming the standard by which domestic and commercial lighting was judged in urban contexts, kerosene/liquid-fuel lamps remained the dominant lighting system for technical and economic reasons. In some respects, kerosene/liquid-fuel lamps and gas light maintained a sort of uneasy equilibrium during the last quarter of the nineteenth century. While gas lighting may have even achieved some dominance in this period, it never really eclipsed the fuel-oil lamp as the principal lighting system in the way that electric lighting eventually would. With the introduction of electric lighting in the 1880s, the phases of transition and equilibrium proceeded quite rapidly and electric systems quickly captured the market for kerosene and gas lighting, becoming the dominant urban lighting source by the 1920s; it was another twenty years before electricity rained supreme in rural areas.[3]

The transition from earlier lighting forms to gas and electric, however, did more than simply mark major technological advances, it represented a fundamental cultural shift in the way in which people interacted with fire and energy. As Margaret and Robert Hazen have noted, "although the need for fire remained unchanged, America's fire habits changed dramatically over the course of the nineteenth century. Technological innovations, the introduction of new fuels and the transition to a consumer-based society conspired to alter forever the ways people handled and thought about the open flame."[4] They summarize this transition as one in which "the user's role was forever transformed from energy producer to energy consumer." Centralized energy delivery systems for gas and electricity altered individuals' independent control over fire, "precluding hands-on manipulation" of most operations involved in creating heat and light. The individual household fire had been a fundamental aspect of daily life, and its loss represented a process whereby "domestic fire users . . . became increasingly insulated from the open flame and the mechanics of fire manipulation." Over the course of century, Hazen and Hazen argue, the "age of the axe was irrevocably transformed into the age of the fuel bill."[5] It comes as no surprise, then, that the hearth and fireplace took on such an important symbolic role within the Colonial Revival and Arts and Crafts movements; the lost of control over energy production was yet another symptom of the industrial and technology revolution.

In order to comprehend the substantial technological and cultural innovations represented by gas light, both from city gas works and self-contained gas systems like the Springfield Gas Machine, a brief examination of the nature of lighting systems used prior to the introduction of gas lighting is necessary. This short discussion is followed by a more detailed examination of gas lighting in general, particularly in terms of the American experience, exploring both urban and rural contexts.

A Brief History of Lighting

The use of various types of oil and rush lamps overlapped with the use of candles and soon became the primary type of lighting in colonial America. The oil lamp, which has taken many forms, represented a major innovation in lighting as it was refined during the late eighteenth century. The grease or "Betty lamp," a very simple device consisting of an open saucer that held the oil and a wick made from "a rag, reed, or twisted flax," was the first lamp to be used in the colonies, brought over by English settlers.[6] The lamp's open wick proved problematic, providing "little light but much smoke and smell" and requiring constant attention; some improvement came from enclosing the wick in a circular tube.[7] Most Betty lamps burned fish oil, but fuels such as chicken grease could be used as well.[8] Smelly and smoky to burn, these fuels gave off weak light. As lighting historian Arthur Hayward has noted, the "light was very feeble, the wick constantly crusting over, and the odor of the burning fish oil anything but agreeable."[9]

Another form of lighting employed during the seventeenth and eighteenth centuries was the rush lamp, which consisted of dried, peeled rushes dipped in tallow. The inside of the rush, a type of reedy plant, served as the wick; a clamplike holder, frequently made from iron held the rush in position.[10] Rush lamps offered some advantages over plain candles: they did not require snuffing, as the plant would burn itself down, and were more economical in that they burned longer and were made of cheaper, more readily available materials. However, rush lamps suffered the same problems as Betty lamps and candles, including poor light quality, smoke, and fire hazard.

Candles, made from a variety of substances such as animal fats, or tallow, and vegetable fats like bayberry wax and beeswax, were a luxury for many years and not a popular type of lighting until well into the eighteenth century. The earliest British settlers did not use candles because of their relatively high cost.[11] Candles also required careful monitoring to stay lit. In general, families tended to use oil lamps for everyday purposes and kept several candles on hand for "special affairs."[12] Tallow candles were the most common type once domesticated animals became available; however, tallow also had important culinary uses, which limited its availability for candle making.

Many improvements to candles, including the introduction of molds, new materials, improved wicks, and candle-making machines, evolved over time. Candle molds allowed multiple candles to be produced at one time in contrast to the laborious process of hand dipping each candle. Spermaceti, a crystalline substance derived from oil found in sperm whales, began to be used in candle making around 1750 and was considered superior to other candle-making materials.[13]

In the nineteenth century, candle makers adjusted the proportion of wax to wick; candles would smoke if they contained too much wax and drip with too big a wick. During the mid-nineteenth century, candle making became mechanized, providing more rapid production of candles at a lower cost.[14] In 1820, Frenchman M. Cambacérès invented the braided wick, which when treated with the fatty acid stearine eliminated the need for snuffing or regular trimming. In addition, people began putting multiple wicks in candles to create more light. Another major improvement came with the introduction of paraffin wax in the late 1840s; a harder wax, paraffin did not melt as rapidly as natural fats.[15] Despite many improvements, however, candles never became a terribly efficient light source. Candlelight remained inherently unsteady and not very bright and candles continued to produce smoke and use potentially dangerous open flames.

A major improvement in lighting came with the advent of whale oil as a lamp fuel, especially when used with several new types of lamps. Although whale oil cost more than previously used fish oils and grease lamp fuels, it burned cleaner and provided a steady, bright light that gave off less odor and smoke than other fuels.[16] Whale oil's growing importance as a lighting fuel prompted the development of new, improved lamps. The cork disk burner, an early type of whale-oil lamp, featured an upright glass lamp with a cork on the lid, through which the wick was fed. In many cases these burners held two wick tubes placed close together, increasing the air draft and causing the flames to burn brighter than two separated wicks.[17] The cork disk burner eventually gave way to the screw thread type, in which the burner was "threaded to screw onto a metal collar on the lamp."[18]

In addition to these glass lamps, several types of innovative metal whale-oil lamps became available, including the Miles patent lamp, the Rumford lamp, the Samuel Rust lamp, and the Argand lamp. The introduction of the Argand lamp in the very late eighteenth century represented a major innovation and improvement in oil lamps—one that would transfer to gas lighting. The lamp's novel design had lasting importance and served as the "principle upon which all [subsequent] oil-burning lamps . . . are constructed."[19] Its chief innovation lay in its use of a circular wick fitted over or into a metal tube. The fuel was stored in a reservoir and supplied to the burner through a narrow tube that fed the wick. The cylindrical burner had an air intake in the bottom and a draft tube in the center.[20] Oxygen could reach the center of the wick as well as its outside, enabling better combustion of the fuel and thus producing a stronger, smoke-free light.[21] Macquer reported in 1783 that "the effect of [Argand's] lamp is very beautiful; its light is very white, lively and almost dazzling, far better than the light of any lamp proposed before, and there is no accompanying smoke."[22] The Argand burner stimulated new invention and experimentation with lamps and lighting.[23]

As whale populations near the East Coast of the United States dwindled, whaling became a long-distance operation and whale oil grew more expensive. One of the new "cheaper" fuels to gain popularity was "burning fluid," introduced in the 1820s and 1830s. Also know in America as camphene, burning fluid typically contained a highly combustible mixture of turpentine and alcohol.[24] Oil industry historian Harold Williamson notes that "camphene, the only major innovation in illumination of native origin during the first half of the nineteenth century, was also the first synthetic oil lamp illuminant utilized in America."[25] Although burning fluid usually consisted primarily of turpentine and/or alcohol, numerous patents exist for some intriguing mixtures. Porters Original Patent Burning Fluid, patented in 1834, proved popular because "it produced a brilliant white light with very little smoke or odor. . . . His secret was turpentine, distilled and purified several times and then mixed with quicklime."[26] Another formula, registered in the 1850s, called for "alcohol, naphtha, white oak bark, slippery elm bark, alkonet root, camphor, saltpetre, and salt."[27] A commentator in the *Franklin Institute Journal* noted that burning fluid produced a flame that "was clear, dense, and brilliant. The light may be made greatly to exceed that from oil, without the escape of any smoke, and there is not the slightest odor of turpentine. The patentee says that mixture is as cheap as spermaceti oil. . . . The friends of temperance will not object to the burning of alcohol."[28] Bright, smokeless, and cheap at sixty cents a gallon, sales of the new fuel exploded; by 1857 Philadelphia manufacturers alone produced over a million gallons per year.[29]

While providing a good, cheap light, burning fluid was extremely combustible and thus one of the most dangerous lighting fuels ever used. Accidents involving the fuel caused hundreds of injuries and fatalities.[30] In 1834, the *Franklin Institute Journal* noted that "the late fatal accidents resulting from the use of such ingredients in lamps will, however, probably put a final stop to the use of these mixtures."[31] An 1853 report in *Scientific American* documented over thirty-three fatal explosions caused by burning fluid lamps in the preceding year. Even with this terrible safety record, burning fluid found wide use until the regular availability of kerosene in the mid-1860s.[32]

Lard and lard oil, by-products of animal slaughter, offered another, much safer substitute for whale oil. Introduced as early as 1820, these yellow substances enjoyed their greatest popularity from 1840 to 1860.[33] Like burning fluid, lard and lard oil cost relatively little and produced a clear, unwavering, and bright flame.[34] Yet, unlike burning fluid, lard and lard oil were very stable and not explosive. The fuels' viscosity, however, presented a major problem: both lard and lard oil congealed at room temperature and would not flow up a wick. Several new types of lamps solved this problem; in one "the fuel was forced into the burner by mechanical or gravitational pressure" and in another "heat was conducted from flame to fuel to melt the latter and keep it fluid."[35] These lard lamps, according to Smithsonian curator C. Malcolm Watkins, were often "crude in appearance and bizarre in function."[36] The more stylish solar lamp could burn both whale oil and lard and became one of the best known lard lamps; it combined an Argand burner and deflector that helped keep the fuel liquid.[37] Watkins notes that the solar lamp "answered all the requirements of 'elegance' demanded by Victorian taste."[38]

The next major innovation in lighting and oil lamps came in the mid-1850s with the introduction of kerosene, a petroleum-based fuel. Watkins describes the introduction of kerosene as "a tidal wave . . . a new discovery in lighting [that] swept aside everything before it, both in form and function."[39] Although kerosene could be produced from the distillation of coal, giving it the common name "coal oil," large-scale production began with the discovery of liquid petroleum deposits.[40] While inventors had known about and experimented with this type of oil since the early 1800s, it wasn't until Edwin Drake drilled his first successful commercial oil well in Pennsylvania in 1859 that the petroleum oil boom really began.[41] Kerosene manufacturers in New York and elsewhere quickly switched from coal- to oil-based production and built new refineries.[42] Initially, factories employed a steam distillation process, but in the mid-1860s cracking or "destructive distillation" came into commercial operation. The refinement of kerosene from crude oil resulted in four types of lighting fuels: benzene, kerosene, naphtha, and gasoline. While these "light fraction" petroleum distillates had few initial uses, their later adoption for gas lighting would create a profitable market for a once largely wasted or discarded fuel.

By the mid-1860s, kerosene or coal oil had become the "almost universal lamp fuel in North America." Based on innovations in burner design in the 1860s, three basic types of kerosene lamps and burners had evolved by about 1870. The most popular burner was the flat-wick model known as the Atwood, a type still popular today. Two other types of burner from that period that remained popular stemmed from the Argand-type burner. The use of kerosene for lighting was almost universal in the last few decades of the nineteenth century and its production expanded rapidly. In the later nineteenth century, distributors transported kerosene in bulk by railroad tank cars and, increasingly, long-distance pipelines.

Retail distribution remained fairly simple with kerosene reaching merchants in barrels and sold to consumers in small cans.[43]

Watkins views the adoption of kerosene and gas lighting as marking a period in which "industrialism took command in the field of lighting" in terms of the production of lighting devices. He explains that the period from 1830 to 1860 was "one of transition between handicraft economy and mass production and distribution. It had been an era when the individual tinkerer applied his talents to inventing the mechanisms of a system which was soon to dispense with his services."[44] Individual ingenuity and craft were giving way to corporate research and development and mass production of standardized lighting devices.

A major departure in lighting systems, one that brought significant improvement in lighting quality and safety, occurred with the introduction of various gas lighting methods in the late eighteenth and early nineteenth century. Gas lighting systems evolved slowly and were slow to gain general acceptance. As they became more familiar with the new lighting fuel, the public developed a healthy fear of gas and its explosive properties, so that its initial application was largely restricted to municipal settings, first as street lighting and then large public buildings. Early gas lighting was also exclusively an urban phenomena because of the expense in creating the distribution infrastructure and the cost of building the gas manufacturing plants. Thus while gas lighting was introduced in America in the first half of the nineteenth century, the majority of the population continued to use lamps fueled with kerosene and other lamp oils throughout most of the century. In fact, the use of kerosene was prevalent well into the twentieth century, and it was the spread of electric lighting that marked its disappearance. Kerosene was used by many rural residents well into the 1930 and 1940s and offered the only reasonable lighting choice for working-class people.

Lighting system designers experimented with several types of manufactured gas, including coal gas, "water" gas, rosin or pine tar oil, benzene, naphthalene, gasoline, and, later, calcium carbonate, as they worked to refine gas lighting in the nineteenth century. Natural gas was introduced in the second half of the nineteenth century but was never a very satisfactory lighting fuel due to its blue flame color. Interestingly, "the petroleum boom that provided kerosene for lighting [also] resulted in the discovery of massive amount of associated natural gas." Historian Christopher Castaneda points out that "natural gas was used primarily for industrial purposes, but where available, its higher heating content made it a superior cooking and heating fuel."[45]

Municipal gas lighting became available in large U.S. cities in the early 1800s. These city gas works produced manufactured gas and later handled natural gas.

From central production and storage facilities, main pipes ran underground along streets, and smaller pipes branched off to serve individual homes and businesses.

Coal gas, derived from bituminous coal, was the first gas used for lighting, beginning in the early nineteenth century; it was the first type of gas to be employed in the United States. Investigations of gas began in Europe in the seventeenth century; various scientists conducted experiments to better understand the nature and uses of gas.[46] By the mid-eighteenth century, numerous chemists were investigating coal gas processing. In his *Chemical Essays* (1767), chemist Richard Watson mentions diverse experiments on the distillation of coal to create "inflammable air."[47]

Inventor and engineer William Murdoch was among the first to entertain "the idea of employing gas to replace candles and lamps." In the 1790s, he explored different kinds of coal for gas production, procedures for removing impurities in gas, and the best method of burning gas.[48] Murdoch installed a coal gas system to light his own house in 1792.[49] He communicated his ideas about using gas as a substitute for oil lamps and candles to his associate and fellow inventor James Watt Jr. in 1796 but did not patent the process at the time. In 1798, Murdoch illuminated part of an engine works in the Soho section of London with coal gas as a demonstration for a few interested friends.[50]

In the first years of the nineteenth century, Murdoch and others began to expand their experimentation with gas and its applications for lighting. In 1802, Murdoch staged a public exhibition of gas lighting at the Soho engine works; a witness reported that it "was of the most extraordinary splendor; the whole front of that extensive building was ornamented with a great variety of devices that admirably displayed many of the varied forms of which gas-light is susceptible."[51] In 1805, still without a patent for his process, Murdoch erected a large gas works to supply the steam engine factory of Messrs. Boulton and Watt (Matthew Boulton and James Watt, James Jr.'s father) and another to serve several cotton mills in Manchester. These mill owners were quick to recognize the advantages of gas light: "relatively inexpensive . . . comparably bright light, yet . . . safer than oil lamps."[52] This major demonstration of industrial gas lighting earned Murdoch the Royal Society's "Count Rumford's gold medal" award in 1808.[53]

Around the same period, Moravian engineer Frederich Albrecht Winsor, who had immigrated to England, performed numerous experiments with gas and obtained the first British patent for gas lighting.[54] Winsor lit the Lyceum Theatre in London with gas in 1804 and dubbed himself the "discoverer, inventor, and patentee of gas lighting." Gas industry historians Samuel Hughes and Henry O'Connor describe Winsor as a tireless proponent of gas lighting who rebutted criticism that gas was unsafe and prone to explosion. In 1810, he and a group of investors incorporated the "Gas-light and Coke Company," obtaining a royal charter in 1812. The new company hired engineer Samuel Clegg, whose devices, including the system for wet-lime purification, the hydraulic main, the governor or regulator for adjusting the delivery of gas into the mains, and the gas meter in its earliest form, eventually overcame many of the major difficulties with employing gas lighting.[55]

The typical coal gas works of the period consisted of a "retort-house, coal-store, purifying house, engine and exhauster, station-meter and governor houses, valve and regulator room, workshops, dwellings, and offices, with the ordinary appliances for testing the quality of the gas, the yard and chimney, together with the necessary plant for the manufacture of the sulphate of ammonia." The retort house contained the ovens or chambers used for the distillation or carbonization of coal; a coal-store, a building used to store raw coal, stood attached to the retort house. The purifying house contained several types of equipment to clean impurities from the gas, including a condenser to remove heavy impurities, a washer and/or scrubber to remove ammonia, and either wet-lime or later dry-lime purification systems that removed sulfureted hydrogen, carbonic acid, and other sulfur compounds. The purified gas then flowed through the station meter, which measured the output or quantity of gas produced, to a gasometer, or gas holder. The station meter was generally housed in a building with the governor, a device that lay between the gas holder and delivery mains to regulate the pressure of the gas supplied to customers.[56]

Households and businesses received gas via a system of mains and pipes; meters measured the consumption of each address. Samuel Clegg patented the first meter, a "gauge or rotative gas-meter," in about 1815; John Malam improved upon his design in 1817. Malam's machine eventually formed the basis for the air pump used on portable gas machines like the Springfield.[57]

In the mid-1860s, municipal gas systems began to manufacture water gas, which was produced by "blowing steam across glowing hot coke in a sealed vessel."[58] This process added hydrogen to enrich the gas, raising its heating and lighting potential. Castaneda notes that "to [further] increase the heating content of [water] gas, atomized oil was sprayed into the gas mixture to produce 'carbureted water gas,' which had a heating content of about 550 BTUs and could be used for lighting and heating."[59] By the 1870s, water gas was the most common form of municipal gas.

Gas lighting caught on quickly in England, particularly in cities. By 1807, coal gas streetlights lit the Pall Mall area of London, and by 1813, Westminster Bridge sported its own gas lights, as did the streets of St. Margaret's parish. By 1823, four gas companies supplied gas to the north side of London, and by 1829, no fewer than two hundred gas light companies served customers in various towns across the United Kingdom.[60]

Early gas lighting was generally rather inefficient, mainly due to the variable quality of the gas, the fluctuating pressure in the gas lines, and the use of crude and unreliable burners. Early burners, little more than pierced holes in iron pipes, easily became corroded or clogged, affecting the quality and economy of the light.[61] A variety of burners, including the single jet, batswing, fishtail, and cockspur, eventually appeared, greatly improving gas lighting.[62] Each of these burners had applications for various types of uses and fixtures, but none produced the same quantity of light as the Argand-type burner would later. The Argand burner, operating on the same principle as the earlier Argand oil lamp, represented a major improvement in gas lighting quality. However, it consumed large quantities of gas and didn't burn brightly enough to compete with nascent electric lighting.[63] The introduction of the incandescent mantle in 1885 at last brought gas lighting to a level of intensity that could compete with electric light, if only for a brief moment.[64]

While the cheapness of kerosene impinged on the growth and popularity of gas lighting in the second half of the nineteenth century, manufactured gas remained one of the best sources of lighting available in Europe and the United States. By the mid-1800s, it had become a principal source of urban and industrial lighting. Gas produced in both municipal and self-contained plants provided a bright, steady light that enabled people to work as well by night as they did by day. This aspect of gas lighting made it welcome on the domestic front too. Brighter gas light reduced the eyestrain suffered by those who read by oil lamp, candlelight, or rush lamp. As time went by, delivery systems for gas light saw great improvement, and new uses for gas fuel, such as the gas stove, continually arose.

Like earlier lighting sources, gas posed certain dangers when handled incorrectly, but many deemed it safer than most oil lamps fueled by liquids that could explode or get knocked over and cause a tragic fire. However, establishing a city's gas works and running pipes to individual structures entailed an expensive and time-consuming operation. It also cost individual businesses and households to hook into the system and refit their buildings for gas lighting. Thus gas lighting, like many new innovations, appeared first in the homes of the upper and, to a lesser extent, upper middle classes. "For much of the nineteenth century," writes gas industry historian Christopher Castaneda, "gas lighting literally illuminated the differences between wealthy and poor; it was an expensive urban luxury." The general populace, who at the time almost universally lit their houses with kerosene, more likely would encounter gas lighting in stores or factory buildings and in urban areas with gas street lighting. Castaneda explains that "public gaslights most obviously changed life for many people, even those not wealthy enough to own private gas lamps."[65]

The proponents of gas lighting regularly touted the advantages of gas over the other types of lighting available in the nineteenth century. *The Gas-Consumer's Guide*, written by gas industry expert William Richards, includes observations on the advantages and utility of gas lighting. Richards lauds coal gas for its "superiority" of quality and "cheapness of the light obtained from it" in comparison to other lighting sources. Coal gas's other advantages, he continues, include its "convenience, cleanliness, brilliancy, manageability, and safety." Gas, according to the author, required no preparation, was quickly lighted, and produced light that could be regulated easily; it also saved time and labor when compared to oil or candle light. In terms of safety, Richards argues, gas offered the safest means of illumination, because it used stationary fixtures that could not be knocked over like candles or oil lamps. He points to the "daily accidents from explosions of camphene and kerosene lamps," to tout the "superior safety" of gas. European insurance companies, Richards notes, actually lowered premiums for buildings lighted with gas.[66]

In his study of the industrialization of lighting, historian Wolfgang Schivelbusch also articulates the principal advantages of gas lighting. The first of these, he writes, was brightness: "Higher intensity was not the only thing making gaslight bright. The larger size of its flame was also a factor."[67] The second advantage of "the gas flame, apart from its brightness was its uniformity. The light intensity of a flame burning around a wick varied enormously." And the "third significant new feature of gaslight, after its brightness and uniformity," notes Schivelbusch, "was that it could be regulated."[68] As gas proponent Frederick Accum wrote in 1815, the "size, shape and intensity of the gas-flame may be regulated by simply turning a stop-cock which supplies the gas to the burner."[69]

Richards also proclaims gas as the "most economical mode of obtaining artificial light." Across the country, on average, gas cost "private consumers" about $3.50 per one thousand cubic feet. Interestingly, Richards notes that it actually cost about $1.50 per one thousand cubic feet to produce coal gas, "leaving a pretty wide margin of profit to the gas companies." The economic advantages of municipal gas lighting proposed by Richards are likely overstated and were certainly challenged by gasoline systems like the Springfield Gas Machine. A further advantage of gas, he wrote, lay in its flexibility for uses beyond lighting such as heating, cooking, and baths.[70]

While electric lighting eventually won the battle as the principal source of lighting, gas remained a major light and heat source in both industrial and domestic settings well into the 1920s. The American Gas Association reported in 1920 that 21 percent of the gas used in the United States went for illumination (67,176,000,000 cubic feet). Domestic uses other than illumination made up 54 percent of the total gas consumed (172,740,000,000 cubic feet), while factories used another 21 percent (69,320,000,000 cubic feet).[71] Thus some thirty years after Edison's launch of electric power generation, gas was still lighting and powering many households and factories.

The Gas Lighting Revolution in America

Following its introduction in the early nineteenth century, gas became an important light source in the United States, particularly in urban settings, by the time of the Civil War. Castaneda argues that the introduction of gas lighting in the United States represents a clear example of technology transfer, in this case, from England and Europe to the United States. American entrepreneurs, he continues, made adaptations to the technology when necessary, and added several important systems that demonstrate adaptive engineering and technological skill. "The early gaslight era," Castaneda explains, "was part of an industrial, social, and technological revolution in American society."[72] This process was slow at first, as a host of technical problems had to be resolved and individual houses and businesses had to be connected to gas works via underground pipes.

Smithsonian curator C. Malcolm Watkins credits David Melville of Newport, Rhode Island, with the first American domestic use of "illuminating gas."[73] Melville obtained a patent on the process in 1810.[74] He apparently installed gas lights in a local textile mill and eventually added street lights outside his home, which many viewed as merely interesting novelties.[75] In announcing his gas lighting demonstration, Melville noted that

> gaslights will be found extensively useful in any situation where much light is required, most especially in the growing manufactories of our country, in as much as they are less expensive than those from tallow and oil, and free from the inconvenience and danger resulting from sparks, and frequent snuffing of lamps and candles—circumstances tending greatly to diminish the hazard of fire and proportionately to diminish the rate of insurance on buildings and material subject to that devouring element. [Gas light is] a discovery that holds out so brilliant a prospect of extensive and permanent usefulness to the country.[76]

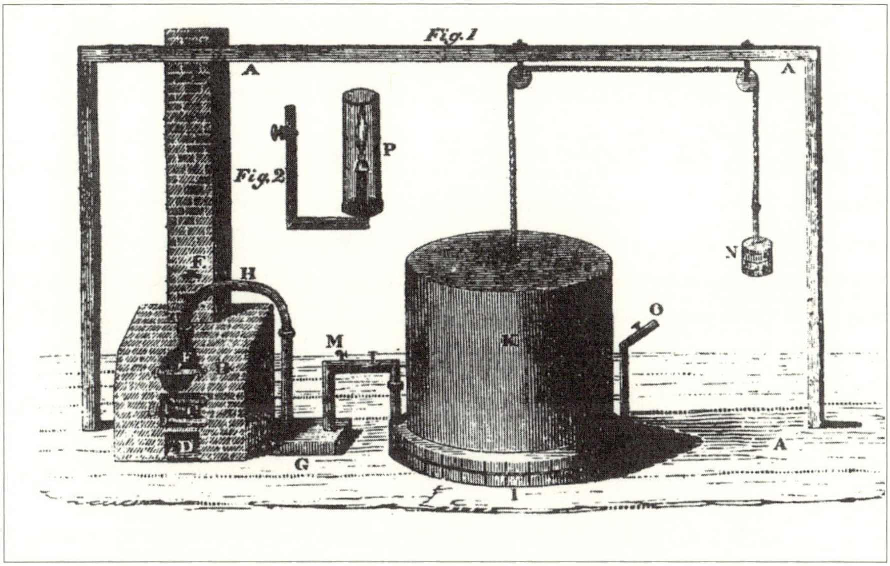

Figure 1.1. David Melville's gas machine. From Thomas Cooper, *Some Information Concerning Gas Lights* (Philadelphia, 1816).

In one of the first major public demonstrations of gas light in America, Rembrandt Peale used gas to illuminate his Baltimore art museum in 1816.[77] The Peale Museum, constructed in 1813–14 as the first purpose-built museum structure in America, drew huge crowds to see the famous mastodon skeleton excavated by Peale's father Charles Wilson Peale. Another of the museum's sensational exhibitions involved gas lighting; a Baltimore newspaper announced "'Gas Light without Oil, Tallow, Wick or Smoke' to be given in the Museum and Gallery of the Fine Arts in Holliday Street." Interestingly, the gas lights reportedly drew more attention than the enormous skeleton.[78] The success and popularity of the gas light installation prompted Peale and others to establish a coal-gas works to provide street lighting in Baltimore; they incorporated the Gas Light Company of Baltimore in 1817.[79] Other northeastern and East Coast cities were quick to jump on the gas light bandwagon.

Following Baltimore's example, Boston established a gas light company in 1822–23, although the city's first gas works was not completed until 1828. An editorial in the *New York Evening Post* described a demonstration of gas light in New York in 1824: "The exhibition of gas lights last evening at 286 Walter Street, attracted a great number of ladies and gentlemen, notwithstanding the unfavorable weather. . . . Six new burners . . . rendered the display considerably more brilliant than [a] preceding occasion."[80] By 1828, part of Broadway was lit with gas lamps, and in 1830 the Manhattan Gas Light Company was organized.[81] Of course, not everyone thought gas light a positive addition to the life of the city. The *American Gas Light Journal* reported that "a lady resident . . . steadily refuses to have the Manhattan gas inserted into her house, because, she says, it is the breath of the devil."[82] These often irrational "fears of explosion and poisoning," argues Schivelbusch, "sprang from the [legitimate] uneasiness people felt at being directly connected to such a dangerous industry"; this uneasiness was further heightened by the perceived infiltration of the gas corporation into the private home. Nevertheless, by 1852, gas lighting reportedly had surpassed the use of oil lamps as New York City's principal lighting system. Following closely on the heels of New York City, Philadelphia built its first municipal gas works in 1834.[83] Beginning in the Northeast and gradually moving west and south, other cities did likewise: for example, Pittsburgh in 1835, Mobile in 1836, and Charleston in 1846.[84] Watkins notes that for the first few decades, gas was confined principally to street lighting, theaters, museums, and other public gathering places. Midwestern states began to introduce gas lighting about the same time as the southern states; Louisville established a gas light company in 1838, Cleveland in 1846, and Chicago in 1849. Smaller cities, such as Lexington, Kentucky; Dubuque, Iowa; and Wheeling, West Virginia, began building municipal gas works in the 1850s. Washington, D.C., opened an operating gas works in 1855. By 1860, over 183 American cities boasted municipal gas works, including a total of 381 gas companies capitalized at over $47 million.[85]

The extensive network of pipes necessary to deliver gas to buildings within a city slowed the spread of the new technology in many locations. The initial cost and regular maintenance of these systems was an expensive undertaking. In fact, after twenty years of operation, Baltimore's gas works had installed only two miles of underground pipes.[86] By mid-century, however, cities like Chicago were rapidly expanding their network of gas mains. In 1854, Illinois governor John Reynolds noted that in Chicago five and a half miles of "large gas-pipes have been laid under ground in this city the last year, and the total of the smaller pipes laid throughout the city is thirteen miles."[87] While distribution within cities had become more widespread by mid-century, municipalities found it too difficult and costly to justify running pipes to individual homes and businesses in suburban or rural areas.

The extensive network of pipes necessary to deliver manufactured gas to buildings also represented the reach of the corporation into the American household. This type of corporate control was also occurring in other business sectors, such as transportation, as railroad corporations gained power and control over "individual entrepreneurial freedom."[88] "The Gilded Age," points out historian Alan Trachtenberg, "marked a significant increase in the influence of business in America, corresponding to the emergence of the modern corporate form of ownership." By the 1870s, he continues, [the corporation] had become commonplace, changing the face and character of American capitalism." The public learned early to question and fear this new monopolistic corporate influence in their lives. In fact, by mid-century "the gas industry was already stigmatized by the monopolistic charters granted to firms." "From the start," notes historian Susan Strasser, "bankers and merchants controlled the gas companies everywhere." Similarly, Schivelbusch sees a direct correspondence between "the concentration and centralization of energy in high-capacity stations [gas and electric plants]" and the "concentration of economic power in the big banks."[89]

While corporate influence on the home and family were commonplace, writes Trachtenberg, "less visible, and often buried beneath the same street, were mechanical intrusions into homes: gas and water lines, plumbing, electricity. Home and flat came under control of an intricate apparatus, in most cities owned by private interests that exacted utility fees for services."[90] Thus, writes historian Ruth Schwartz Cowan, "Americans learned what it means to become embedded in a set of technological systems in the years between 1870 and 1920."[91] As Schivelbusch explains, "With a public gas supply, domestic lighting entered its industrial—and dependent—stage. No longer self-sufficiently producing [their] own heat and light, each house was inextricably tied to an industrial energy producer." He continues: "This loss of domestic autonomy is part of the larger dissolution of the 'total household.' A market and exchange economy, based on the division of labour, absorbed ever more activities and functions originally perfected by individual households."[92]

For American homeowners, it "seemed that industries were expanding, sending out tentacles, octopus-like, into every house. Being connected to them as consumers made people uneasy. They clearly felt a loss of personal freedom." People gazing at gas lights had lost their link to "primeval fire" and, "if anything, they were thinking of the gas bill." Commenting on the continued use of simple paraffin lamps in upper-class drawing and living rooms, Schivelbusch explains that "it seems to have been a reaction to the industrialization of lighting. By keeping their independent lights, people symbolically distanced themselves from a centralized supply." "The traditional oil-lamp or candle in a living-room, he adds, "expressed both a reluctance to be connected to the gas mains and the need for a light that fed on some visible fuel."[93] The portable or self-contained gas lighting system addressed this need for control over lighting and distancing the household from the intrusive tentacles of corporate influence.

Portable/Self-Contained Gas Generating Systems

People who lived in rural areas, well beyond the reach of city mains, also wanted the benefits of gas lighting in their homes and businesses. The invention of self-contained gas generator systems in the early to mid-nineteenth century offered an eventual solution to this problem. As noted above, the self-contained gas lighting system also answered consumers' desires to control their own lighting, particularly for those who hoped to keep their picturesque country homes and estates free of industrial and corporate influence. While gas production and lighting originated in England and Europe, America contributed the self-contained or "isolated" gas machine system to gas lighting technology.[94]

As early as the 1810s, inventor Samuel Morey had begun working on a self-contained gas machine system. Morey's "American Water Burner," patented in 1817, passed steam across "heated turpentine," resulting in a mixture of hydrogen and turpentine that burned easily and produced a bright, clean flame. While Morey's invention didn't ultimately succeed as a portable or self-contained system, it proved important for later developments in municipal gas production for creating water gas.[95] In general, coal-gas production did not lend itself to small portable machines, particularly for residential uses. The process required expensive and difficult to operate machinery. Moreover, these systems posed a significant danger, as they involved open flames, high temperatures, and, often, pressurized materials.

Not until closer to mid-century did portable gas machine systems become available on a commercial basis, the earliest of these machines using rosin, turpentine, and other oils. Worthington notes that C. R. Woodworth and Company's "Portable Gas Works," which used coal and rosin oil, typified the commercially produced domestic gas machine. In order to produce a supply of gas, Worthington writes, "the user would, by burning coal, heat a closed iron cauldron to a glowing red. Rosin oil was then released into the chamber where it instantly vaporized on contact with the hot iron. From the chamber the vapor then passed through a bath where impurities were removed and went from there to a gas holder much like its larger municipal counterpart." Worthington explains further that the system supplied only limited amounts of gas and required frequent recharging and regular maintenance to remove carbon buildup inside the kettle. Even with these drawbacks, the "value of the better light far outweighed the inconveniences, and the Woodworth machine and others like it proved the idea workable."[96]

Another portable machine of the period, the Gas Apparatus of the Maryland Portable Gas Company of Baltimore, went into production circa 1852. The system brought the "luxury [of gas light] home to every man," and was, the company claimed, as easy to use as an "ordinary cooking stove."[97] Like the Woodworth machine, the Gas Apparatus utilized rosin oil heated by coal in a retort. The rosin oil evaporated into gas, which passed through a purifier and into a gasometer for storage. The company's single retort model produced and stored about three hundred cubic feet of gas. Burners of the period used about two cubic feet of gas per hour, so this volume provided some 150 hours of light using one burner.

In 1858, the U.S. government installed one of the Maryland Portable Gas Company's machines at the Soldiers' Home located in northwest Washington, D.C.[98] The system at the Soldiers' Home represented an innovative technology that clearly improved the quality of life for the residents of the home. Its location about four miles north of the urban core of Washington meant that the only other lighting choice was kerosene lamps. The system operated on an as-needed basis; when the gasometer began to run out of gas, someone needed to "fire up" the retort to produce gas and refill the gasometer. The apparatus was likely located in a separate gas house sited some distance from Scott Hall (now the Sherman Building) and its neighbor the Lincoln Cottage. Although Washington, D.C., had a municipal gas company by 1855, the system operated until about 1872 or 1873, when the gas house was razed following the conversion of the complex to municipal gas service.

During the Civil War, the Lincoln Cottage became the summer home for President Lincoln and his family. Based on evidence uncovered during restoration work in 2008, the cottage had a number of gas fixtures in each room. As such, the system provided some small amount of light during what were certainly some of the darkest days for the president.

Experimentation with gas-making machinery reached a high pitch by midcentury as inventors searched for the best fuels and means of turning them into burnable gas. The discovery of petroleum in the 1850s and early 1860s provided a variety of new fuels for experimentation. The refinement of crude oil for kerosene produced both gasoline and naphtha as byproducts. Inventors secured hundreds of patents in the 1860s and 1870s for portable gas machines that used these petroleum-based fuels.[99]

It appears that gasoline was used to fuel portable gas machines beginning in the early 1860s.[100] An article in the 1864 *Mining and Petroleum Standard and American Gas-Light Journal* discusses the introduction of Drake's Gas-Light Machine, which used gasoline to produce air-gas for lighting. The article acknowledges the past use of other volatile oils for generating gas but pointed out a principal objection: "In the winter when light is most wanted, the low temperature of the atmosphere condenses the vapor for the oil, there by depriving the gas of so much carbon that but little light would be given." The article continues: "This apparently insuperable difficulty is now overcome by the use of a volatile oil called Gasoline, being an early distillation from Petroleum."[101]

Gasoline gas machines operated by evaporating gasoline to produce an air-gas mixture burned in standard gas burners and fixtures. Numerous types of gasoline gas machines, using what were known as hot or cold processes, received patents during the final three decades of the nineteenth century. Worthington reports that by 1872 "no fewer than 100 companies were manufacturing isolated gas making machinery."[102] Cold process systems, like the Springfield, operated by forcing air across the surface of liquid gasoline or other petroleum distillates like naphtha, causing the liquid to vaporize. These machines experienced problems with evaporation rates and the fluctuating richness of the gas. In order to evaporate all of the gasoline or other hydrocarbon fuel, some designers utilized a hot process approach. This type of system employed fire and heat to speed up the evaporation of liquid fuel into a gas; the machines varied in design but many used some type of radiator to indirectly warm the fuel to hasten evaporation.

Some of the earliest portable or self-contained systems utilized the hot process to vaporize the fuel. One such system, patented by Oliver P. Drake in 1853, used a "blower forcing air through a carburetor that contained benzole," relying "on heat to vaporize the distillates." Benzole was extremely volatile and safety was a major issue with this system, particularly given the use of heat for evaporation. Among the earliest gas machines to be commercially available was the Benzole Gas Apparatus, produced by A. M. Mace of Springfield, Massachusetts, beginning in 1855.[103] Much like Drake's system, the Benzole Gas Apparatus consisted of "revolving disks of wire gauze set in parallel with the screw pump which draws in air into the liquid chamber where it is agitated and diffused among the liquid . . . and produces illuminating gas." In order to keep the benzole at the ideal temperature of 70 degrees Fahrenheit, one could simply light a "jet at the bottom of the tank, [and] the fluid in the chamber above would be raised to the proper temperature."[104]

Another early hot process system, a carburetor designed by Charles Cunningham in 1855, utilized gasoline instead of benzole. Cunningham's machine combined a meter-wheel air pump with a drum-type carburetor that was driven off the same weight that turned the air pump. The pump forced air through the carburetor, which had a revolving wheel of wicklike material much like a modern humidifier. The air picked up the gasoline vapor from the surface of the drum and then flowed to the distribution system. Because of the small size of the drum and carburetor, the system failed to evaporate all of the liquid efficiently. In order to improve the evaporation rate, the system included a heater that raised the

Figure 1.2. Advertisement for Drake's Solar Gas Machine. Author's collection.

temperature of the gasoline in the carburetor either through a direct heat source or via a radiator system that circulated hot water in a pipe through the carburetor.[105] Like other early machines, the Cunningham carburetor combined the air pump and carburetor in one machine, necessitating its placement outside the building to avoid introducing gasoline into the structure.

Inventor and former Gilbert and Barker employee Hiram S. Maxim patented an "improved gas-machine" in 1868 that "is so arranged that the production of gas will be entirely automatically regulated, and that the volume of gas, as well as its pressure, is under automatic control." Maxim's invention used the hot process. He wrote in his patent application that "the flame thus arranged in the furnace heats the tank and causes the evaporation of the gasoline therein and in the cylinder."[106] With the incendiary combination of open flame and gasoline, it is a small wonder that insurance companies found these machines problematic.

Early hot process gas machines had several design problems that rendered them dangerous. The machine posed a danger principally because it held a supply of liquid gasoline and gasoline vapor (in the gasometer or storage tank) and used an open flame for heating. As noted in the contemporary newspaper reports on gas machine accidents, this combination in one small mechanism held the potential for explosive consequences. An article on gas machines in the *American Gas Light Journal and Chemical Repertory* noted that "in certain machines . . . it is sought to aid the volatilization by heat, the air being blown into a vessel of oil, heated beneath by a flame; but, however effectual this process may be in maintaining regularity in the quality of the gas, it will never become a popular favorite, owning to the grave objection of the use of fire, by which the possibility of accident is greatly increased."[107] Due to the amount of gasoline stored in the hot process machine, the unit typically sat outside of the building it lit; doing otherwise created a major safety and fire hazard that made insurance companies very wary of these systems.

While locating the gas machine outside the dwelling improved safety, it was both inconvenient for the owner operator and a maintenance problem. Many of these machines used water to provide the seal for the gasometer and in the air pump, and thus freezing water presented a real problem for operating the machine in winter months when the demand for good lighting was greatest. In terms of operation, "hot process" systems typically required close and regular attention to make the air-gas. Relighting the flame for the heating process was required each time the system was to be run, and thus they were not in any way automatic production systems.

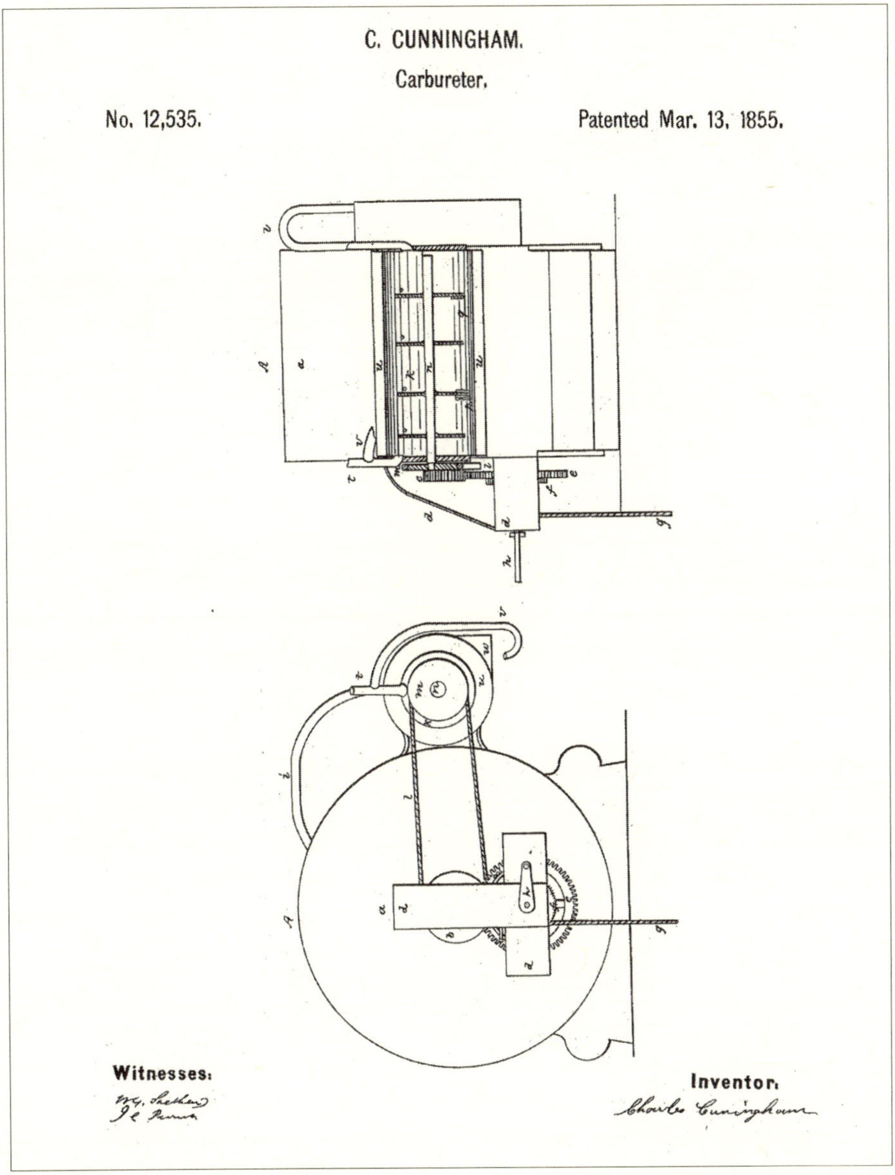

Figure 1.3. Patent No. 12,535 for carburetor by C. Cunningham, March 13, 1855. U.S. Patent and Trademark Office.

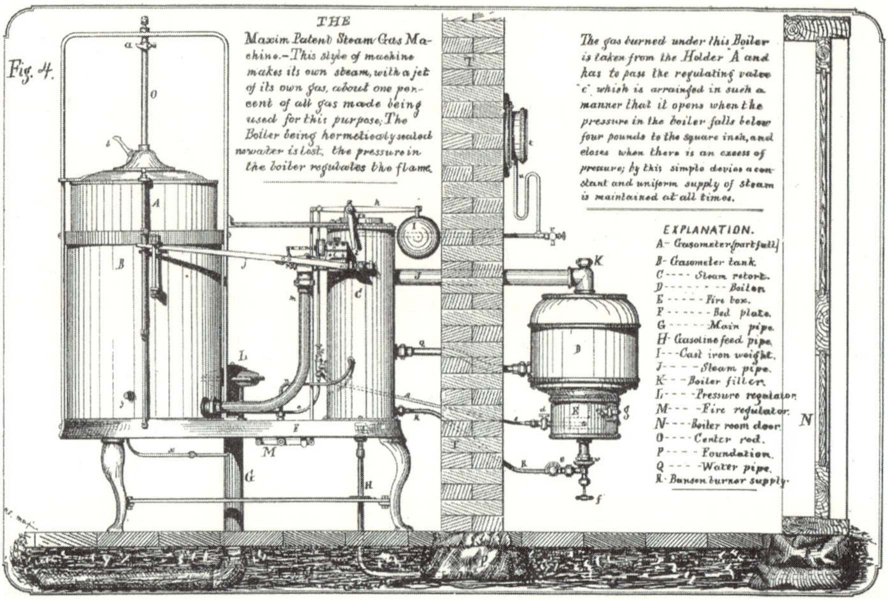

Figure 1.4. The Maxim Steam Gas Machine, Maxim Gas Machine Company, New York, 1873. Library of Congress.

Cold process gasoline systems introduced in the 1860s quickly eclipsed heated systems, like the Maryland Portable Gas Machine. For example, several Gilbert and Barker customers reported replacing their "old rosin" systems with the Springfield Gas Machine because of the time involved in operating the rosin-oil system. Customer Mathew Baird complained that "with the old rosin works they [his staff] were obliged to make gas at least twice a week" and noted that he was "much pleased with the change" to the Springfield system.[108] Cold process gasoline gas machines, like the Springfield, also eliminated many of the greatest safety concerns of the hot process systems but generally suffered from problems caused by fluctuating evaporation rates and incomplete evaporation of the fuel.

Typically, the early cold process systems had air pumps that moved air across and/or through liquid gasoline in order to pick up gas vapor.[109] In some cases, weight-driven mechanisms powered the air pump, as seen in Drake's Automatic Gas Machine from the early 1860s. In others, like Prichard's "Gas Machine" (1868), water pressure "condense[d]" the air and forced "it through the carbureting fluid."[110] Typically, a gasometer stored "carbureted" air-gas for use as needed. Like hot process machines, many early cold process systems required attention during the gas-making process to fill the gasometer. Most of these machines also contained quantities of liquid gasoline and thus needed to be placed away from the building. Like the hot process systems, this made the machine more onerous to operate in the winter due to freezing and inconvenient to operate in bad weather to produce gas.

An 1868 design by William Foster Jr. and George P. Ganster of New York, attempted to eliminate the problems with gasoline storage by adding a tank or reservoir for the gasoline that was buried outside the building.[111] This arrangement supplied the gas machine, located in the building's basement, with gasoline via a small pipe. Thus the machine in the basement held a minimum amount of gasoline at any one time. However, even this small amount of gasoline in the building presented a hazard, leading insurance companies to quickly eliminate these systems from their lists of approved and insurable devices. Drake's Automatic Gas Machine, patented in 1862, also employed a gasoline storage tank or reservoir that fed the combination air pump and generator unit. In 1864, the *Mining and Petroleum Standard and American Gas Light Journal* touted Drake's new gasoline fueled system: "Mr. Drake has a beautiful mechanism to perform the duty of making the gas, and it is now perfected as a successful and complete thing." Customers could have this type of unit installed in the basement of the house, like the Foster and Ganster machine, or, more typically, in a gas house or underground vault.[112]

Both hot and cold process gasoline systems failed to catch on initially, as customers shied away from the unfamiliar fuel as the press regularly reported horrific accidents. The *New York Times* carried many stories of calamities involving gas machines. A September 1869 article, for example, reported on a "patent gas machine containing gasoline located on the first floor of the premises [that] took fire through carelessness . . . setting the whole building on fire." In another case, a *Washington Post* headline from March 1878 read "A Gasoline Blow-Up. Terrible Explosion and Loss of Life." The story provided gruesome details about the resulting deaths of a city councilman and ten-year-old girl.[113]

The Springfield Gas Machine, patented in several stages between 1867 and 1869, also employed cold process technology and utilized a generator unit and air pump. The Springfield system, made and sold by the firm of Gilbert and Barker Manufacturing, had several novel features that set it apart from its competitors. These included a stacked-pan generator that greatly expanded the evaporation surface of the generator; the placement of the generator unit, and thus all of the liquid gasoline, in a brick vault away from the structure to be lit; and an adjustable burner that allowed for the regulation of the air-gas mixture at the fixture. The air pump, placed in the basement of the building to keep it from freezing, was connected to the gas generator via iron pipes. The Springfield configuration employed a remarkably simple design, that made it easy to operate and generally effective at providing good quality gas and light. Moreover, insurance companies considered it safe enough to insure.

In the mid- to late 1870s and early 1880s, numerous inventors continued to work on portable gas machine systems that employed both the hot and cold processes. The Excelsior Gas Machine Company of Warren, Massachusetts, went to great lengths in their 1872 catalog to explain the many difficulties and problems of "cold process" carbureting machines like the Springfield, touting their "modern and improved system of gas making" that applied "external heat to the retort or vaporizer" to produce gas automatically. "By the application of heat to gasolene automatically supplied in small quantities to a suitable retort," they explained, "the liquid is reduced to a very finely-diffused vapor." Thus only a small amount of gas was actually subjected to the heat, and was vaporized under pressure. The pressurized vapor was funneled through a small jet, and this set up an induction of air that then mixed with the gasoline vapor; the resulting air-gas was stored in a small gasometer.[114]

Like the Excelsior firm, the Maxim Gas Machine Company of New York also made much of the failings of cold process systems in their 1873 catalog. Maxim argued that "air-blowing" machines, like the Springfield, were not safe, economical, or reliable because they could not control the relative proportions of gas vapor and air, particularly due to changes in the rate of evaporation with changing temperature. These machines could not use up all of the gasoline, Maxim maintained, because the lightest fractions evaporated first, leaving heavier, less volatile, liquids that would not easily evaporate. The solution offered was the Maxim Steam Gas Machine, which consisted of a steam boiler, retort, and gasometer. The steam boiler was separated from the retort by a brick wall; the unit delivered steam to the retort to heat the gasoline and vaporize it. The retort was fed with small amounts of gasoline from a storage tank, so the specific gravity of the gasoline did not decrease due to the evaporation of the lightest fractions. The system purported to automatically regulate the flame to the boiler, using a steam regulator valve, so that changes in steam pressure were translated into an increase or decrease in the boiler unit. Thus the appropriate pressure and temperature were maintained to offer uniform and consistent evaporation of the gasoline. Similar to the Excelsior machine, the Maxim system utilized air induction, that is, it sent the pressurized gasoline vapor through a narrow jet which also drew in large quantities of air. This system, the manufacturer boasted, eliminated the need for the cumbersome air pump and weight used with most cold process machines.[115]

Frank W. Ofeldt of New Jersey also experimented with steam pressure to drive his "Gas-Apparatus for Carbureting Air" and an aerometer (a device to measure and maintain air pressure and density) to eliminate the need for a bulky gasometer to store the gas.[116] Thomas B. Fogarty of New York patented an "automatic" gas machine system in 1876 that also vaporized gasoline in a heated retort. Fogarty claimed that his machine was novel and deserving of a patent due to its automatic nature and its use of "joints impervious to flame."[117]

Several other types of portable gasoline gas systems appeared in the later nineteenth century. Perhaps the simplest of all was the "hollow-wire" lighting system developed in the 1890s. Its name is thought to have been chosen to capitalize on the growing popularity of electric light, particularly the flexibility provided by wires. The hollow-wire unit consisted of a gasoline storage tank and a hand-operated air pump, making it extremely simple and inexpensive compared to systems like the Springfield. The pump pressurized the system and the liquid gasoline (not gasoline vapor) flowed to lighting fixtures via small brass tubing (the hollow wire). The fixture contained a generator tube, where the liquid gas was vaporized, and an incandescent-mantle burner. The generator had to be heated to begin the evaporation process; once lit, the heat of the burner warmed the generator and vaporized the gasoline. Once pumped up to fifteen to twenty pounds of pressure, the system reportedly would work for several days.[118] The "hollow tube" system's use of liquid gasoline within the home posed an obvious danger and marked a clear departure from cold process systems like the Springfield Gas Machine, which sought to eliminate all of the gasoline from the house. Despite the obvious dangers, the low price of the hollow-wire system, which cost about a third of the typical gasoline gas machine, and its relative ease of installation and use made it attractive to many rural and suburban homeowners.

The air-gas produced in portable gas machines, whether hot or cold process systems, required open flame burners, which could result in unsteady light. Several inventions, including the Welsbach incandescent gas burner introduced in the 1880s, significantly improved the flame quality. This burner produced a steady light that did not radiate as much heat as "other high-power gas lamps" and used gas much more economically. However, the Welsbach burner utilized very fragile cotton mantles that broke easily; moreover, the glass cylinders that surrounded the filament were prone to cracking and breaking the mantle.[119] Even with these drawbacks, manufacturers quickly adapted the incandescent burner for use with self-contained gas machines, resulting in immediate and significant improvements to light quality.

Despite the relative accessibility, efficiency, light quality, and ease of operation of gasoline gas machines, they presented a potential hazard due to the explosive nature of gasoline and required careful handling. This was particularly true for hot process systems that used heat to speed evaporation of petroleum-based fuels, but even cold process machines posed a danger to consumers who knew little of the properties and dangers of gasoline. Many gas machine catalogs contained specific warnings to consumers of the dangers of gasoline. For example, the Maxim catalog included a "Caution," disclosing that "in what has been said of the safety of the Maxim Gas Machine, there is no intention whatever of deceiving or misleading any person as to the really dangerous nature of gasoline."[120]

Gasoline gas machine systems were also employed to enhance the quality of municipal coal gas, which produced notoriously poor light. For this application, companies sold just the gas generator, which was installed on the gas company's pressurized line. Thus the coal gas was forced through the carburetor or generator unit and hydrocarbon vapor was added to the coal gas. Catalog references indicate that both individual building owners and small municipal gas utility companies used this method to enrich their gas and improve the quality of light. John A. Bassett patented another interesting and seemingly dangerous approach to carbureting municipal gas. His "Improved Apparatus for Carbureting Gas" consisted of a small burner with a reservoir of gasoline that could be attached to a standard wall fixture. After lighting the coal gas and heating the burner, the user opened a stop cock and the reservoir fed a small stream of liquid gasoline into the heated burner; the gasoline immediately evaporated as it entered the burner and increased "the illuminating power of gas used for lighting."[121] It seems likely that this device was not popular with insurance companies, which in the 1860s, after a series of losses of property and life, began to develop specific underwriting rules related to gas machines. In virtually all cases, fire insurance companies insisted that gasoline gas machines be installed in vaults or gas houses at least fifty feet from the building to be lit and favored cold process machines that didn't mix heat and flame with any quantity of gasoline.

With the safety of gasoline gas machines a regular concern for customers, manufacturers introduced another fuel type, acetylene gas, in the closing decades of the nineteenth century. Throughout the 1890s, municipal gas provided the bulk of urban lighting, although electricity was gaining in popularity; rural residents still relied on kerosene lamps or portable gasoline machines for lighting. The use of acetylene gas for lighting was known well before the 1890s, but the cost of producing the calcium carbide fuel used to make the gas was prohibitively expensive. When mixed with water, calcium carbide produces pure acetylene gas, which provided a steady, white, open flame that was better than that provided by many other fuels. For rural consumers, the acetylene generator provided a clear improvement over fuel-oil lamps.[122]

The advent of electricity, ironically, led Thomas L. Willson to invent an electric furnace in 1892 that dramatically lowered the cost of making calcium carbide fuel, and this paved the way for its use in portable gas machines. The relatively simple and inexpensive acetylene gas machine system caught on quickly during the 1890s as a means for lighting America's farmhouses.[123] For instance, in 1895, acetylene gas lighting machines received just ten patents; by 1900, the U.S. Patent Office had issued over 150 patents for various types of acetylene gas lighting systems. This level of invention continued for at least the next five years.[124] One acetylene gas system catalog boasted of its value over gasoline gas and kerosene, stating that "until recently no practicable substitute had been found for the smoky, flickering, uncertain gasoline gas jet and the almost universally used, untidy, odoriferous, dangerous, oil-spreading kerosene lamp."[125] In general, Russell notes, the acetylene light did not remain popular for long as a source of domestic lighting because its fuel was "messy and corrosive" and the acetylene gas had an unpleasant odor.[126]

One of the advantages of the acetylene gas machine compared to the gasoline gas machine was that it reportedly produced a reliable, brilliant, and clean light. In addition, acetylene machines were said to burn their gas more efficiently than gasoline machines and took up less space.[127] Catalogs for acetylene gas units generally emphasized the much lower initial cost of the acetylene system and argued that acetylene lighting was cheaper overall when considering the intensity of the light produced.[128] For example, in 1906 a one-hundred-light acetylene system produced by the Sunlight Gas Machine Company cost $325.00 (FOB New York), while a one-hundred-light iron/steel Springfield unit cost $650.00 (based on an 1882 price list).[129] While the initial cost of the unit was considerably less, however, the literature produced by gasoline gas manufacturers argued that the overall cost of operation was higher for acetylene gas than gasoline gas. In Gilbert and Barker's 1906 catalog, the company reported that "one gallon of gasoline costing sixteen cents will supply a Springfield Incandescent Burner giving the light of twenty-five candles for 111 hours. Fourteen pounds of Carbide of Calcium costing fifty six cents will supply a burner giving the light of twenty-five candles for 111 hours. Acetylene light costs three and one-half times as much as gasoline."[130] This cost analysis was countered by the acetylene system manufacturers, who reported that "it requires for 1000 candlepower for one hour: 313 feet of common gas, or 62 1/2 16-candlepower incandescent lamps, or 20 feet of acetylene gas." Thus city gas at one dollar per thousand feet cost thirty-one cents while the twenty feet of acetylene gas cost only twenty cents (or one cent per foot).[131]

Acetylene systems had some notable disadvantages, including the maintenance they required, as they needed regular recharging, and the cleaning of waste material they produced.[132] The waste consisted of a sludge of slaked-lime precipitate that accumulated at the bottom of the generator; Worthington notes that one "over enthusiastic manufacturer" called the sludge removal "a safe, clean, pleasant job."[133] On the positive side, the lime sludge was a useful byproduct on the farm, where it could be used to lime agricultural fields, make whitewash for buildings, and make plaster and concrete. Not surprisingly, company catalogs painted rosy pictures of life with acetylene systems, characterizing them as "easy to clean" and not requiring "constant attention."[134]

Gas Lighting Fixtures and Burners

Both municipal and portable gas lighting systems used specifically designed fixtures to burn gas and produce light. These fixtures typically consisted of a body, usually ornamental, a burner, and a globe. The indoor varieties of fixtures included brackets, pendants, and chandeliers or gasoliers.[135] Brackets extended horizontally from the wall, and pendants and chandeliers or gasoliers hung from the ceiling. Pendants had only one or two burners, while chandeliers or gasoliers had more than two burners.[136] A less common fixture stood upright, often on top of a newel post, essentially functioning as an immovable lamp. Gilbert and Barker refer to this type as "Figures for Newels."[137] Lighting historian Loris Russell notes that by the 1880s, the wall bracket was the most common fixture; the simplest of these included a "brass pipe projecting from the wall, with a stopcock at the wall end and a jet at the other." Double- and triple-jointed versions of the wall bracket also became very popular, as they allowed the position of the light to be adjusted.[138]

The lack of portability was one of the major disadvantages of gas light fixtures over kerosene.[139] Manufacturers of gas lighting systems responded in the 1870s and 1880s with portable gas lamps that were connected to the fixed distribution pipe with flexible rubber hose. William Richards, author of the 1871 *Gas-Consumer's Guide*, writes that "in occupations where a portable light is required . . . an apparatus similar to a candlestick, to which the gas is conveyed by a flexible tube, supplies the want."[140] These flexible tube systems worked with many types of fixtures and gas types, including the Springfield Gas Machine.

In terms of light quality, the most important part of the fixture was the burner. The gas burner sat within the larger, often decorative, fixture, and came in several types: single-jet burners, flat-flame burners, Argand burners, multiple flat-flame burners, regenerative burners, and incandescent burners. One of the most important considerations for good burners, Richards notes, was that "the orifices for the egress of the gas should be perfectly regular in size, and sufficiently large, so as to permit it to issue with very feeble pressure or force." If the burner holes were too small, the light was diminished by the increase in pressure of the gas; if the holes were too large, the flame might smoke.[141]

The single-jet burner was the oldest form of burner as well as the simplest. An early example, the rat-tail burner, was in use by 1808. This burner consisted of a plain, upright body with a small, round opening for the gas. According to gas industry expert William Gerhard, rat-tail burners were used on a limited basis as they gave off little light.[142]

The flat-flame burner included two varieties, the bat-wing or batswing, in use by 1816, and the fishtail, in use by 1820.[143] Flat-flame burners had a slitted opening which created a fanlike emission of gas. While they did not require a chimney to protect against smoking, they did need a glass globe to protect the flame from drafts. The batswing burner had a dome-shaped top with a thin slit across it, while the fishtail had two jets that came together to produce a flat flame.[144] Customers generally preferred these burners for domestic use over the earlier single-flame burners because of their increased brightness. Richards called the batswing burner the "best adapted for all out-door lights" as it did not require a glass chimney and was not affected by "every breath of wind." The fishtail, Richards notes, was the most defective burner in some respects because it wasted gas and was subject to smoking with an increase in gas pressure.[145] Multiple flat-flame burners were used mainly to obtain brighter street illumination or to light large indoor spaces, such as stores or great halls. Multiple burners, made of "several concentric rings of round burners, or three, five, or more flat-flame burners," produced a very bright flame.[146]

Another important gas burner was the Argand type, consisting of a "hollow ring connected with the gas tube and perforated on its upper surface with a series of fine holes, from which the gas issued, which formed a hollow, round flame." Like the Argand oil lamp, these burners also required a glass chimney. This type of burner supplied air to the flame more efficiently, resulting in better combustion. Several inventors made improvements to the Argand burner over time. William Sugg altered the original burner by having the gas delivered at a very low pressure, ensuring an even distribution of gas. The Gleason Manufacturing Company created a "noiseless" Argand burner with a volumetric gas governor, a device that maintained an even flow and pressure of gas. Another improvement came with the Siemens precision burner, which had a flat disk in the center of the flame that enhanced combustion and increased luminosity.[147] Writing in the *Gas-Consumer's Guide* in 1871, Richards reported that "when a quantity of light is required in enclosed places, as shops, warehouses, etc., Argand burners are beyond comparison superior to all others."[148]

Friedrich Siemens's invention of the first regenerative burner in 1879 marked an important development in the evolution of gas burners. This burner offered better combustion and a greatly increased power of illumination, which resulted from raising the temperature of the gas before ignition, increasing the air supply before combustion, or both. Improvements by Lundgren and Sugg enabled more extensive use of the burner in dwellings in the final years of the nineteenth and early twentieth centuries.[149]

The last burner type to be introduced, the incandescent gas burner, represented the greatest advance in gas lighting technology. In this type, the flame heated a refractory substance, or mantle, that then glowed with light.[150] The mantle, a cotton fabric woven into a netlike bag, was produced by impregnating the cotton with a mixture of thorium dioxide and cerium dioxide and then heating it; the cotton burned away and the metal oxides fused together to form the solid mesh. In crafting the modern mantle in the late 1880s, chemist Carl Auer von Welsbach discovered that this mixture of metal oxides gave off a white light and produced a strong mantle.

In addition to vastly improved light quality, these burners used gas in a very efficient manner.[151] One report claimed that "a single home burner using water gas and a Welsbach burner provided light equal to 22 to 35 candlepower," a great improvement over the 15 to 17 candlepower generated using flat-flame coal-gas burners.[152] An advertisement for the Welsbach incandescent gas burner in the *Journal of Gas Lighting* (1890) boasted, "Electric Light Surpassed. All the advantages of Electric Light and none of its drawbacks"—a clear attempt to discredit electric light before it really got established.[153] Schivelbusch notes that new technologies, like electric light, often spurred innovation that kept the old technology viable, if only for a short period. Historic lighting expert Roger Moss suggests that without the introduction of the Welsbach burner, electric light would have spread much more rapidly.[154]

In addition to burners, gas lighting fixtures included various types of chimneys, globes, and reflectors. Globes protected the flame from drafts and the house from fire. Made of glass, they met customers' aesthetic demands with their decorative style and ability to diffuse light in an attractive way. Most, but not all, burners required glass chimneys and globes to eliminate drafts; the many different forms and shapes depended on the buyer's aesthetic preferences and the types of burner they protected. Richards reported that the glass chimney proved indispensable for use with the Argand burner. In all cases, glass around the flame could "detract considerably from the light." Many fixtures had reflectors that threw back, or reflected, some of the light that otherwise would have been lost.[155]

CHAPTER 2

THE GILBERT AND BARKER MANUFACTURING COMPANY

The Springfield Gas Machine, a lighting system now almost wholly forgotten, was the product of the inventive minds of two remarkable men, Charles N. Gilbert and John F. Barker. In the years immediately following the American Civil War, each man became aware of the existence of portable or self-contained gas machines for lighting buildings beyond the reach of municipal gas systems and with the problems inherent in their current designs. Barker, a skilled machinist, designer, and inventor, eventually crafted a remarkably simple yet effective gas machine unit. When he retired from the firm in 1911, he held over fifty patents on gas machine equipment. Gilbert, principally a businessman and entrepreneur, saw the sales potential for a safe gas machine system that worked well.

The combination of men like Gilbert and Barker was an essential ingredient in the industrial and technological revolution of the nineteenth century. As technology historian Ruth Schwartz Cowan has noted, "Inventors invent: they create things that have never been created before," while "entrepreneurs innovate and diffuse: they figure out how to enter the marketplace to make money with the things that inventors invent." She continues: "Inventions turn into innovations when they fall into the hands of entrepreneurs who are willing to provide the funds or the hard work or the managerial skill that is necessary to turn an invention into an innovation."[1]

While the combination of inventor and entrepreneur was essential, so was timing, and, in the case of Gilbert and Barker, timing was everything. Cowan and others have pointed out that the "years between 1870 and 1930 were the golden age of entrepreneurship," as economic support was growing and regulation was almost nonexistent. Successful inventors and entrepreneurs such as Gilbert and Barker, notes Cowan, "were people who had the vision to understand the potential use of new technologies coupled with the political and economic skills to create and sustain large, complex organizations."[2]

Public sentiment was also important for the acceptance of the new portable or self-contained lighting technology under development by Gilbert and Barker and others. At the most basic level, there was a clear demand by the public for gas lighting in rural areas, particularly the expanding suburban landscapes around cities. Some suburban homeowners even piped their homes for gas light in anticipation of the arrival of gas mains or machines. For example, Gilbert and Barker customer and speaker of the Connecticut House of Representatives Charles Ives noted in 1868 that "hoping that I might in some way before long light my new house with gas I had it piped when it was built, and for the last eight years I have been on the look out for some invention that would enable me to make my gas-pipes available and enjoy once more the luxury and increased safety of a gas illuminator; I was unable to satisfy myself however until I learned about the apparatus which your company manufacture." Ives was not alone in this approach, as many builders recommended this course of action to their customers for suburban and country homes.

As discussed in chapter 1, the rise of the corporation, particularly large, monopolistic gas companies, engendered a level of public fear and apprehension about connecting to a centralized supply of power. The central gas plant's reach into the home, with its tentacle-like gas mains and supply pipes, was viewed by many as an unwanted intrusion in the private sphere; the portable or self-contained gas machine provided an alternative to the tyranny of corporate control. The "introduction of technological systems [such as utilities]," Cowan notes, "brought politicians, landlords, industrialists, and managers of utilities into the equation."[3] Gas machine companies reacted to these concerns with targeted marketing campaigns touting their systems' "independent" production of light. While some companies just hinted at the "freedom" provided by their units, others boldly advertised "Every House Occupant his Own Gas Manufacturer."[4] Gilbert and Barker customer Nathaniel Lamson addressed the corporate world of gas light in a partially tongue-in-cheek testimonial. Explaining that he provided light to his home and business and a block of stores on the main street, he wrote that "each occupant has a meter, and pays his gas bill to me quarterly, as he would to a 'grinding, heartless' coal gas corporation, if he lived in the city. I own all the stock in the company. I am President, Treasurer, Secretary, Superintendent, Director, Meter Inspector, etc. My tenants sometimes jokingly call me a 'soulless corporation,' yet as I furnish them gas at only half the price they would have to pay for coal gas, they are very well satisfied."[5]

Inventor John Francis Barker, the son of Albemarle and Abigail Barker, was born in Newton, Massachusetts, on December 16, 1839, the youngest of the couple's seven children. When John was four years old, his father suffered a serious injury in a fall that left him permanently paralyzed. The former blacksmith reportedly loved to read and passed the time in his wheelchair reading to his younger children, who had to attend the local public school because of the family's reduced means. In 1848, at the age of nine, John Francis went to work on a farm. By twelve he could no longer attend school, because his family depended on his income for survival. In 1852, John got a job with Brown and Company in the village of West Newton, Massachusetts, driving a two-horse express carriage between West Newton and Boston.[6] The story of Barker's early years is a familiar one in terms of American inventors of the period, notes Cowan, who writes that "many of the nineteenth-century inventors for whom biographical information exists were the sons of middle-class families who had for various reasons experienced periods of poverty."[7]

Barker's older brothers learned the machinist trade as apprentices in Otis Pettee's shop in Newton. John's brother Horace eventually left the Pettee shop and ventured into Boston, where he found employment with a steam-fitting establishment. In 1854, Horace opened his own small shop in the steam-fitting and machine trade in Lowell, Massachusetts. The business, known as H. R. Barker Machine Shop, flourished and John joined the firm shortly after it opened. Within two years, the fifteen year old had acquired a "completed set of tools, and had a gang of men under him."[8] The brothers' early business association would

Figure 2.1. John Barker. Photograph by H. C. Moore. Springfield Picture Collection of the Connecticut Valley Historical Museum, SPC-04-0090.32-08.

continue in one form or another throughout their lives, with Horace investing capital in John's gas machine firms.

While Barker's on-the-job training was typical for nineteenth-century inventors and tinkerers, by the early twentieth century, "the typical . . . engineer was a man with a bachelor of science degree, the product of school, not shop training."[9] Trachtenberg explains that many entrepreneurs and inventors like Barker "had begun their working lives as craftsmen, mechanics with a knack for invention, and risen to wealth and status as a result of their mechanical skill and entrepreneurial expertise."[10]

After working for Horace for four years, John moved to Philadelphia to take a similar position in the late 1850s. Although little is known of his time in Pennsylvania, he clearly continued to hone his skills as a machinist and supervisor. In 1862, he received an offer to take charge of the pipe work in the Water Shops of the U.S. Armory at Springfield, Massachusetts, from Major General Alexander Dyer.[11] Barker apparently met Charles N. Gilbert when both lodged in a boardinghouse on State Street in Springfield.[12]

Charles N. Gilbert was a midwesterner, born in Chicago in 1843 and raised in Wisconsin. The son of Nelson and Emily Gilbert, he attended preparatory school from 1855 to 1857 and college from 1858 to 1860 at Beloit College in Beloit, Wisconsin.[13] After his father died in 1860, when Gilbert was seventeen years old, he and his family left Beloit.[14] He apparently moved back to Chicago sometime after 1860 and there became involved in the manufacture and design of portable gas machines.[15] He may have worked with the principals of the firm that would become the Combination Gas Machine Company in 1868. The company moved to Detroit after the Chicago fire of 1871 and went on to manufacture the Combination Mixing Regulator, for which Gilbert and Barker Manufacturing served as the exclusive East Coast agents. In the early twentieth century, the Detroit firm would buy the rights to the Springfield system.

By about 1865, Gilbert had moved to Springfield, Massachusetts, where he set up the New England Portable Gas Works Company.[16] It appears that he had obtained patent rights to a gas machine while in Chicago and brought these with him to Springfield.[17] Gilbert apparently "saw a promising future in a new type of gas machine to safely burn gasoline." He quickly hired John Barker as superintendent of the newly established firm. Barker, it turned out, had a very personal interest in gas machines; he had been seriously burned during an explosion and fire at the U.S. Armory in Springfield caused by a defective gas machine.[18] Because of his accident, Barker insisted that Gilbert and Barker make safety one of its the overriding concerns. "No machine was ever permitted to be installed," company

Figure 2.2. Charles Gilbert. From *Gilbarco News,* May 1940. Springfield Picture Collection of the Connecticut Valley Historical Museum, SPC-04-0200-07.

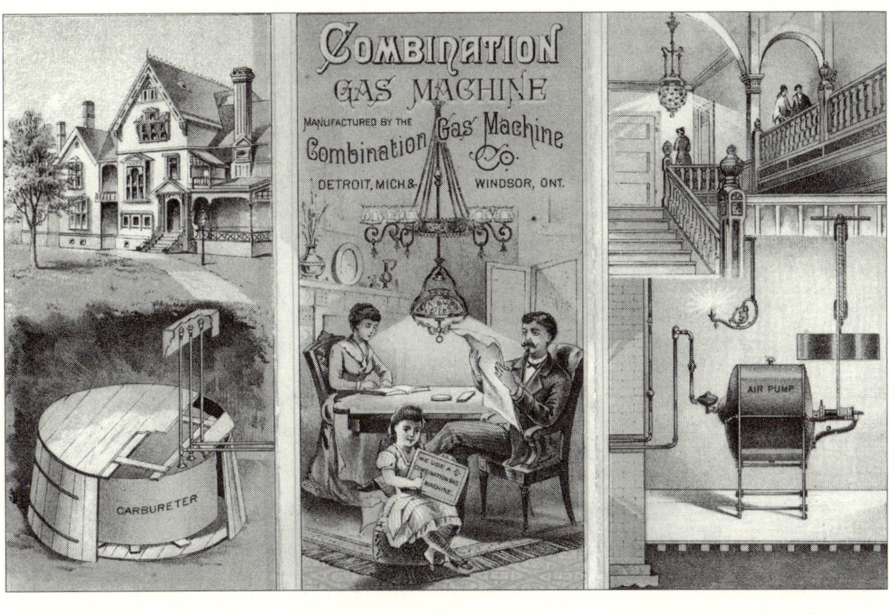

Figure 2.3. Brochure for the Combination Gas Machine, Combination Gas Machine Company, Detroit, n.d. Author's collection.

literature later asserted, "until it had passed the most rigid inspection and was known to be absolutely safe."[19] Barker was not alone in his safety concerns. In the late 1860s and early 1870s, U.S. newspapers regularly reported on horrific explosions and fires caused by gas machines, keeping safety on the minds of American consumers.

A New England Portable Gas Works Company catalog indicates that the firm began manufacturing its first lighting system, known as the Springfield Gas Machine, sometime in 1866.[20] An October 1866 report by an employee of the credit reporting firm R. G. Dun and Company notes that the New England Portable Gas Works Company "owns a right . . . to make gas [machines] under a patent issued to someone whose name I have not learned in Chicago." The employee identifies Gilbert, "whom I know nothing of," as a partner of this new firm, along with "Edward Southworth, rich, S. G. (C) Bemis, Ephrim W. Bond [president of Massachusetts Mutual Life Insurance,] worth $50m, perhaps twice that amount, [H. A.] Gould (of Chapin & Gould), worth $50–100m, Otis Childs, worth $15m, John Hamilton, . . . & one or two others whose names I have not [heard]."[21] These local Springfield investors, the Dun report continues, apparently believed in the company's potential for success despite "the fact that only a few short years before a gas converter in the Springfield Armory at Watershops pond had blown up." This referred, of course, to the accident that had injured John F. Barker and others.[22]

About this time Gilbert and Barker hired draftsman and inventor Hiram S. Maxim to help them perfect a safe and efficient gas machine.[23] Like Barker, Maxim, a Maine native, came to Springfield to work at the armory. After leaving the company in the late 1860s, Maxim went on to patent several other types of gas machines, which were sold through his Maxim Gas Machine Company. None had much success in comparison to the Springfield Gas Machine.[24] He focused his energies exclusively on hot process systems after leaving Gilbert and Barker, regularly berating the cold process machines of his former employer.[25]

The founders of the New England Portable Gas Works Company initially set it up as a copartnership that would later "be organized into a corporation under the general law." The R. G. Dun report estimates the capital for the company as "about 125m, the chief part of which is the cost of the patent right. The balance is the working capital." Given the standing of the partners, the report also concludes that as long as the firm remained a partnership in which the partners bore personal liability, it would remain "good beyond question."[26]

The New England Portable Gas Works Company, however, found the design of their first gas machine, which combined an air pump and gas generator in one unit, impractical and unsafe.[27] Due to the danger posed by storing gasoline within an occupied building, the machine had to be placed in an outbuilding well away from the building to be lighted.[28] Its placement in a separate gas house presented problems because the water in the air pump could freeze, owners found the placement away from the building inconvenient in bad weather, and the machine's iron and zinc construction was subject to rusting and corrosion.[29] Not surprisingly, the R. G. Dun representative wrote in early September 1867 that "insurers are down on [New England's] original apparatus for lighting houses. So they have small sale." On September 24, 1867, R. G. Dun recorded that the company "had closed, they are the same as the Springfield Gas Machine Co."[30]

The newly formed Springfield Gas Machine Company, established on July 10, 1867, and capitalized at $125,000, consisted largely of New England's previous investors and principals.[31] The new company included E. W. Bond, president; H. Dickinson, treasurer; and E. W. Bond, Edward Southworth, H. A. Gould, C. A. Jessup, J. F. Barker, Otis Childs, and C. N. Gilbert, directors. C. N. Gilbert served as agent and J. F. Barker as superintendent.[32] The company produced a redesigned machine that utilized an "improved arrangement" that separated the system's two principal components. The air pump was placed in the cellar of the building to be lit and the gasoline gas generator in an underground vault outside and away from the structure.[33] R. G. Dun noted in September 1867 that the company "now claim[s] to have got a better thing." The Dun reporter cautioned about the company's meager working capital of fifteen thousand dollars, recording that "it has made no money," but noting that "among the owners are some rich men."[34]

This "better thing" owed its existence to a new design and patent that Gilbert, Barker, and E. N. Ives had assigned to the New England Portable Gas Works Company in January 1867.[35] Although they modified and improved this design several times, it formed the essential "improved arrangement" that the Springfield Gas Machine Company and its successor firms would tout and protect over the years. The firm's reorganization may have been a way to distance itself from its failed original product in order to avoid tainting the new design by negative association, and/or as a way to reestablish financial credibility. By the time Springfield published its 1874 catalog, the company enjoyed increasing successes with their new design. The principals could look back on and refer to their first failed machine as a way to demonstrate the firm's major innovations to the system.

The Springfield Gas Machine Company's first major hurtle was to patent the improved cold process design of the gas machine. However, the U.S. Patent Office was reportedly "at this time running four months behind." Barker apparently appealed to his former boss and friend, General Dyer, by then the chief of ordnance in Washington, D.C. Barker traveled to Washington to argue his case, reminding Dyer of the danger of gas machines currently in use at places like the U.S. Armory in Springfield. It was, he reasoned, "in the public benefit to have his patents granted at once." After the commissioner of patents sent a letter to the chief of the chemical department, the patents, valued at $110,000, were soon granted.[36]

A 1906 company catalog noted in retrospect that the "first Springfield Gas Machine [a one-thousand-light system], having its generator in the ground, was installed at the Water Shops of the United States Armory, at Springfield, Massachusetts" in 1867.[37] By September 1868, the company's improved design was showing strong sales. R. G. Dun reported that the firm was "considered good and improved. Business well managed."[38]

A major factor in the company's success was an 1867 rating by the New York Board of Fire Underwriters that determined the Springfield system and air-gas units of similar design were as safe as coal gas.[39] The firm's 1867 catalog noted this endorsement and offered customers a copy of "the report upon Gas Machines, Carbureting Apparatus, Gasoline, &c., adopted by the National Board of Fire Underwriters in September last."[40] As discussed in detail below, the New York Board of Fire Underwriters' rating meant that buildings using the gasoline gas system were insured at the same premium as those with municipal coal gas; use of either system resulted in considerably lower premiums than the use of kerosene lamps.[41]

The Springfield Gas Machine Company's manufacturing plant and general offices were first located at the corner of Spring Street and Lyman Street in Springfield, Massachusetts; their initial building, fronting Lyman Street, was reportedly quite small.[42] The company manufactured its "improved apparatus" in both galvanized iron and in more expensive copper, and sizes ranged from a small ten-burner unit to a huge one-thousand-burner system (table 2.1).[43] The 1867 catalog calculates the Springfield system's installation costs as about half that of portable coal-gas or rosin-oil works. In addition, the system provided gas for lighting at the relatively low cost of approximately two dollars per thousand cubic feet. In what would become Springfield's standard marketing pitch, the company touted its system as ideal for "all those who are beyond the reach of Coal Gas Mains," particularly manufacturers, "owners of country residences, proprietors of town halls or blocks of stores, church committees, and all others with property to light."[44]

The firm also marketed a special coal-gas carburetor to municipal coal-gas customers to enrich the gas and "improve your light." As the catalog notes, "Parties now using a gas of *poor quality* can by this means, obtain a light *fully equal to the best Cannel Coal Gas*."[45] In addition to lighting products, the company supplied "good gasoline" to its customers in metallic tanks and barrels, claiming that

TABLE 2.1
Price List for Springfield Gas Machines, 1867–1868

No.	Number of burners rated to supply	Galvanized iron price ($)	Copper price ($)
1	10	225	250
2	25	275	300
3	50	350	400
4	100	550	600
5	150	700	750
6	200	850	900
7	300	1,125	1,200
8	500	1,550	1,700
9	1,000	2,600	2,800

SOURCE: Handwritten notes in Gilbert and Barker 1867–68. Springfield Gas Machine Company, Circular of the Springfield Gas Machine Company.

"we have constantly on hand a stock of such [gasoline] as is nice, it being distilled expressly for us."[46] Due to gasoline's newness, its variable quality, and the general difficulty in purchasing it at this time, marketing gasoline played a large part in the successful sale of the machines.

Although the details are a bit unclear, Barker apparently briefly left the firm, or at least his position as superintendent, in late 1868 or early 1869 to start a store in New York City in order to sell the Springfield Gas Machine Company's products to a wider market. Run in partnership with Charles N. Gilbert, the store opened in about February 1869. In September 1869, it apparently became a sales office of the new Gilbert, Barker and Company.[47]

The firm sold several hundred units of its new machine between 1867 and 1869.[48] The company advertised its "celebrated machines" in a variety of publications, including the *New York Times*, *Appletons' Journal of Literature, Science and Art*, and *Scientific American*.[49] While R. G. Dun deemed the new machine "a good thing" in 1868, Dun rated the company itself with some caution. Dun's representative observed that Springfield had little operating capital, having used up most of the initial fifteen thousand dollars, and "had not made any money." Despite these financial difficulties, the company reportedly had "good men in the concern."[50]

The unprofitable Springfield Gas Machine Company stayed in business for only about two years, when the newly formed firm of Gilbert, Barker and Company bought it out on June 28, 1869.[51] The new company purchased "of the Springfield Gas Machine Company all their Patents, Stock, Machinery, Tools, etc." In addition to principals Charles N. Gilbert and John F. Barker, the firm included John's brother and former employer Horace Barker of Lowell, Massachusetts, and W. S. Gilbert of Cohoes, New York.[52] In September 1869, the R. G. Dun representative recorded that Gilbert, Barker and Company had succeeded the Springfield Gas Machine Company, noting that "G&B have not much means. B's brother of law furnishes the means. Have never done very well, but have a good thing [product], and if it is properly managed will make money."[53] It seems likely that sale of the firm to the Gilbert and Barker families brought needed capital, particularly from Barker's older brother Horace, whose machine shop in Lowell had done very well.

Gilbert, Barker and Company remained in business until March 7, 1870, when it was incorporated, "principally of members of the late firm," and "styled the Gilbert and Barker Manufacturing Company."[54] Trachtenberg notes that this period saw a "decisive decline of proprietors, family businesses, simple partnerships" and that "increasingly, the instrument of success proved to be more effective organization, restructuring of enterprises into corporations in which financing and sales along with production fell under control of a single entity." He continues: "In a period of expanding economic growth and more or less unstable money supply, the advantages of incorporation were manifold, for it permitted a number of people to pool their capital and their efforts under one name, as a single entity."[55] H. R. Barker served as president of the new corporation, John F. Barker as treasurer, and Charles N. Gilbert as general agent.[56] The new company noted in their catalog that they would endeavor

> by the employment of skillful mechanics and fidelity in details of construction, to fully sustain in the Machines that we shall in future manufacture, the enviable reputation that the "Springfield Machine" now holds.
>
> We shall keep on hand, for sale at the lowest market rates, an abundant supply of the best quality of Gasolene for Gas Machines. We shall also manufacture Metallic Barrels and Tanks for storing Gasolene.[57]

Gilbert and Barker Manufacturing Company published a small sales catalog in 1870–71 that explained the operation of its gas machine, provided illustrations of the equipment, offered testimonials from satisfied customers, contained a price list for the machines, and included a list of houses and businesses then using the Springfield system. As promised, the firm's gas machines continued to follow the design of the machines produced by the company's previous incarnations. The systems were made entirely of copper by this time; a later catalog indicates that the firm abandoned iron and zinc in favor of copper in 1867 and 1868.[58] In 1870–71, the Springfield system ranged in size from a No. 1, rated to power twenty

TABLE 2.2
Price List for Copper Springfield Gas Machines, 1870

No.	Number of burners rated to supply	Capacity (in gallons)	Price ($)
1	20	100	350
2	35	150	450
3	65	200	600
4	100	250	800
5	150	300	1000
6	200	350	1,200
7	300	475	1,500
8	500	700	2,000

SOURCE: Gilbert and Barker 1870. Gilbert and Barker Manufacturing Company, *Springfield Gas Machine*, 1870–71.

burners, to a No. 8, rated at five hundred burners (table 2.2). The catalog notes that larger machines could be specially ordered and were priced accordingly.[59]

The Springfield Gas Machine quickly grew in popularity and by 1870 was already in use in a wide variety of businesses, including mills, factories, hotels, and stores, and in private dwellings across New England; Gilbert and Barker's 1870–71 catalog lists some 133 domestic and 54 commercial installations (see the online appendix to this book, http://www.utpress.org/download). Based on testimonials and recommendations solicited from current customers, the company established a solid reputation for quality and reliability, which it touted to potential customers. While the R. G. Dun and Company representative who visited the firm during this period remained cautious because of the company's small operating capital, he noted in August 1871 that "they are doing quite a large business and apparently making money."[60]

Gilbert and Barker clearly saw their success tied to the Springfield's design as a cold process gas machine and articulated its advantages over coal-gas and rosin-oil systems in the 1870–71 catalog:

> Our machines have many advantages over the process now so generally used of making gas from coal or oil by fire heat and the use of retorts (regular coal-gas works):
>
> *First*—Their first cost is much less.
>
> *Second*—They do not necessitate so great an expenditure for setting, or for out-buildings, for coal sheds, retort-house, etc.
>
> *Third*—They do not require one-tenth the labor and attention, or the services of a gas engineer to manufacture the gas, but are very nearly automatic in their operation, always ready, and supplying so much or little gas as may be required, and so fast as used.
>
> *Fourth*—The gas manufactured is perfectly pure; there are no foreign elements to be expelled by means of purifiers, washers, or any like apparatus.
>
> *Fifth*—The expense of keeping such machines in order is nothing.
>
> *Sixth*—All the unpleasant odor usually attending the manufacture of coal or oil gas is avoided.
>
> *Seventh*—They do not offend the eye or deface a place by their unsightliness.
>
> *Eighth*—The expense per thousand [cubic] feet for the gas manufacture is less—for in the manufacture of gas from coal or oil by retorts in the common way, the general expenses (for labor, for repairs upon retorts, etc.) are so heavy, that when only a limited amount of gas is made, the total cost per thousand feet is vastly increased over merely the cost of the material used.[61]

It appears that the Springfield Gas Machine was slightly more expensive than many other gasoline gas machines on the market at the time, although it was comparable to or less expensive than the more elaborate coal-gas and rosin-oil retorts and hot process gasoline systems. For example, the Underground Gas Machine Company sold its twenty-five-burner system, a cold process machine similar to the Springfield, for $300.00 and its one-hundred-burner machine for $650.00 (see table 2.2). Like the Springfield, the price included everything needed for the system but did not include delivery and installation. The Springfield cost more than the Underground systems and similar machines, as elaborated in Gilbert and Barker advertisements, because of its copper, rather than iron, construction.

The Underground Gas Machine of this period also cost less because it was buried directly in the ground and not placed in a vault. While the company points out that this eliminated the cost of constructing a brick vault, Gilbert and Barker quickly responded that iron systems buried directly in the ground experienced severe corrosion and maintenance problems, unlike the copper Springfield Gas Machine installed in a brick vault.[62]

By 1870, the firm had opened a sales office in New York City at numbers 70, 72, and 74 Wooster Street (between Spring Street and Broome Street). The 1870–71 catalog notes that the New York office shared space with the Archer and Pan-Coast Manufacturing Company, an established firm described as "manufacturers and wholesale dealers in gas fixtures." Sometime between 1870–71 and the 1872 publication of an article in the *American Gas-Light Journal and Chemical Repertory*, which refers to Leander A. Bevin as Gilbert and Barker's agent, the firm acquired a sales office at 554 Broadway.[63]

An insert in the 1870–71 catalog indicates that the company was concerned about competitors who might infringe on its patents. Several hundred firms manufactured gas machines by this time, making the business extremely competitive. Gilbert and Barker warned readers that "our improved PLAN OF SETTING a gas apparatus is patented, and the placing of an air pump, driven by a weight, within the house, operating in combination with any gas generator, detached from the same, and located in a vault outside the building, is an infringement upon our patents. Manufacturers of machines, and those about to purchase, are warned against such infringement, as we shall prosecute all trespassers upon our rights to the full extent of the law." The company clearly intended to go to court to challenge those who appeared to infringe on its patents. In fact, the 1874 catalog includes the full text of the opinion in the company's suit against Oakes Tirrill for "infringement of a patent granted to J. F. Barker and C. N. Gilbert, on the 3rd August, 1869, for an improved apparatus for carbureting air."[64]

An examination of the catalog for Oakes Tirrill's Underground Gas Machine Company of New York reveals that his system closely resembled the Springfield. Tirrill apparently copied the design of the generator and air pump and, more important, the separation of the air pump, located in the basement of the house, and the gasoline-filled generator, buried some distance from the house.[65] In his opinion in the case, the judge notes that "neither the process nor the chief parts of the apparatus are claimed to be new." He continues that the "chief feature of the improvement is in the placing of the carburetor under ground, in a vault, separate from the building to be lighted . . . while the . . . [air pump] is placed in an apartment in the building." Although the Underground Gas Machine was buried directly in the ground rather than in a vault, the judge upheld the validity of the patent for Gilbert and Barker's "improved arrangement" and found that the defendant had infringed on the patent.[66]

While protecting its existing patents in a very competitive environment was essential to its long-term success, the firm relied on other business strategies as well. Gilbert and Barker's continuing attention to the safety and ease of use of its systems, in both new product development and the refinement of current systems, led to the firm's growing prosperity in the second half of the 1870s. Attention to quality, safety, and innovation consistently distinguished the company in the ever-expanding field of gas lighting, even with the advent of new light sources such as acetylene gas machines and nascent electric generating systems.

In terms of safety, the company's continuing endorsement by the New York Board of Fire Underwriters and its system's subsequent rating with the National Board of Fire Underwriters proved a major marketing and sales advantage. For example, in the National Board of Fire Underwriters' report on gas machines, published in 1876, the board acknowledged that the design of the Springfield system, which separated the air pump and generator, made it the safest of the many types of gas machines then available. The report asserted that

> it has ever been the policy of the Board to refrain from giving preference to any particular process or apparatus but those machines which, by the durable material used in their construction, the personal care and supervision given on the part of the manufacturers to insure safety in the setting of their apparatus, the secure fitting of pipes, the intelligent provision they make to avoid condensation, and the precautions shown to insure safety, should have our fullest endorsement, and especially should this be said of the proprietors of the machines who do not sell rights or allow unskilled or irresponsible persons to put up machines.[67]

Significantly, the report continues that "it is not the experience of the Committee that more than one or two such companies exist."[68] While the report doesn't specifically name individual companies, it does include a drawing of the ideal gas machine installation virtually identical to one from the Springfield Gas Machine catalog. This strongly suggests that Gilbert and Barker was one of the "one or two such companies" referenced in the report. Not coincidentally, perhaps, report author D. A. Heald used a Springfield Gas Machine to light his personal residence near New York City.[69] Because the media regularly reported the recommendations of the Fire Underwriters, Gilbert and Barker benefited greatly from the board's thinly veiled endorsement. An 1873 *New York Times* article, for example, described the instructions for safe installation and use of gas machines recently issued by the Special Committee on Gas-Machines of the New York Board of Fire Underwriters.[70] The Springfield system conformed with these guidelines in every respect, providing further evidence of its safety.

By 1874, the company had grown into a prosperous and well-known concern and consequently expanded its manufacturing facility. An April 1874 R. G. Dun report records that "by their statement of March 1, 1874, they showed a surplus of over $30,000"; the firm's capital stock had also increased to twenty-eight

TABLE 2.3
Gilbert and Barker Manufacturing Annual Financial Statements, 1876–1886

Assets	1876	1883	1885	1886
Real estate	$0.00	$22,017.28	$26,750.00	$0.00
Buildings	$0.00	$0.00	$0.00	$4,273.28
Land and water power	$0.00	$0.00	$0.00	$26,750.00
Machinery	$2,452.65	$8,862.29	$4,976.60	$0.00
Quick assets	$0.00	$0.00	$0.00	$115,292.85
Cash and debts receivable	$61,083.40	$80,728.07	$58,882.62	$0.00
Manufactures, material, and stock in process	$26,464.83	$41,892.46	$31,768.36	$0.00
Miscellaneous	$6,212.41	$4,317.25	$10,456.86	$12,325.28
Total assets	**$96,213.89**	**$157,817.35**	**$132,834.44**	**158,641.41**
Liabilities				
Capital stock	$28,000.00	$40,000.00	$40,000.00	$40,000.00
Debts	$36,631.82	$78,987.63	$42,103.05	$49,360.38
Balance—profit and loss	$23,057.82	$17,774.32	$34,450.17	$50,132.79
Reserve for depreciation	$8,424.25	$21,061.40	$16,281.22	$19,148.24
Total liabilities	**$96,213.89**	**$157,817.35**	**$132,834.44**	**$158,641.41**

SOURCE: Reconstructed from R. G. Dun records, 1876–86 (Massachusetts, vol. 42: 44, 436), R. G. Dun and Company Collection, Baker Library, Harvard Business School).

thousand dollars. The Dun report characterized Gilbert and Barker as "respected, prompt, & perfectly good for obligations." The firm's stockholders were listed as J. H. Parsons and Company of Cohoes, New York; H. R. Barker of Lowell, Massachusetts; and J. F. Barker and C. N. Gilbert of New York.[71] Over the next few years, R. G. Dun continued to describe the firm as a solid, profitable enterprise, that was "doing well and making $" and "considered sound and safe."[72] The company's 1876 financial statement indicates that the firm had assets of $96,213.29 and an annual profit of $23,057.22 (table 2.3). The firm's success during this period is all the more remarkable when put in the context of the national economic picture of the late nineteenth century. Trachtenberg explains that "a severe Wall Street crash in 1873 triggered a round of bankruptcies and failures in the United States, six thousand businesses closing in 1874 alone, and as many as nine hundred a month folding in 1878." In fact, the next twenty years saw a "perilously uneven business cycle . . . affecting all sections of the economy."[73]

When the company issued its 1874 catalog, it had moved its New York office and salesroom to 238 Canal Street, opened sales offices in Philadelphia (12 N. Seventh Street) and Boston (117 and 119 Court Street), and contracted with agencies in Baltimore, Cincinnati, Chicago, and Milwaukee to sell its products. It seems possible that the company's expansion into the Chicago market followed the Combination Gas Machine Company's relocation to Detroit in 1871. The company also may have taken advantage of the expanding urban marketplace that resulted from rebuilding after the Chicago fire of the same year.[74] In 1875, the firm added a sales office on the West Coast, establishing a relationship with Nye and Acheson of San Francisco. An advertisement in the *Santa Cruz Sentinel* reported that Nye and Acheson had "accepted the agency of the celebrated Springfield Gas Machine" and that they were "sole agents for the Springfield Gas Machine for California and the Pacific Coast." The ad also boasts that "over 1,000 testimonials attest the superiority of the Springfield Gas Machine," suggesting that the company enjoyed brisk sales.[75]

With sales flourishing, Gilbert and Barker's 1874 catalog presents an expanded product line, including a hydraulic blower initially patented in 1871. For businesses like mills with a regular supply of water, this unit could replace the weight-driven air pump. The firm's 1874 product line consisted of ten sizes of the gas machine system, produced solely in copper, ranging in capacity from 15 burners (No. 1) to 750 burners (No. 10) (table 2.4). In a testimonial in the 1874 catalog, one commercial customer acknowledges the Springfield as "apparently the highest-priced machine, yet we judged it would prove the cheapest in the long run, on the principle that 'the best is always the cheapest.'" The 1874 catalog also reports that

TABLE 2.4
Price List for Copper Springfield Gas Machines, 1874

No.	Number of burners rated to supply	Capacity (in gallons)	Price ($) (air pump)	Price ($) (gas generator)	Price ($) (machine complete)
1	15	70	138	162	300
2	25	100	160	240	400
3	40	150	225	300	525
4	75	200	320	380	700
5	100	250	400	450	850
6	150	300	500	525	1,025
7	200	350	620	630	1,250
8	300	500	890	760	1,650
9	500	750	1,150	1,100	2,250
10	800	1,200	1,850	1,650	3,500

SOURCE: Gilbert and Barker Manufacturing, *Springfield Gas Machine,* 1874.

"we furnish the best quality of [gasoline] at market rates,—no charge for barrels," and also manufactured both iron and copper tanks for storing gasoline.[76]

With gas machine explosions and fires regularly reported in the news, safety and quality continued to be the major selling points for the Springfield unit. The 1874 catalog notes that of the more than seven hundred units sold to date, none had been reported to have caused fires or damage to the buildings that they lit. The company once again touts the "report of the New York Board of Fire Underwriters upon this matter" and offers to "mail it to any one who may wish it." The firm's 1870–71 and 1874 catalogs both contain various testimonials from insurance underwriters in the northeastern and mid-Atlantic states who report that they personally used the machine, acknowledge its safety and quality, and would insure a building with a Springfield system at a lower premium than if lit with kerosene.[77]

The *Commerce, Manufactures and Resources of Springfield, Massachusetts* (1883) remarks that the company's "claims for patronage are based upon its correct principle of construction, its safety, reliability in operation, the mechanical excellence of its manufacture, and the durability and intrinsic value of the metals composing it."[78] These claims were further supported and acknowledged by a special award from the American Institute exhibition in 1872, diplomas for "continued excellence" at the American Institute expositions in 1874 and 1875, and a special award at the 1876 Centennial Exhibition in Philadelphia.[79] The American Institute wrote in 1872 that "we find the machine to be a most perfect automatic gas making apparatus, adapted alike for large buildings as well as small ones. It is simple in construction and effective in operation, requires no skill and very little attention to keep it in complete working order, reliable both summer and winter. We pronounce this machine to be by far the best in the market, and would recommend for it the highest award you can bestow."[80] The special award presented to the Springfield Gas Machine at the Centennial Exhibition was one of only forty-nine awards given to exhibitors in Group XIV, which included "every kind of apparatus for heating, lighting, ventilation."[81] The company's displays and awards at major expositions provided important exposure for its gas machine system and contributed significantly to its marketing successes. The 1876 Centennial Exhibition, for example, drew over nine million visitors in 159 days. These "ritual celebrations," writes Trachtenberg, celebrated the machines of the age and provided "fervently optimistic prophecies of abundance" throughout the Gilded Age, a period of general economic instability.[82] In this sense, inventions like the Springfield Gas Machine were viewed as engines of progress and improvement.

Year-round reliability and ease of operation were very important selling points for the Springfield system, as was the durability of the machine due to its copper construction. An 1872 article/advertisement in the *American Gas-Light Journal* echoes the American Institute award, reporting that the Springfield "is built entirely of copper" and "is used under any and all circumstances that coal gas could be used, and are [*sic*] giving universal satisfaction, alike when exposed to the cold of New England and Canada winters and to the tropical heat of the summer sun of the extreme south."[83]

The list of homes and businesses with a Springfield system in the 1874 catalog indicates that the company marketed its machine to a broad manufacturing and commercial clientele as well as to homeowners across the country and in several foreign countries (see the online appendix to this book, http://www.utpress.org/downloads). The 1874 catalog emphasizes the use of the system for lighting large mill and manufacturing complexes and in the "mechanical arts" and for brazing, soldering, and plating applications. Gilbert and Barker note that it used "this gas at our own factory for doing all the heating and soldering required in the manufacture of our gas machines."[84]

The Springfield Gas Machine also lit a variety of large tourist hotels, libraries, hospitals, prisons, colleges, schools, stores, and even Howe's Cave in Schoharie County, New York. Large installations at tourist destinations like the tourist hotels of New England, Florida, and even Bermuda, served as wonderful advertisements for the company, attracting upper-middle- and upper-class clients who could afford a Springfield Gas Machine system for their own home or business. Small cooperative groups also used the system to light multiple houses in neighborhoods. For example, in Greenfield, Massachusetts, one machine lit "six dwelling houses and two street lamps" on a residential street; each house had an average of twenty-four burners.[85]

The overwhelming success of the Springfield Gas Machine system and its parent company clearly drove the competition to adopt the principal design feature patented in 1867, that is, the separation of the gas generator in a pit or vault away from the building to be lit and placement of the air pump in the structure's cellar or basement. In fact, this was the only gas machine arrangement that fire insurance companies would permit. While the courts initially upheld the company's patent right, it appears that other firms continued to use the arrangement and challenge Gilbert and Barker's right to a patent on it.[86]

In the late 1870s, Gilbert and Barker filed two new lawsuits in U.S. circuit court for patent infringements: against the Walworth Manufacturing Company in the U.S. Circuit Court of Massachusetts and against Coleman's Pneumatic Gas Works Company in the U.S. Circuit Court of Ohio. In both cases, Gilbert and Barker alleged patent infringement on the design of the Springfield Gas Machine system, specifically its "improved arrangement."[87] Neither circuit court upheld Gilbert and Barker's patent on this aspect of their machine. Gilbert and Barker reported the circuit courts' negative decisions in its 1878 catalog, as well as its pending appeal to the U.S. Supreme Court, noting confidently that "our patent will here be sustained." The Supreme Court heard the case in 1879, whereupon the chief justice wrote an opinion that upheld the lower court's ruling.[88] As the court reporter for the *Washington Post* wrote, "This decision opens up the manufacture of machines for making illumination gas from gasoline in which the generator is enclosed in an underground vault."[89]

By 1880, the Walworth Gas Machine had become a serious competitor to the Springfield system. Like Gilbert and Barker, Walworth also had the approval of the National Board of Fire Underwriters in terms of the safety of its system; as late as 1911, an insurance policy for John S. Ames's Langwater property specifically permits the use of both the Springfield and Walworth machines.[90] The Walworth unit, like the Springfield, placed its generator away from the building to be lit.

The Walworth catalog reads much like Gilbert and Barker's; Walworth touted its machine's quality, safety, and dependability and reported that "we use no fire about our apparatus. It is safe and reliable at all times, winter and summer." Walworth generators were constructed of both copper and iron, as was the Springfield by 1880, although the actual design of the generators differed somewhat.[91]

Typically, Gilbert and Barker was already moving ahead with product innovations even as the Supreme Court heard its appeal of the Walworth case.[92] While the loss of its patented "improved arrangement" must have inflicted a blow on the company, Gilbert and Barker countered with new products, like the water-powered air pump and mixing regulator, which increased the efficiency, safety, and ease of operation of its systems.[93] The company also strove to further refine its product line, introducing a new generator that could be directly buried in the ground, eliminating the need for a brick vault and resulting in lowered installation costs.[94] Many of these new components were exclusive to the Gilbert and Barker product line and enabled the company to maintain its strong sales and customer interest at a time when the Supreme Court decision potentially opened the flood gates on the manufacture of copycat gas machine systems. This type of resilient, innovative, and entrepreneurial drive is at the heart of the firm's success and longevity.

Although faced with increasing competition, the company's gas machines sold well; sales doubled between 1874 and 1878. These sales figures no doubt improved with the introduction of another innovative product to the line in 1878: an automatic water-powered air pump that eliminated the need to wind the weight.[95] Although the company continued to sell the weight-driven pump for customers without access to a pressurized water supply, the new pump became the favored installation because it made the system "entirely automatic" and virtually maintenance free. The company made a point of emphasizing the advantage of "our latest improvement" over other gas machine systems, notably the Walworth Gas Machine and Underground Gas Machine, which relied solely on weight-driven pumps.[96] This important innovation helped distinguish the Springfield system from its many competitors.

In the early 1880s, the company added another component to the gas machine system that further distanced it from its competitors. A new mixing regulator insured a consistent mixture of gas and air, allowing the use of "ordinary open coal gas burners"; this component eliminated a major complaint of gasoline gas machines, namely, the uneven quality of the gas and thus the varying quality of the light. The 1883 *Review of Commerce, Manufactures and Resources in Springfield* reported that the company held the "sole monopoly of these mixers, there being no other instrument of the kind in the country."[97] The report goes on to note that the mixers were generally used for larger installations, such as those in summer hotels, factories, and public buildings. The importance of the firm's initial mixing regulator and later its sale of the Combination Mixing Regulator cannot be overemphasized. These components revolutionized the portable gas machine's ability to provide dependable and high-quality light. The mixing regulator also affected the cost of operating the machine by maintaining a stable air-gas mixture and eliminating the unnecessary use of gasoline. At the same time,

TABLE 2.5
Price List for Springfield Gas Machines, 1882

Type	No.	Number of burners rated to supply	Capacity (in gallons)	Price ($) (air pump)	Price ($) (gas generator)	Price ($) (machine complete)
Copper	1	20	120	125	175	300
	2	30	180	150	250	400
	3	40	220	200	300	500
	3 1/2	50	250	250	350	600
	4	75	300	300	400	700
	5	100	350	350	500	850
	6	150	400	400	650	1,050
	7	200	500	500	750	1,250
	8	300	750	650	1,000	1,650
	9	500	1,000	900	1,350	2,250
	10	750	1,500	1,400	1,800	3,200
	11	1,000	1,900	1,800	2,200	4,000
Iron	1	20	120	100	150	250
	2	30	180	125	175	300
	3	40	220	150	250	400
	3 1/2	50	250	175	275	450
	4	75	300	200	350	550
	5	100	350	250	400	650
	6	150	400	300	450	750
	7	200	500	400	600	1,000
	8	300	750	500	750	1,250
	9	500	1,000	700	1,000	1,700

SOURCE: Gilbert and Barker Manufacturing, *Gilbert & Barker M'f'g Co's Price List*, 1882.

the regulator increased the safety of the system by preventing dangerously rich and explosive mixtures.

Around 1882, the company added a less expensive version of the Springfield Gas Machine system in iron, clearly trying to broaden the market for its product by competing with less expensive rivals. The firm's price list indicates that the larger iron systems cost significantly less than copper systems. The price of copper systems ranged from $300.00 for a No. 1 system rated at twenty burners to $4,000.00 for a No. 11 system rated at a remarkable one thousand burners (table 2.5). The new iron systems ranged from $250.00 for a No. 1 to $1,700.00 for a No. 9, the largest unit produced in iron or steel.[98]

TABLE 2.6
Price List for Walworth Gas Machines, 1880

Type	Number of burner rated to supply	Capacity (in gallons)	Price ($) (machine complete)
Fully buried generator (iron)	20	65	300
	40	110	375
	60	150	475
	80	200	600
Generator (copper) in brick vault	100	250	750
	150	250	900
	200	400	1,100
	300	400	1,500
	400	500	1,800
	500	500	2,000

SOURCE: Walworth Manufacturing, *Walworth Gas Machine*, 7, 16.

The Springfield Gas Machine continued to cost about the same as the competition's models throughout the 1880s, although Gilbert and Barker's iron version eventually cost less than some other systems. For example, the Walworth Gas Machine cost slightly more than the smaller Springfield units and slightly less than the larger Springfield systems (table 2.6).[99] Thus Gilbert and Barker clearly priced its iron machines to match or undercut competitors with similar systems.

By the 1880s, customers could purchase iron or steel machines in several new configurations, with a choice between a partially buried "flat" gas generator (extending into a small brick shaft for access to controls) and a fully buried gas generator (operated via pipes and controls rising to the surface).[100] These innovations significantly reduced the installation cost of the system by limiting or entirely eliminating the need to construct a complex brick vault. Several other companies had marketed partially buried or fully buried systems for a number of years; in fact, the Underground Gas Machine used a fully buried configuration as early as 1871.[101]

The sale of gasoline also continued to be a significant part of Gilbert and Barker's business. This is emphasized in the 1883 *Review of Commerce, Manufactures, and Resources in Springfield*, which notes that "an important element of the business of this company is supplying gasolene for use in its machines, which trade alone amounts to several thousand barrels annually." Even with increased competition and the loss of their patented "improved arrangement," the company's sales continued to increase in the late 1870s and early 1880s, and by 1883 there were reportedly over three thousand Springfield Gas Machine units in operation.[102] In April 1883, the R. G. Dun and Company representative noted

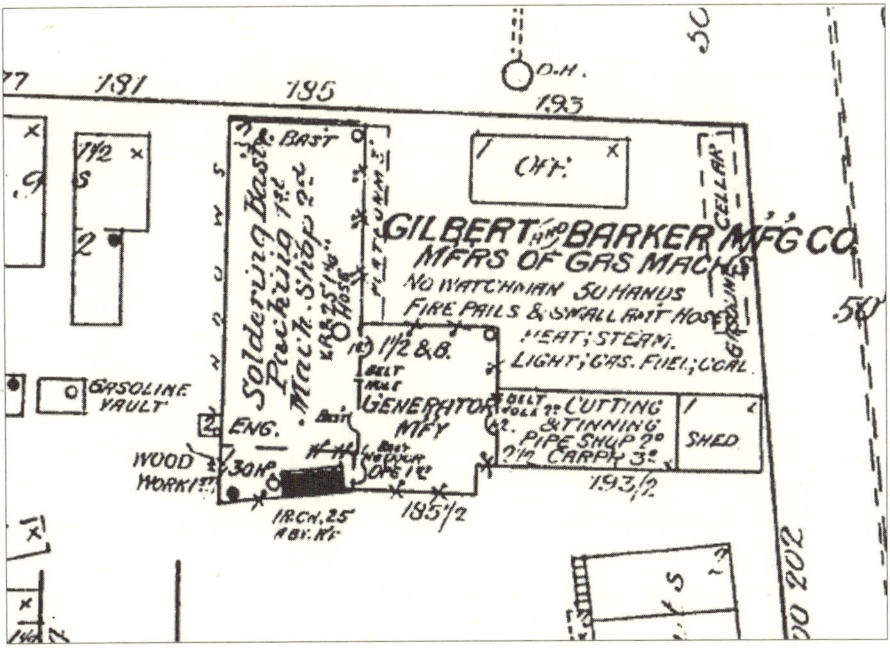

Figure 2.4. Sanborn Fire Insurance map showing Gilbert and Barker factory. Sanborn Library. Sanborn Fire Insurance Company, Springfield, Massachusetts, 1886.

of the firm: "It is generally understood that this company are doing a successful business and their credit is never called into question." After a follow-up visit in October 1883, the Dun's reporter asserted that as a "money making concern [Gilbert and Barker] is first class credit."[103]

Gilbert and Barker further expanded its manufacturing plant on Lyman Street as success of its products grew. By 1883, the plant included a "three story brick building 40 x 70 feet in dimension, with a one and one-half story 'L' 40 x 40, and a three-story frame building 20 x 80 feet, with boiler and engine room attached."[104] This description is confirmed by the 1886 Sanborn Fire Insurance map of the company's factory property, which indicates that the three-story brick building contained soldering operations in the basement (most likely for assembling the machines), a packing area on most of the first floor with a small woodworking shop at one end, and a machine shop on the second floor; a 30-horsepower engine is shown at the east end of the building. The one-and-a-half story "L" attached to this building contained the firm's generator manufacturing operation. The three-story frame building held the cutting and tinning operation, related to creating parts for the air pump and generator, on the first floor, the pipe shop on the second floor, and the carpenter shop on the third floor; a small one-story shed was attached to the end.[105]

The works reportedly had "special and general iron-working machinery, [with] employment being furnished to about fifty skilled artisans." The Sanborn map shows a gasoline vault to the south of the three-story brick building, which almost certainly contained the gas machine used for soldering the gas machine units and barrels. A large, below-ground "gasoline cellar," apparently used for storing bulk gasoline for sale, stood at the corner of Spring Street and Lyman Street, some distance from the other buildings. Finally, a small one-story building housed the company's office fronting Lyman Street.[106]

The company continued to have sales offices in New York City (in a new location at 75 Maiden Lane) and Boston (a new location at 51 Union Street) and sales agencies in Philadelphia (George W. Hulme, 12 N. Seventh Street) and San Francisco (A. F. Nye and Company, 609 Market Street), clearly taking advantage of the concentration of commercial and business activity in these cities as well as the many businessmen with country homes and estates that worked in these firms. The advantage of their East Coast market is underscored by Walter Licht, who writes in *Industrializing America* that "New York and Philadelphia more than quadrupled and doubled their populations, respectively, during the period and continued to lead the nation in industrial output."[107]

Gilbert and Barker had retailed gasoline to its customers from the start, selling bulk gasoline and storage barrels, shipping gasoline via rail and drovers. This proved an important business strategy for the company, one that would have a lasting impact on its later development.[108] The Underground Gas Machine Company and the Excelsior Gas Machine Company, neither of which advertised the sale or supply of gasoline to their customers, simply went out of business as the market for gas machines evaporated.[109] It was Gilbert and Barker's expertise related to handling and storing gasoline that proved critical for its eventual entry into the market for service station equipment as sales of the Springfield system declined.

Gilbert and Barker Manufacturing Company was, by 1884, one of the largest retailers of gasoline in the United States. Gasoline, the first of the new petroleum byproducts, became commercially available by about 1863. Gasoline was distilled or refined from naphtha, another fuel used in early gas machines. Naphtha itself resulted from the first distillation of crude petroleum to produce kerosene and other heavy lubricating oils. The further distillation of naphtha or the light fractions produced benzene, refined naphtha, and gasoline.[110]

Oil industry historians Williamson and Daum note that most refiners derived little profit by producing refined naphthas prior to the Civil War. After using benzene, the heaviest distillate, for burning oils, they would throw out or give away the other distillates, including refined naphtha and gasoline. When the Civil War cut off supplies of turpentine from the South, northern companies began to use benzene as an alternative solvent in many products, chiefly paint and varnish, stiffening agents for hats, and furniture polish. Following the war, the lighter byproducts of naphtha, refined naphtha and gasoline, suddenly found uses as solvents. The rapidly growing production of kerosene from crude, however, soon produced a huge surplus of naphtha byproducts. With supply outstripping demand, these byproducts had little or no value, and some unscrupulous producers used them to cut or dilute the kerosene that they sold as lamp oil. The resulting mixture greatly increased the risk of explosion and fire, exacerbating well-placed fears about using these new petroleum fuels for lighting.[111]

By 1861, manufacturers had discovered a whole new market for naphthas and other light fraction byproducts: supplying devices, like the Springfield Gas Machine, that evaporated volatile oils into air-gas. Initially these machines used mostly the heavier benzene and naphtha, but these fuels did not evaporate as completely or quickly as desired. Moreover, they had a tendency to condense, affecting the quality of the light and potentially ejecting liquid fuel from the burner. Joshua Merrill, an employee of the Downer Kerosene Oil Company whom Williamson and Daum call one of the "outstanding technological innovators of his generation," appears to have first developed the procedures to refine gasoline from petroleum naphtha.[112] Thereafter, gasoline, the lightest and most volatile of the naphtha group, became the fuel of choice for portable gas machines. Given gasoline's higher volatility and lower boiling point, compared to benzene and naphtha, many considered it less subject to condensation and thus a better choice for evaporative gas machines. Gilbert and Barker reported that gasoline "is highly volatile, having such strong affinity for air that when brought in contact with a current of [air] at normal temperature, the entire mass passes readily to a vapor, leaving no residuum behind."[113]

In 1863, Joshua Merrill began commercial production of gasoline at the Downer Kerosene Oil plant in Boston, reportedly "for the largest manufacturer of air-gas machines." This referred not to Gilbert and Barker, who had yet to begin work on their machine, but perhaps to the maker of the Drake Automatic Gas Machine, patented in 1862. At first, shipping the volatile product posed some problems and required the use of "the stoutest beer barrels" to limit evaporation during delivery. The production of gasoline, however, grew rapidly with the success and spread of air-gas machines across the country in the 1860s. Williamson and Daum note that while the use of gasoline for air-gas machines didn't absorb the "entire output of light fractions from petroleum refining, [it] did provide numerous small refiners . . . a profitable specialty business." As the decade ended, the firm of Charles Pratt and Company of New York, later an affiliate of Standard Oil of New Jersey, had "reputedly become the largest refiner of gasoline in the world."[114] Thus it is not surprising that Pratt purchased Gilbert and Barker in 1884. The several thousand barrels of gasoline reportedly sold by Gilbert and Barker in the late 1870s had grown to an amazing twenty-five thousand to fifty thousand barrels, or 1,050,000–2,100,000 gallons, by the late 1880s.[115]

By 1873, John D. Rockefeller was using Standard Oil stock to buy up as much oil refining capacity as possible. Licht reports that in "1878, the firms who had joined with him accounted for more than 80 percent of the refinery capacity in the nation." In combination with his secret deals with railroad carriers to ship oil at lower rates than his competitors, Rockefeller essentially controlled the price of crude, the cost of transportation and the price of refined products.[116] This stranglehold on the industry was formalized in January 1882 with the creation of the Standard Oil Trust.

In August 1882, the Standard Oil Company of New Jersey was incorporated by H. M. Flagler, Paul Babcock Jr., James McGee, Thomas C. Bushnell, and the trustees of the Standard Oil Trust. The new company sold chiefly kerosene, along with naphtha and benzene and a range of lubricating oils; for example, their Bayonne, New Jersey, refinery reported annual production of 77.41 percent kerosene, 7.37 percent naphtha, 7.54 percent tar, and 7.68 percent loss. In 1882, the combined production of oil in all of Standard's plants reached a staggering 15 million barrels, or 630 million gallons, and continued to grow rapidly.[117]

With production on the rise, Standard Oil focused its efforts on distributing and marketing its products, organizing several new companies for wholesaling and beginning retail operations. The company's executives worked hard to increase the sales of the firm's many petroleum products; for example, they began to distribute "heaters, stoves, lamps, lanterns, irons, and other utensils, usually at little or no profit, as a means of expanding the consumption of Standard Oil kerosene and gasoline." In 1883, the company formed the American Wick Manufacturing Company to fabricated wicks specifically for kerosene burners. As part of this strategy to expand the demand for their petroleum products, the Pratt Manufacturing Company, a Standard Oil of New York subsidiary, purchased Gilbert and Barker in 1884. Pratt acquired 306 of the 400 shares of Gilbert and Barker for $83,612.00. With the purchase of Gilbert and Barker, Standard Oil acquired "one of the largest marketers of gasoline, selling as it did some 75,000 to 100,000 barrels, or 3,150,000 to 4,200,000 gallons, per year, and a versatile manufacturing company producing articles associated with the storage and use of petroleum products."[118]

Reportedly selling his personal interest to Standard, Charles N. Gilbert retired from the business at age forty-one, and John F. Barker was made president.[119] One wonders if Gilbert's decision to retire wasn't influenced by the Standard Oil purchase; perhaps he disagreed with the sale or simply decided to take his share of the profit and pursue other interests.[120] The operation of the firm remained in the hands of John F. Barker and his management associates. The change in stock ownership did nothing to slow the growth of the company; in fact, R. G. Dun rated it "a prosperous concern always apparently easy in money matters and its credit is never brought into question." The company's profits increased from $34,450 in 1885 to over $50,000 in 1886, prompting Dun to note the company's continued "high standing and credit" (see table 2.3).[121]

Gasoline sales only increased with Gilbert and Barker's purchase by Pratt Manufacturing and Standard Oil. Gilbert and Barker sold gasoline in forty-two-gallon wooden or steel barrels and generally it shipped by rail and/or express delivery company. For example, bills of lading for shipping gasoline from Gilbert and Barker's Boston office to the Ames family in North Easton indicate that it cost about forty-two cents per barrel on the New York, New Haven and Hartford Railroad Company in 1912.[122] The company charged a refundable deposit on the barrels. Gasoline in wood barrels cost slightly more than in steel barrels; in 1911, prices stood at twenty-one cents per gallon in wood and twenty cents per gallon in steel.

After its acquisition by Pratt, Gilbert and Barker continued to expand its efforts to sell gas machines to a variety of industries, running ads in trade journals such as the *Paper World* urging industrialists to "light your paper mill with a Springfield Gas Machine." The canning industry also adopted the Springfield system, which allowed it to use a low cost fuel for soldering cans in canning factories. In addition to new commercial applications for the system, the company again expanded its domestic market by adding household appliances like stoves and hot water heaters to its product line. These were exactly the types of products produced by Pratt Manufacturing and perfectly met Standard Oil's desire to create demand for its oil products. As noted in a *Springfield Sunday Republican* article, with a "gas machine a family may light and heat the house, cook their food, wash their clothes and have hot water in every sleeping-room and bath-room."[123] This new line of appliances and the new uses for the machine beyond just lighting boosted sales of the system to owners of country houses and estates.[124]

An article on Gilbert and Barker in *Picturesque Hampden,* a county business almanac and history, indicates that by 1892 the company had sold and installed nearly twenty thousand Springfield Gas Machine systems. The publication notes that "a modern dwelling is no more complete without a machine of this character than it would be without a modern furnace, as compared with stoves for heating or without a system of plumbing as compared with the ordinary pump." By 1892, the company's factory and works on Lyman Street had expanded to over thirty thousand square feet, included a powerful steam plant, and employed one hundred men.[125] The company's work force had doubled between 1886 and 1892, further attesting to its rapid growth. The Sanborn map of 1896 illustrates the continued growth of the company's facilities, showing a 25-horsepower steam plant, or engine, in the east end of the firm's three-story brick building, which now had a new attached shed for storing lumber. The gasoline vault shown on the 1886 map has a one-story structure over it that included a furnace; the building has the notation "gasolene experimenting" and clearly served as a research and development facility for gasoline products. With the exception of the first floor of the three-story brick building, now labeled as "packing and painting," and the shed addition to the three-story wood frame structure, now labeled "pipe and lumber storage," other operations remained the same. The office received the largest expansion of any part of the facility during the period, doubling in size with an addition to its eastern facade.[126]

Gilbert and Barker obviously increased both its work force and facility in response to the widening variety of its clients. New manufacturing concerns that used the Springfield system beginning in the mid-1880s included the Hampden Watch Company, the Overman bicycle company, the Edison Machine Works, and, ironically, the Edison Lamp Company. The company also continued to expand its product line, becoming agents for appliances sold by a variety of Standard Oil–owned companies. In 1892 alone, the company reportedly sold four thousand stoves in New England. Gilbert and Barker also sold "Vapor Gas Stoves," a line of stoves very popular in the "western part of the country." Having sold over fifty thousand barrels, or 2.1 million gallons, of gasoline in 1892, one commentator effused that their "brand of gasolene for gas machines, known as 'PURE SPIRITS,' has acquired a reputation seldom enjoyed. It is manufactured expressly for them and they constantly guard its purity and quality."[127] The company never failed to remind its customers in its literature that "our 'PURE SPIRITS' Gasolene is the best. Use no other." Not surprisingly, by the turn of the century, Gilbert and Barker served as a dealer for "Pratt's Patent Prepared" gasoline, a product of Charles Pratt and Company's refinery.[128]

In the late 1880s and early 1890s, a lawsuit was brought against the Standard Oil Trust for its alleged stranglehold on petroleum production and sales, as well as its control over related manufacturing establishments. The public saw the company as a "vicious, venal menace" and a "dangerously powerful monopoly." In 1892, the Supreme Court of Ohio found that the Ohio Standard affiliate of the Standard Oil Trust was a monopoly controlled and managed by the Standard Oil Trust; the court forbade Ohio Standard to remain a part of the trust. As a direct result of this decision, Standard dissolved the trust and reorganized its combination of interests under the control of Standard Oil Company of New Jersey or Jersey Standard. Jersey Standard bought the stock of Gilbert and Barker from the Pratt Manufacturing Company during the reorganization in 1892, so that Gilbert and Barker became a direct affiliate of Jersey Standard. The changes in Standard Oil seem to have done little to lessen its considerable influence in the petroleum industry. The firm's kerosene sales rose from 9,551,000 barrels in 1895 to over 11,000,000 barrels in 1899; likewise, sales of naphtha and gasoline (primarily for lighting) rose from 2,639,000 gallons in 1895 to 3,270,000 gallons in 1899, a 24 percent increase.[129] Gilbert and Barker's popular Springfield Gas Machine contributed significantly to these expanding sales and the firm's subsequent expansion into the manufacture of fuel-oil equipment further boosted Jersey Standard's sales.

In 1897, Gilbert and Barker added a new type of installation and several new products to the line. The company continued to manufacture and market its three styles of air pump: weight-driven, water wheel–driven, and hydraulic, or suction, blower. The partially buried and buried gas generator types continued to be sold, as did the Springfield Mixing Regulator. An 1897 pamphlet, *Directions for the Care of The Springfield Gas Machine*, illustrated several new products introduced in the late 1880s and early 1890s. The Combination Mixing Regulator, a completely automatic or "self-governing" mixing unit, appeared circa 1888. The new regulator clearly outperformed the company's earlier mixing regulator. For example, customer Henry Kibbe, who had used a mixing regulator with his Springfield system for twelve years and had recently upgraded to an "Automatic Mixing Regulator," wrote in 1906 that the new device "has worked to our entire satisfaction, furnishing uniform gas without care or attention, which we think is a great improvement over our experience in the twelve years noted above."[130]

Gilbert and Baker's "New Feed" Springfield Gas Machine, introduced about 1894, combined the buried gas generator with a large buried storage tank. Used with the hydraulic blower or water-driven air pump, this system required little attention and was marketed as virtually maintenance free. This type of system would automatically shut down and require no attention until needed, making it ideal for seasonal camps and country homes. The company also continued to experiment with and refine their gas burners. The firm advertised its Argand-type mantle burner and later Springfield Incandescent Burner as extremely efficient and effective, producing a light that compared favorably with electric light.[131]

From the late 1880s to early 1900s, Gilbert and Barker marketed its products and services directly to upper-middle- and upper-class country house and suburban home owners, as well as to the major resorts and hotels spawned by a burgeoning tourist industry. In 1888, and again in 1906, the company published a catalog titled *How Best to Light our Country Homes and Resorts,* featuring the country homes of the era's rich and famous.[132]

By 1906, the firm was constructing its air pumps and gas generators entirely of "galvanized Bessemer steel" and the mixing regulator of cast iron. The product line remained essentially unchanged after 1897. The 1906 catalog focused attention on the Combination Mixing Regulator, heralding it as fully automatic, requiring "no attention whatever; once set up right, it may be left entirely to the operation of nature's unchanging laws." Safety continued as the watchword of the company, which asserted that it installed its machines "in exact accordance with the Rules of the New York Board of Fire Underwriters as revised." Domestic uses for the system continued to expand, and by the end of the nineteenth century, the company began producing a separate illustrated catalog of "gas ranges, broilers, laundry iron heaters, gas logs, gas stoves, water heaters, and gas radiators." In addition to the cooperative neighborhood installations reported in earlier catalogs, the company now explained that the system operated in "village plants, most of them regularly incorporated gas companies," which used Springfield machines to produce air-gas for their customers.[133]

By the turn of the century, however, the company clearly began to feel the competition of electric light. The firm had ironically installed systems in the Edison Machine Works, the Edison Lamp Company, and even Thomas Edison's personal "experimenting laboratory" at Llewellyn Park, New Jersey.[134] Electric light enjoyed increased popularity by the first years of the twentieth century and customers gradually began to switch from gas to electric lighting. To mitigate undependable early electric systems, manufactures developed combined gas and electric fixtures so that customers could light their homes with gas when electric sources failed.

While Gilbert and Barker initially offered a product that was cheaper than electricity, the new energy source had advantages that the public quickly recognized. Electricity provided a simple and maintenance-free energy source with minimal installation costs, while gas machines like the Springfield required high initial investments for individual property owners and needed regular maintenance. The automatic and self-contained nature of the Springfield Gas Machine system, however, allowed it to compete with electricity into the 1920s and 1930s, particularly in rural areas not yet served by the electrical grid.

Although the sales of Springfield Gas Machines dropped in the first quarter of the twentieth century, a newspaper article from 1940 reported that "many Gilbert and Barker gas machines are still in use in remote sections of the country. Many of these machines are 30, 40 and even 50 years old."[135] The company celebrated the "old indestructible," a gas machine in the Monson Academy building in Monson, Massachusetts, that operated from 1889 until 1953, a span of time that witnessed extraordinary changes in American technology and society. In 1889, the machine and the two Gilbert and Barker installers arrived at this small Massachusetts village via train and horse-drawn stage. The machine originally lit the entire building and provided fuel for laboratory burners; although electricity lit the building for much of the twentieth century, students continued to use the lab burners supplied by the machine until a fire, unrelated to the Springfield system, destroyed the building in 1953.[136]

As will be explored in chapter 6, while Gilbert and Barker continued to manufacture and market the Springfield Gas Machine until about 1916, the company had begun to make changes by the late 1880s that positioned it to survive the end of the gas machine era. Cowan notes that by the end of the century, "petroleum refiners had come to realize—as had the gas companies—that the handwriting was on (or, rather, in) the wall: electricity was going to erode the market for kerosene fairly quickly."[137] Thus in 1889 the firm began to manufacture fuel-oil burning equipment in an effort to expand the market for Standard Oil's fuel-oil products.

Fuel oil, a low-viscosity, flammable liquid petroleum product, was increasingly used for various types of residential and industrial furnaces. In 1910, Gilbert and Barker introduced the "first practical gasoline pump" for the growing automobile industry.[138] Both of these new product lines no doubt resulted from Standard Oil's efforts to offer new products that utilized and expanded demand for their petroleum products.

CHAPTER 3

The Springfield Gas Machine

From the very beginning, the Springfield Gas Machine Company, later Gilbert and Barker Manufacturing, focused on providing customers, particularly those beyond the reach of city gas works, with good, safe, and relatively inexpensive gas light.[1] This approach, combined with constant innovation and marketing savvy, ensured the company's remarkable competitiveness and success over the years.

The company's basic Springfield system consisted of two principal components: a weight-driven air pump placed in the building to be lit and a gas generator located in a vault some distance from the building. The air pump, powered by a suspended, concrete weight that was wound periodically like a clock, stood in the basement of the house or workplace. This unit forced air through a pipe to the generator placed at a safe distance, at least fifty to seventy feet, from the building.[2] Pressure from the pump moved the air-gas mixture back to the building and distributed it through pipes to light fixtures. Ingeniously, the system operated on demand. When all burners were turned off, the air pump automatically stopped once pressure in the system equaled the pressure of the descending weight; when a burner was opened, the pump, driven by the slowing dropping weight, would restart to maintain constant pressure in the system. This chapter explores the technical development of the ingenious Springfield system and its various components from their initial design and introduction through their refinement as the technology evolved.

It appears that Charles Gilbert and John Barker experimented with several approaches to a portable gas machine design during their first few years in Springfield. The exact sequence of events is a bit cloudy, but a general order of the machine's early development can be gleaned from the company's catalogs, U.S. Patent Office records, and newspaper accounts. Examination of these records makes it clear that Gilbert and Barker, much like Edison and the lightbulb, didn't invent the gas machine so much as improve on existing models to create a commercially viable product.[3]

The company's "first Springfield machine," consisting of a weight-driven air pump and gas generator in one fixed unit, was illustrated in its 1874 catalog. The unit was "constructed in the manner we have described. A, being the air pump, B,

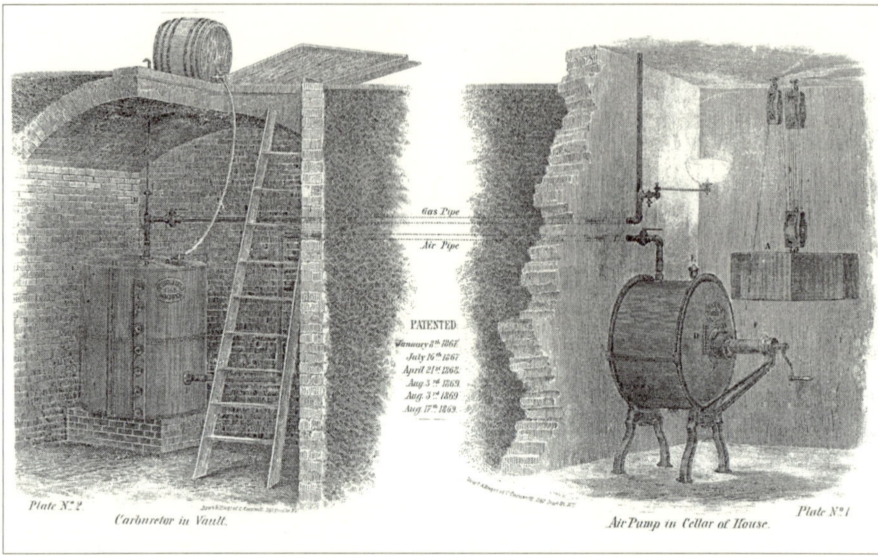

Figure 3.1. Springfield Gas Machine, carburetor in vault and air pump in cellar of house. From Gilbert and Barker Manufacturing, *Springfield Gas Machine,* 1870–71. Library of Congress.

the gas generator, both were set upon one iron frame and connected together."[4] This arrangement, combining and generator and air pump in one unit, was fairly typical of gas machines then in operation. For example, the Empire State Gas Machine, dating to the mid-1860s, was very similar in its design and construction to the first Springfield machine, combining a weight-driven air pump and gas generator in one compact unit.[5] According to an 1867 *New York Times* article, the original Springfield system employed "the old English naphthalizer applied to coal gas." The apparatus "was crude, irregular and uncertain, and refrigerated at the most rapid rate. Underwriters were intolerant of it."[6]

Gilbert and Barker later acknowledge that "so arranged," [the machine] was open to all the objections above enumerated." These objections related to both safety and ease of operation. To avoid having quantities of liquid gasoline in the structure, the machine had to be placed in a building other than the one to be lighted. However, placing the machine in an outbuilding created several operational difficulties. The water in the air pump would freeze during winter months in an unheated building, rendering the machine inoperable. Cold weather could also adversely affect the evaporation rate of the liquid gasoline. The remote placement of the unit also made it inconvenient to tend in bad weather. In addition, reported the inventors, the unit's iron and zinc construction made it subject to

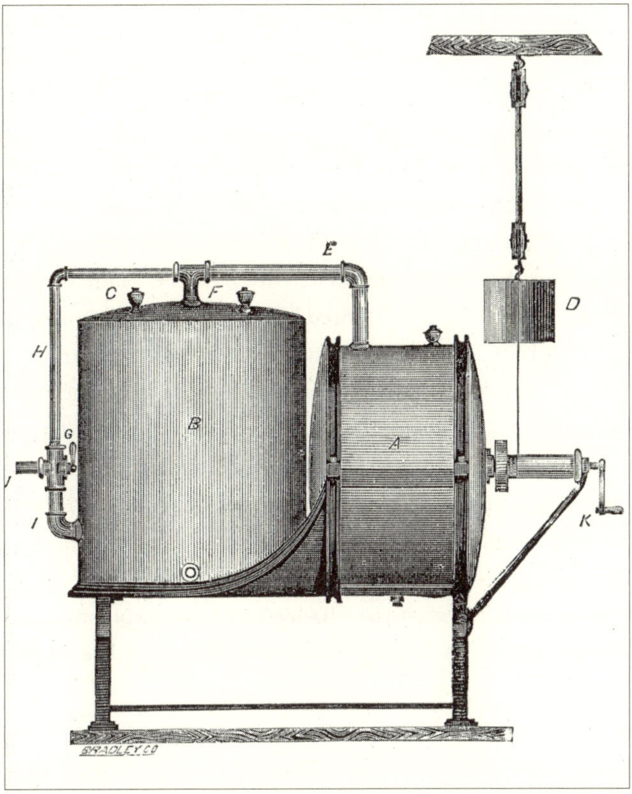

Figure 3.2. The Springfield Gas Machine as originally constructed. From Gilbert and Barker Manufacturing, *Springfield Gas Machine,* 1874. Author's collection.

rusting and corrosion, particularly in an unheated outbuilding. The company found these problems "so formidable . . . that we despaired of effecting its general and successful introduction, until we discovered and perfected our improved arrangement."[7]

The company, reorganized as the Springfield Gas Machine Company, arrived at this "improved arrangement" about 1867 or 1868. Patent information and the company's catalogs indicate that this system used a cold process that relied principally on evaporation rather than heating of the fuel (either naphtha or gasoline); however, several early patents included an auxiliary heating system to speed evaporation in larger units. Importantly, the design of Springfield units from 1867 onward called for separating the gas generator and its bulk naphtha or gasoline fuel from the structure to be lit and containing it in an underground chamber.

Gilbert and Barker's first patent for a gas machine was granted on January 8, 1867. In this early system, a series of stacked evaporating pans or chambers sat within the round, tanklike generator; small, separate window gauges were included to monitor the levels of liquid naphtha within the machine. This arrangement of several stacked evaporator pans, known as a compound gas generator, greatly increased the total evaporating surface of the machine over other systems that used a single evaporator reservoir. According to the patent, the gas generator was installed in a "fire-proof and gas-tight chamber, made of brick" that was "detached from the building to be lighted." The generator chamber was accessed via a manhole, "should it become necessary at any time to repair the generator or readjust the piping." The detached generator chamber could, the inventors noted, be heated "by means of a radiating box or pipe passing through the chamber outside the generator" to warm the generator and speed evaporation. Also outside the building, note the inventors, "we place a fire-proof tank-house, in which is placed a tank for storing the naphtha, and connected with it is a force pump, by means of which we force the naphtha through proper pipes to the reservoir of the generator."[8] This arrangement lessened the chance of fire or explosion and eliminated the principal safety concerns of insurance companies and the increasingly wary public.

A patent by Gilbert and Barker dated July 16, 1867, suggests that the initial arrangement for heating the generator chamber in the previous design was not entirely satisfactory. In this new patent, the inventors added a heating coil apparatus to the generator unit to aid in the evaporation process. This was accomplished by "means of a heated fluid passing from a heater through suitable pipes and radiators placed within the carburetor, so that the naphtha or other hydrocarbon oil contained therein shall be heated by the radiation of the heat from the fluid passing through the pipes and radiators instead of being heated by direct circulation of the naphtha itself." The purpose of adding heat to the system was to counteract the evaporative cooling that occurred as the temperature of the liquid, giving up energy in the change from liquid to gas, grew ever cooler, a reaction that slowed the evaporation process and affected the quantity and quality of the gas. As the inventors noted, the "naphtha is continually growing colder in proportion to the rapidity of the evaporation and the temperature of the surrounding atmosphere, and the fluid, therefore, vaporizes less rapidly; and unless the lost heat can be restored, or in other words, unless the fluid in the carburetor can be kept at a uniform temperature, the quality of the gas made or enriched is variable."[9] The company reported in its 1870–71 catalog that the cooling problem caused more difficulties for large machines in industrial and institutional settings. Gilbert and

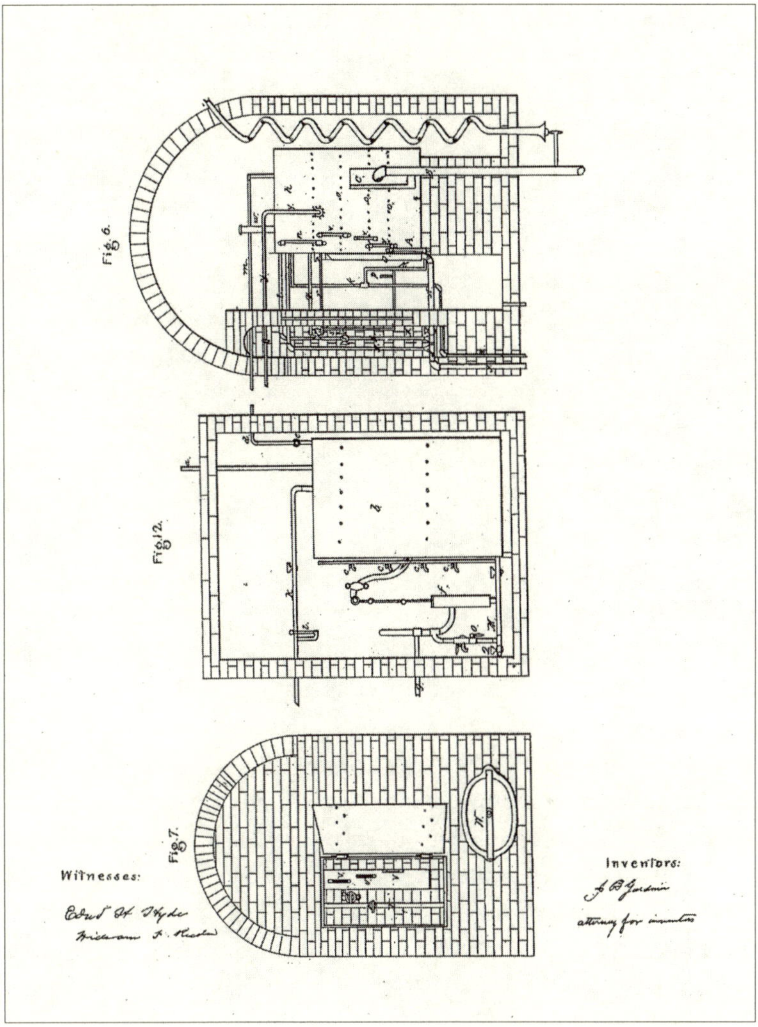

Figure 3.3. Patent No. 61,004 for carburetor by Gilbert, Barker and Ives, January 8, 1867. U.S. Patent and Trademark Office.

Barker's standard compound generator unit for lighting homes and small businesses apparently had enough volume of fuel and evaporating surface for evaporation to suffice.[10]

While the heating process described in the 1867 patent allowed for the maximum absorption of vapor into the air, the subsequent cooling of the warmer air-gas mixture in the iron piping system resulted in some of the naphtha or gasoline reverting to liquid; this liquid could be ejected from the burners and potentially cause fires. To counteract this problem, the patented generator heating system also included a condenser unit to lower the temperature of the air-gas "for the purpose of condensation, so as to remove or send back to the carburetor the surplus naphtha or other hydrocarbon oil, which would otherwise collect in the pipes or be thrown out of the burners."[11] Typically, manufacturers solved this condensation problem by angling the supply pipes from the generator to the building so that the pipe dropped in elevation from the house to the generator; thus condensed liquid fuel would safely drain back into the generator unit. Gilbert and Barker noted that "the whole system of piping should tend to decline toward the main rising pipe, and be free from low places or traps, and be proven with a pressure of air, under a test gauge, to be perfectly air and gas tight."[12]

The heat for the firm's newly devised radiator system came from a "benzine-stove with a gas burner." The heat source warmed a liquid-like glycerin that circulated in the heat coil within the actual generator unit. The inventors note that "to guard against danger from fire, we place the carburetor A in a small apartment by itself, and detached from the building to be lighted, and place the heater B outside this apartment containing the carburetor A, with the pipes . . . passing through the partition from one vessel to the other, so that the two vessels shall each be in separate compartments."[13] An August 1869 patent laid out additional improvements to the heating system for the generator described in the company's July 1867 patent; the company retained the heating system for use with very large systems that were continually troubled by evaporative cooling. The principal modification to the system involved the addition of a reservoir to provide for the "necessary space or room for the expansion of the circulating fluid" in the heater unit.[14] In trial and use of the earlier machine, Gilbert and Barker discovered that the heated fluid needed more space to expand than that provided by the circulating pipes alone. The company's 1870–71 catalog notes that for large machines it "make[s] provision for heating the carburetor, by means of a copper pipe running through each chamber, through which either steam or hot water is circulated. We build our machines arranged either with or without this heating pipe."[15] By 1874, however, the catalog no longer mentions this system, suggesting that Gilbert and Barker had eliminated the heating component; in fact, the 1874 catalog boasts that "no fire or heat is used" to run the machine.[16]

The firm made several other improvements that it retained in later versions of its gas generator, including the addition of valves to drain fluid from one evaporator chamber or pan to another and a new design for a single gauge to monitor the fluid levels in each of the evaporator chambers. The valves served to move heavy,

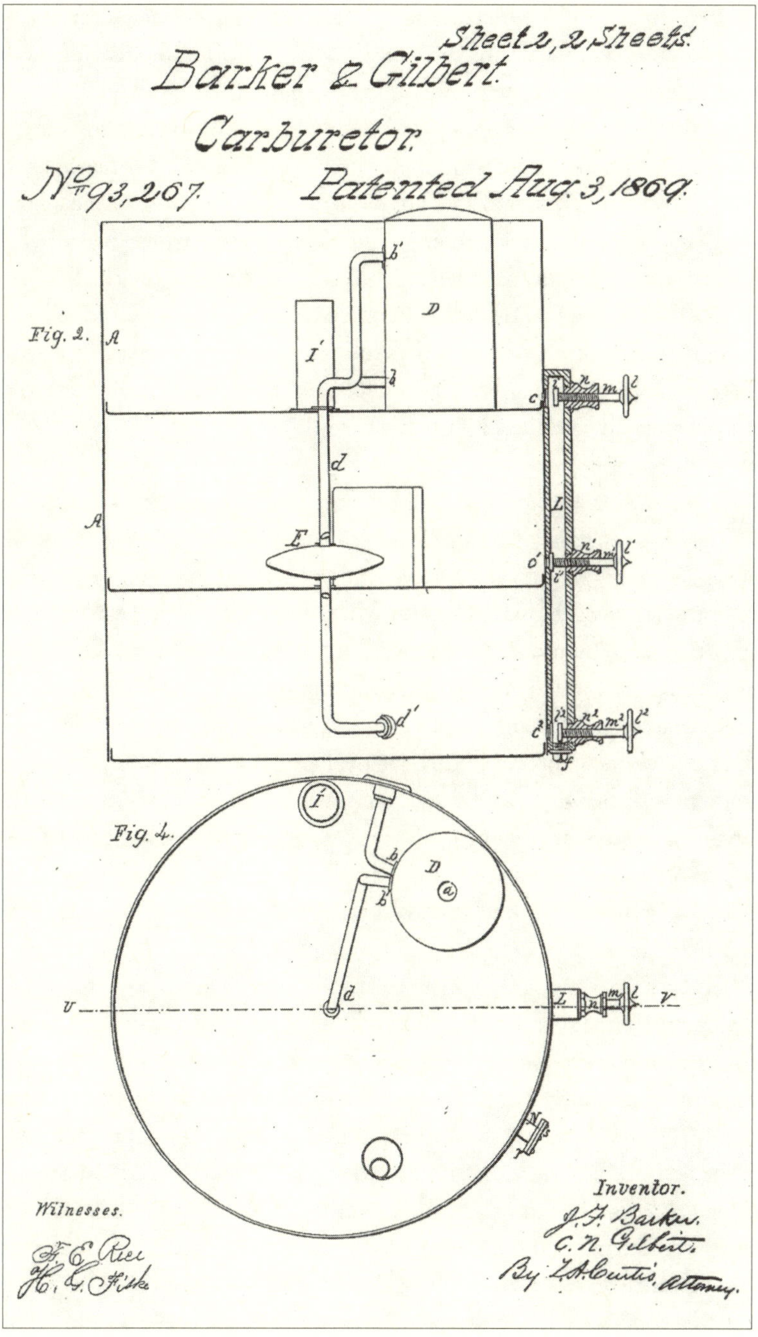

Figure 3.4. Patent No. 93,267 for carburetor, sheet 2 of 2, by Barker and Gilbert, August 3, 1869. U.S. Patent and Trademark Office.

less volatile oils or residual hydrocarbon fuels that did not evaporate as quickly to the chamber at the bottom of the unit. In this position, the heavier fuel mixed with pure air coming from the pump and thus evaporated more quickly and completely. Additionally, in machines with heating units, the lower pans sat closer to the heated coils, thereby increasing the evaporation rate of less volatile fuel; these valves also allowed for the draining and cleaning of the entire machine as needed.[17]

Using this approach, the company argued, the system would use all of the liquid gasoline and would always have a good, rich mixture of air-gas, and draining the heavier, less volatile fuel from the machine would not be required. The August 1869 patent first mentions the use of gasoline in the generators; all previous patents referred to the use of naphtha as a fuel. Since gasoline is the lightest and most volatile petroleum byproduct, its use also would improve the evaporation rate of the fuel in the machine.

In August 1869, Barker and Gilbert received a patent for an air pump device for "driving air through a carburetor, to the burners." This unit consisted of "the arrangement and construction of a shaft of what is known as a meter-wheel, with its connecting parts, whereby the wheel, when made to revolve, during its ordinary operation, by means of a descending weight, shall be made to continue to revolve while said weight is being wound up, by means of a spring, in connection with suitable ratchets and pawls."[18] The invention was borrowed from earlier municipal gas plants, which used meter wheels to measure gas usage; the gas pressure itself turned the wheel and drove the recording device, with each revolution representing a fixed quantity of gas. The inventors noted that a meter wheel–type air pump, which reversed the process—driving the shaft of the wheel in order to pressurize and pump air—was already a standard device for forcing air through gas machine systems.[19]

Gilbert and Barker improved the standard meter-wheel, air pump system by adding a spring and ratchet mechanism. The patent notes that the spring kept the pump turning even during the winding process, which eliminated the inconvenience and even danger of gas lights going out during winding. In previous systems, the inventors warned, "the wheel would instantly stop its rotary motion as soon as the operation of rewinding the cord was commenced, and the supply of gas at the burners would be cut off, as the wheel would cease to force any air through the carburetor during the operation of rewinding the cord, and the lighted gas would be instantly extinguished."[20] The early models of the company's air pump consisted of a copper shell or case that enclosed the wheel of the pump.[21] The 1874 catalog explains that

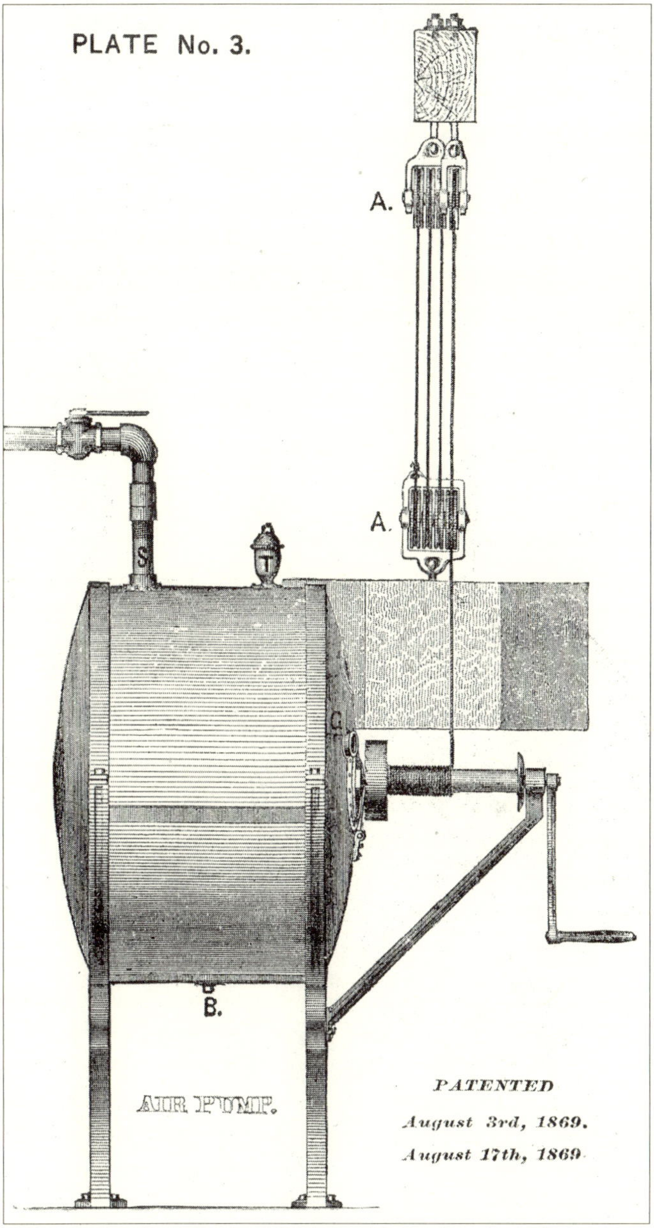

Figure 3.5. Springfield Gas Machine air pump. From Gilbert and Barker Manufacturing, *Springfield Gas Machine,* 1874. Author's collection.

> the wheel of the pump is divided longitudinally into four spiral compartments. . . . When the pump is set in place, ready for operation, it is filled with water. . . . Air enters through the openings, filling all those compartments or such portion of them as are above the water line; as the wheel, actuated by the weight, revolves, the lips of these several compartments pass under the water, thus sealing the air within and preventing its escape, except through the opening in the rear of the pump and the air pipe. . . . As one compartment is emptying, another is filling. . . . Thus . . . a constant pressure of air through the air pipe, gas generator, and upon the burners in the house is maintained.[22]

The 1870–71 catalog describes the firm's "Air-Forcing Apparatus," designed "especially for large mills, factories, and establishments where water power is ob-

Figure 3.6. Air-Forcing Apparatus, Springfield Gas Machine. From Gilbert and Barker Manufacturing, *Springfield Gas Machine*, 1870. Library of Congress.

tainable." This system consisted of a small water turbine that drove a rotary fan blower; a gearing system and belts connected the turbine to the fan blower. A two- to three-inch water supply pipe, the company reported, could "operate this wheel and furnish gas for a thousand burners."[23]

By 1871, customers could use a hydraulic blower to power their system instead of the weight-driven air pump. The blower, like the air pump, usually sat in the basement, but it could also be placed in the wheel pit of a mill or at the most convenient supply of water.[24] The company describes the hydraulic blower as consisting "of two copper vessels, one inverted in the other, to which are attached pipes. . . . The injection pipe . . . brings water from the flume, or whatever the source of supply may be, to the blower. This pipe is fitted with the nozzle . . . through which the water rushes, falling through the larger pipe. . . . In so doing,

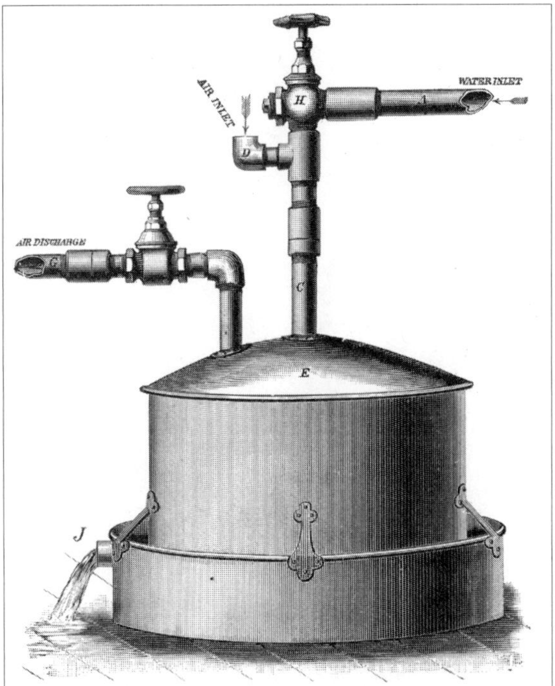

HYDRAULIC BLOWER.
Patented June 6, 1871.

Figure 3.7. Hydraulic blower, Springfield Gas Machine. From Gilbert and Barker Manufacturing, *Directions for the Care of the Springfield Gas Machine,* May 15, 1897. Author's collection.

air is drawn in at the suction pipe. . . . Both water and air fall together into the inverted vessel . . . where the air is confined. The water passes off through the discharge pipe. . . . Through the [distribution] pipe . . . AIR is conveyed to the gas

Figure 3.8. Water-powered air pump, Springfield Gas Machine. Gilbert and Barker Manufacturing, *Directions for the Care of the Springfield Gas Machine*, May 15, 1897. Author's collection.

generator."[25] "With this apparatus," Barker wrote, "a constant and steady flow of air through the carburetor is obtained."[26] Thus, given a continuous flow of water, this air pump could operate with little attention; the elimination of all moving parts made this a largely maintenance-free component.

In 1878, the company introduced its water wheel–powered air pump and began touting it as "our latest improvement in . . . air pumps" and later as "one of the most important distinctive features of the Springfield Gas Machine."[27] With this device, driven by a small stream of pressurized water directed onto an overshot water wheel attached to the side of the cylindrical pump case, Gilbert and Barker achieved a major innovation in gas machine systems.[28]

It appears that the firm bought the rights to an 1875 patent for this technology, an "Improvement in Automatic Air-blasts for Gas-machines," from Robert Hewson of California sometime between 1875 and 1878.[29] An 1890 flyer for Gilbert and Barker's Water Wheel Air Pump explains that the "novel feature of an automatic air-blast is controlled by us under Letters Patent to Robert Hewson. . . . No one, therefore, can construct or use an automatic air-blast run by water-power without infringing our rights under said patent."[30] Hewson wrote, "What I claim . . . is, not a water-wheel of any special construction," but one that cleverly employed a system of gears and levers. Hewson patented both the use of the water wheel to turn the air pump and the gas holder with shut-off valve that in combination allowed the machine to operate virtually "automatically."

Gilbert and Barker eventually patented their own version of this pump, similar but not identical to the one illustrated in the company's 1878 catalog. This Gilbert and Barker design, patented in November 1879, included a rather complicated lever and valve system to control the air pressure generated by the pump; the 1890 patent version of the pump eliminated this system altogether.[31] This new version of the pump employed a simple shut-off valve, very similar to the float valve in a toilet tank, which used a small counterbalance arm to adjust the pressure in the system. Interestingly, the air pump illustrated in the company's 1878 catalog appears essentially the same as the one illustrated in the 1890 patent. It is unclear whether the company actually built the version patented in 1879 and why it took them until 1890 to patent the model that it began selling as of 1878. The acquisition of Hewson's earlier patent may have eliminated Gilbert and Barker's need for a patent of its own before then; the 1890 patent also seems to include some minor improvements on Hewson's patent.

In general operation, as the water wheel–driven air pump pressurized the system, an expanding air reservoir or air holder located on the top of the pump would rise and move a lever or rod that would shut off the water to the wheel. When someone turned on the gas at the burners, the pressure in the system dropped and the reservoir would slowly descend and activate the lever to turn on the water, thus starting the air pump. Notes on an installation drawing for the Springfield system at Hill-Stead, a country house in Farmington, Connecticut, indicate that the top section of the air holder, called the float, held water to provide a seal and moved up and down to accommodate the system's changing air pressure. The lower pump chamber or meter wheel originally used water for the seal but later contained "parafine [*sic*] oil to a hight [*sic*] which will come up to about 3/4 of oil gauge."[32] The paraffin oil allowed for an airtight seal around the blades in the pump chamber and didn't freeze like water.

This air pump had many advantages over weight-driven pumps. The manufacturer reported that "it requires no attention; never runs down; is ready for use at all times; and, as it runs under a light power and low speed is more durable, and furnishes a more uniform power."[33] The automatic operation of this new pump and its extra reservoir of pressurized air-gas enabled the whole system to respond better to changes in the number of burners in use without any manual adjustments or change in quality of the light. For example, if someone lit six extra lights in a system already in use, the pressurized air-gas reserve in the holder would help compensate for the initial drop in pressure to the system and the potential drop in light quality or intensity; the use of this extra volume of air would also trigger the pump to start up and maintain system pressure.

The air pump, whether weight-driven, hydraulic blower, or water-driven, supplied pressurized air to the gas generator. The company describes its gas generator as

> well adapted for its work . . . and so arranged as to give the greatest possible evaporating surface to the action of the air. Strictly speaking, it is a compound gas generator . . . made up of a number of evaporating pans or chambers placed in a cylinder, one above another. . . . These chambers are divided by frames [over which] a suitable capillary material is stretched. The main gas pipe leading to the building to be lighted, is attached at K, the cocks . . . are for shutting off the pressure of air or gas, whenever desired. Supposing the generator to be set in position in the vault, and the pipes connected, it is then filled with gasolene through the rubber tubing and pipe from the barrel . . . according to the size of the generator. In opening the two way cock N, to allow the gasolene to run into the generator, it will be seen that the vent pipe is opened at the *same time*, thus affording a free escape for the air or gas that is displaced by the fluid, and

> preventing any strain upon the generator, or backward pressure of the gas upon the air pump in the cellar of the house.
>
> The gasolene, as it enters the generator, fills the uppermost chamber first, to the top of the overflow tube. . . . This tube allows of its passage to the next chamber below, this in turn filling, afterward the one below this, and so on. . . .
>
> The filling cock . . . then being shut, the apparatus is ready for use. Air, forced by the pump or hydraulic blower, enters the generator at B, passes over the fluid and through the meshes of fibrous capillary material, now thoroughly saturated with gasolene, back and forth through the subdivisions of this chamber . . . then up through the tube C, to the next chamber above, winding through this in a similar manner . . . and so on, until finally becoming thoroughly impregnated with the vapor arising from the gasolene, it is delivered a richly carbureted air gas, through the gas tube K, to the burners of the building lighted.[34]

The Springfield Gas Machine generator's unique design of stacked evaporator pans that articulated from one level to the next produced a large evaporating surface in a relatively compact unit; the use of a wicklike capillary material further improved its evaporative capacity. For example, the evaporating surface of a No. 1 Springfield system with four evaporator pans, each 34 inches in diameter, produced 62 square feet of evaporating surface. A No. 10 Springfield system with eight evaporator pans, 90 inches in diameter, produced 1,063 square feet of evaporating surface.[35] The size of the evaporating surface in the machine was directly related to the quality of the air-gas and the efficiency of a cold process gas machine. Systems with less evaporating surface could not produce a rich enough mixture, particularly as temperatures decreased due to evaporative cooling during operation.

By 1874, the company's catalog indicates that it had entirely eliminated the use of the auxiliary heating coil system and that even for the "largest establishment . . . the entire care required in making gas . . . is simply to keep the generator supplied with gasolene."[36] It appears that increasing the system's evaporative surface, adding the valve system to drain older fuel to the bottom chamber, and utilizing a new specially designed "adjustable" burner at least partially solved the problems associated with evaporative cooling and the variable mixture of the air-gas. Gilbert and Barker also significantly increased the evaporative surface of the machine between the late 1860s and mid-1870s by simply enlarging the diameter of generator and its evaporator pans. Even modest increases in the diameter of the

Table 3.1

Generator Specifications

Number	No. of burners	Capacity (gal.) (sq. ft.)	Evaporating surface	Diameter (in.)	Height (in.)	Evaporating pans (no.)	Size of air inlet (in.)	Size of gas outlet (in.)	Width when cased for shipping (in.)	Height when cased for shipping (in.)	Weight when cased for shipping (in.)	Length (ft.)	Width (ft.)	Depth (ft.)
1	15	70	62	34	27	4	¾	1	38	31	267	6	4 ½	7 ½
2	25	100	98	43	28	4	1	1 ¼	47	32	343	7	5	7 ½
3	40	150	133	48	29	4	1 ¼	1 ¼	52	32	440	7	5 ½	7 ½
4	75	200	177	51	35	5	1 ¼	1 ½	54	39	510	7 ½	5 ½	8
5	100	250	221	55	36	5	1 ½	2	59	39	580	8	6	8
6	150	300	266	60	36	5	2	2 ½	64	39	718	8	6 ½	8
7	200	350	310	60	42	6	2	2 ½	64	46	819	8	6 ½	8 ½
8	300	500	440	70	43	6	2 ½	3	74	46	1,213	9	7 ½	8 ½
9	500	750	665	85	45	6	3	4	89	53	1,680	10	9	9
10	750	1,200	1,063	90	58	8	4	5	94	65	2,210	11	10	10

SOURCE: Gilbert and Barker Manufacturing, *Springfield Gas Machine*, 1874.

machine resulted in significant expansion of the evaporative surface due to the multiple evaporator pan arrangement.

The Springfield Gas Machine system's fabrication entirely in copper also contributed to its quality and success. Gilbert and Barker reported that it began producing the machines in copper in 1867 after finding it more durable than the original zinc galvanized iron construction. The 1870–71 catalog reports that "our carburetors are made entirely of brass or copper, with copper pans within—and copper heating pipe (when used)." The dampness of the vault and impurities in the water used in the air pump corroded the zinc galvanized iron more easily so that early machines apparently didn't last more than a few years. Copper fabrication extended the life of the machine and increased its reliability but made it more expensive than other machines of similar design. Playing up its choice of top-quality materials, Gilbert and Barker advertised the Springfield system as lasting far longer and running more efficiently than its competitor's machines, saving customers money in the long run.[37]

About 1880, the company offered its gas generators in both copper and galvanized iron, although in both models, the air pumps continued to be manufactured entirely of copper. As of 1888, the firm built its gas generators exclusively of "heavily galvanized iron with every seam not only riveted like a steam-boiler, but massed in solder"; it continued to make its air pumps from copper. By 1906, the firm manufactured all of its gas generators using the new, more corrosion-resistant Bessemer steel.[38]

Gilbert and Barker offered a broad range of machines designed to suit installations in small and large buildings alike (see tables 2.1, 2.2, 2.4 and 2.5). Individual systems came in a number of configurations, allowing for a variety of installations. The Springfield could light any number of buildings, such as a house and its various outbuildings or a large mill complex. Customer Mathew Baird of the Baldwin Locomotive Works in Philadelphia noted that "the Gas Machine you put up at my country residence, near Darby, a year ago last May furnishes me with an abundant and good light, not only for my dwelling, but also for my stables, billiard room, laundry, and other buildings."[39] The dimensions of each model sold varied according to the required capacity of the pumps and generators (tables 3.1 and 3.2). The diameter of the piping needed to connect the various models also differed according to the number of buildings involved, the number of burners installed, and the building's distance from the generator (table 3.3).

Perhaps the most important refinement to the Springfield system and to gasoline gas lighting machines in general came with the invention of the mixing regulator. Gilbert and Barker patented an "Apparatus for Mixing Aeriform Fluids" on November 30, 1880. It appears that the company marketed their mixing regulator

Table 3.2
Pump Specifications

No.	No. of burners	Length (in.)	Diameter (in.)	Length in frame (in.)	Diameter in frame (in.)	Height in frame (in.)
1	15	20	16½	30	20	32
2	25	20	21	32	25	38
3	40	25	25	36	28	42
4	75	29	32¼	43	36	55
5	100	40	32¼	54	36	55
6	150	33	42	52	46	64
7	200	44	42	63	46	64
8	300	41	54½	58	61	68
9	500	54	54½	75	61	68
10	750	65	64½	85	70	71

[a]Two of 4,000 lbs. each.

SOURCE: Gilbert and Barker Manufacturing, *Springfield Gas Machine*, 1874.

		Dimension of Propelling Weight				
Propelling weight (lbs.)	Size of wire cord (Roebling's)	Diameter (in.)	Height (in.)	Air outlet (in.)	No. of feet of cord for each foot of space	Gallons of water when in use
450	29	22	14	¾	8 7/10	10
675	28	26	16	1	8 7/10	15
1,000	27½	30	18	1¼	8 7/10	26
1,450	27½	34	20	1¼	8 7/10	48
1,650	27	36	20	1½	8 7/10	77
2,000	27	40	20	2	8 7/10	96
2,800	26	40	27	2	8 7/10	144
4,000	25	46	30	2½	8 7/10	240
5,000	24	48	34	3	8 7/10	372
8,000[a]	25	46	30	4	8 7/10	525

Table 3.3
Piping Specifications

Greatest number of feet to be run	Size of pipe (inches)	Greatest number of of burners to be supplied
20 feet	⅜	2
30 feet	½	4
50 feet	⅔	15
70 feet	1	25
100 feet	1 ¼	40
150 feet	1 ½	70
200 feet	2	140
300 feet	2 ½	225
400 feet	3	300
500 feet	4	500

SOURCE: Gilbert and Barker Manufacturing, *Springfield Gas Machine*, 1874.

commercially by at least 1883.[40] The mixing regulator, essentially a machine that weighed the air-gas as it entered its chamber, opened a valve from the air pump to add fresh air to lighten the mixture or shut the valve to reduce the volume of air and make the mixture heavier. The machine produced a mixture of 15 percent gasoline vapor and 85 percent air, the ideal ratio for burning air-gas according to the company.[41] Gilbert and Barker found the mixture of air-gas from the generator too rich in the weeks and months after refilling it with fresh gasoline, particularly in larger gas machines used in industrial and commercial settings, and not rich enough as the fuel depleted over time. Both of these situations resulted in poorer quality light and the need for additional maintenance and adjustment. The former situation posed a danger as supersaturated air-gas could condense in the pipes and shoot out from the burners as liquid fuel. As Barker notes in his patent of 1880,

> the "gas" . . . as it comes from the carburetor, is generally too rich in carbon to be burned economically or to produce the best illuminating effect . . . without dilution with common air. To effect such dilution an air-supply pipe usually leads from the air pump or blower to the pipe which takes the gas from the carburetor to the burners, and the adjustments of the port in such air-supply pipe is commonly effected by hand manipulation.

> A prominent defect of difficulty incident to such a hand-manipulated apparatus is readily illustrated: Given gas of a certain richness in carbon, and the port in the air-supply pipe so adjusted that proper proportions of gas and common air are being supplied to one hundred lighted burners, now if fifty of those burners be turned off the supply of common air will be altogether disproportionate to and in excess of the supply of gas, with the result that the light will grow dim and perhaps go out altogether.[42]

Barker intended his regulator to provide the proper mixture of air and gas under all operating conditions and in reaction to changing operating scenarios (i.e., the addition or subtraction of burners while the system was in use or changes in the operating temperature). As he continues in the patent, "One prominent advantage . . . is that when this improved apparatus is once adjusted to furnish proper proportions of gas and air those proportions are substantially maintained in the illuminating-fluid supplied to the burners, whether their number be greater

Figure 3.9. Springfield Mixing Regulator. From Gilbert and Barker Manufacturing, *Directions for the Care of the Springfield Gas Machine,* May 15, 1897. Author's collection.

or less, and no further adjustment of the apparatus is required till the quality of the gas supplied from the carburetor substantially changes."[43]

As eventually constructed, the Springfield Mixing Regulator combined Barker's original patent design of 1880 with several other versions of the machine by Barker, Edwin Burdick, and Charles Hovey, the latter two as assignors to Gilbert and Barker Manufacturing.[44] The last patent design by Hovey most closely resembles the device pictured in Gilbert and Barker's 1885 pamphlet *Directions for the Care of the Springfield Gas Machine*.[45]

Hovey's Springfield Mixing Regulator essentially relied on the same type of expanding gas-holder tank as earlier versions. This floating gas holder allowed the system to maintain consistent pressure to the burners by opening and closing "a mixing valve . . . according to the number of burners that may be used." An "air-regulating valve" allowed the adjustment of the size of the opening from the air pump inlet into the mixing regulator. Hovey's addition of this single mixing valve to the system allowed for much simpler operation than the original Barker design with its complicated system of levers and valves. As Gilbert and Barker state in its 1885 instruction manual, "How frequently the valve . . . requires to be changed is purely a question of circumstances; do it whenever the lights indicate that it is necessary." For example, "when the lights become thin and pale it indicates a too large proportion of air, and that the gas needs enriching by reducing the quantity of air. To do this, turn the Regulating Valve . . . to the right."[46] Once adjusted, the mixing regulator would then adjust the main valve as necessary to maintain the mixture as set with the Regulating Valve, regardless of the number of burners in use. The system came with a test burner, located on the gas outlet pipe just beyond the mixing machine, for monitoring the quality of the gas mixture and to allow easy adjustment of the Regulating Valve, thus balancing the system to create the best, cleanest burning flame.

While the Springfield Mixing Regulator denoted a real advance in terms of maintaining good quality light regardless of the number of burners in use, it had several drawbacks. The system used water to provide the seal for the air-gas holder float device, which made the regulator prone to rusting and corrosion. More important, the regulator required regular adjustment and maintenance to maintain the proper air-gas ratio necessary for good, clean combustion. This proved particularly true when the generator had just been refilled and provided a very rich mixture and as the generator emptied and provided a much less rich mixture. It remained for several inventors in Detroit to create the "Combination Mixing Regulator," which eliminated the need for a water holder and automated the operation of the adjusting valve, in the late 1880s.

By 1890, Gilbert and Barker had become the "sole agents for the Eastern United States" to sell the Combination Mixing Regulator manufactured by the Detroit Heating and Lighting Company of Detroit, Michigan (formerly the Combination Gas Machine Company).[47] Gilbert and Barker catalogs and the labels on several extant Combination Mixing Regulators identify it as "Manufactured for Gilbert & Barker Mfg. Co."[48] The Combination Mixing Regulator was based on an 1888 patent design by Edward Frost, later modified by Edward Frost and Benjamin Middleditch of Detroit in 1897.[49] The machine consisted of a "very substantial cast-iron case in which is suspended a strong sheet-metal can B, closely sealed, and containing nothing but air. The balance-beam E, to which this is

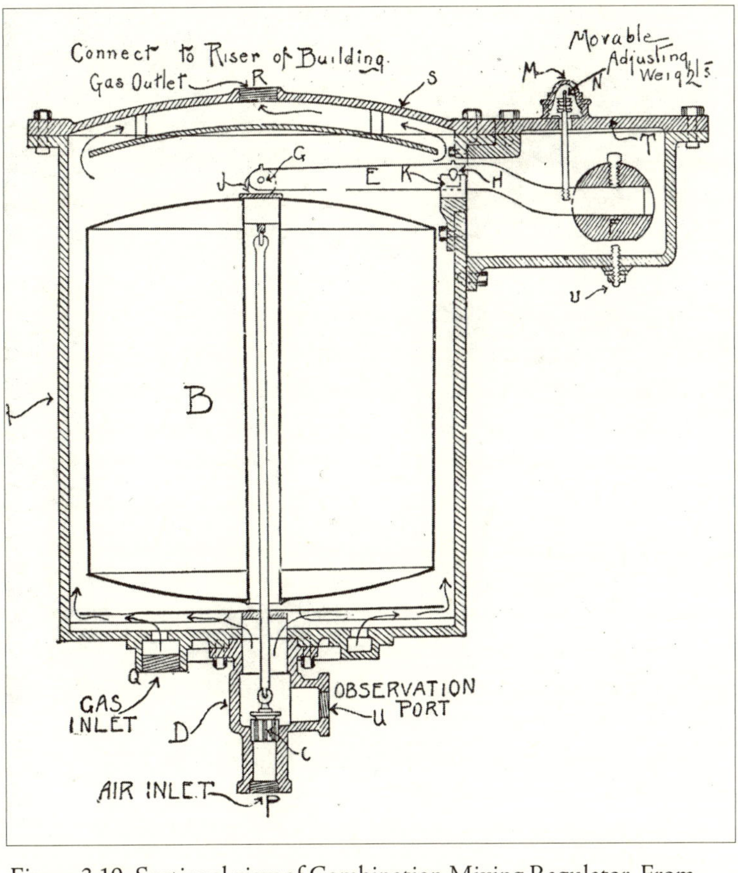

Figure 3.10. Sectional view of Combination Mixing Regulator. From Gilbert and Barker Manufacturing, *How Best to Light Our Country Homes and Resorts,* 1906. Author's collection.

hung, is sustained by pin H on agate bearings K. Since the counterpoise F exactly balances the weight of the metal in the can B and its attachments, that factor is eliminated, and we have only to consider the weight of the air in the can, which is a constant quantity."[50]

In operation, air-gas entered the gas holder and, based on the density of the air-gas, would raise or lower the interior sheet-metal can (or float) filled with pure air. The company explains: "One thousand cubic feet of gasolene gas of 20-candle power contain seven gallons of gasolene, weighing nearly forty pounds. The air contained in the can . . . will, therefore, float in any gasolene gas which may be admitted into the shell around it, and as the gas tends to become rich or dense, and so cause smoke, the can rises and opens the port, and so admits air and dilutes it. On the other hand, if the gas becomes thin and impoverished the can falls and closes off all or a portion of the air." Thus the machine automatically adjusted to changes in the density or weight of the air-gas coming from the generator. For example, after refilling the generator with gasoline, the air-gas tended to be richer in gasoline vapor and thus denser. This denser mixture would raise the air float in the unit, opening the valve and admitting additional fresh air to create the desired blend of 15 percent gasoline vapor to 85 percent air. As the firm's advertising materials indicated, this mixing regulator "required no care whatever, just leave it alone. When the quality of the gas is continuously poor, it is safe to assume that the generator needs replenishing."[51]

Prior to the invention of the mixing regulator, Gilbert and Barker and other firms that produced gas machines addressed this issue of changing air to gas vapor ratios by increasing the evaporative area of the generator itself and by developing specialized, adjustable burners. These burners could compensate, at least partially, for fluctuations in air and gas levels and could be operated in the room being lit, eliminating the need to go to the cellar to adjust the system at the air pump. Over the course of the Springfield Gas Machine's forty-plus-year history, Gilbert and Barker developed and produced a wide range of gas burners for both home and industrial applications. Beginning with simple, single-flame burners, the company experimented with numerous approaches to developing adjustable burners. By the end of the nineteenth century, the firm had adapted the newest incandescent burner technology for gas machine use.

John Barker patented the firm's first improved gas burner in 1870. In the patent description, he noted that his invention was a device "for regulating the flow of carbureted air or gas from the burner to its point of combustion." This flat-flame burner contained "a series of perforations . . . made into the lower part of the burner, so that when the burner is made or set for the combustion of carbureted air or gas of any certain quality the flame may be increased or diminished by turning the shell either up or down." He continued: "My invention is designed to obviate all difficulty in this respect [the problem of varying richness of the gas mixture], as the burner is set or made to let out at the tip the minimum quantity of gas that will produce a good flame, and as the gasoline remains longer in the generator and becomes weaker in its illuminating qualities, the outertube or shell may be turned so as to let out more gas and increase the flame without liability of smoke."[52] Barker patented several other burners over the next ten years, including two that were designed to increase the flame size and strength of the light produced by using radial arms with a series of small gas orifices.[53] In 1881, Barker patented another adjustable flat-flame burner that used a small regulating valve operated by an adjusting screw that controlled the flow of gas to the tip of the burner.[54] Thus the burner could be finely adjusted as the air-gas mixture in the system decreased in richness over time.

The introduction of the mixing regulator in the 1880s largely resolved the problem of controlling the varying richness of the air-gas mixture and systems

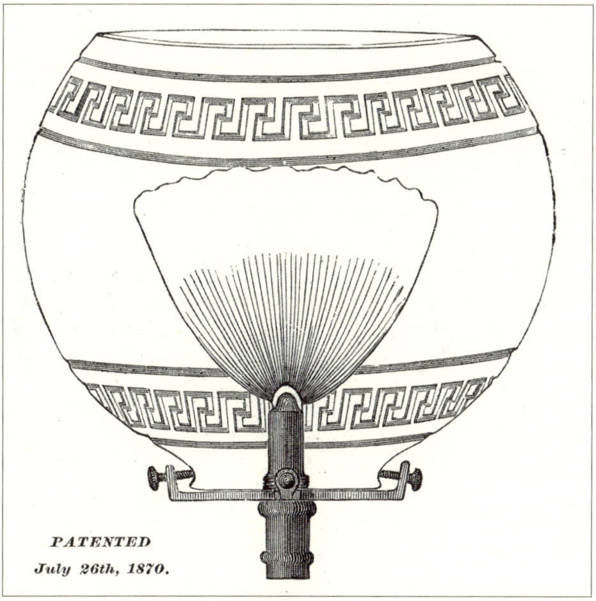

Figure 3.11. Barker's patent adjustable burner. Gilbert and Barker Manufacturing, *Springfield Gas Machine*, 1874. Author's collection.

using the mixing regulator could employ standard coal-gas burners without encountering problems. Customers with systems that did not have mixing regulators could continue to use the specialized burners for gas machines or add a mixing regulator to their older system. Both the standard burners and the special air-gas burners developed by Barker worked with "ordinary gas-fixtures . . . with all the improvements that have of late years been made upon them: extension lights for dining room; portables, with flexible hose and ornamented shades, for Libraries, Figures for Newels, &c."[55]

The most important innovation in gas burner design was the introduction of the incandescent gas burner in the 1880s. In general, these revolutionary new burners applied the heat of the flame to a refractory substance, or mantle, which then glowed with light. The Welsbach Incandescent Gas Burner was introduced in the late 1880s and advertised as producing "a given amount of light with less than one-half the gas used in the ordinary burner." It produced "an extremely steady light" and gave off "only about one-third the heat of ordinary burners." An 1888 advertisement for the Welsbach burner in the *New York Times* specifies that "it can be used with the Springfield or any other gas machine." Gilbert and Baker clearly saw the efficacy of this type of burner for lighting and introduced their own Springfield Incandescent Gas Burner in the late 1890s.[56] Used along with the mixing regulator, these burners provided a steady, bright light that for several decades helped gas lighting compete successfully against electric light.

The Springfield Incandescent Gas Burner combined the enhanced light quality of the incandescent system with the adjustability of Gilbert and Barker's previous gas machine burners.[57] As Barker wrote in his patent, "My improvements [are] to provide for the regulation of the gas to a constant quality at the burner under varying conditions of quality and pressure in the mains."[58] The burner was adjusted, notes the company's catalog, "by revolving it on its screwed base. Turning to the right enriches: turning to the left reduces the flame. . . . gas is thrown into the pillar of this burner through the annular space around the needle point A. This gives a rotary or cyclonic movement to the gas and more perfectly draws the air up the Bunsen tube."[59] The company reported that the all-brass burner operated safely because "all parts are screwed together and are tight. The gas cannot leak." The burner was very easy to adjust by simply turning the screwed base by hand or with the available "burner wrench." Gilbert and Barker's design of the burner also provided "a bypass so that when the burner is turned clear down so that the needle valve is shut completely, sufficient gas still passes through the needle to maintain a small blue flame on the crater under the mantle. By this means we are enabled to turn the gas down low and leave it when not in full use."[60]

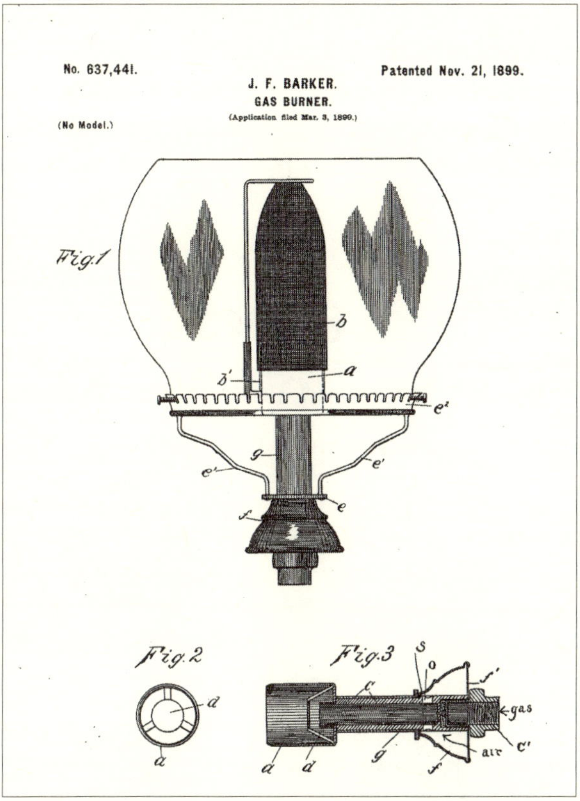

Figure 3.12. Patent No. 637,441 for gas burner by J. F. Barker, November 21, 1899. U.S. Patent and Trademark Office.

As such, this burner could function as a pilot light of sorts, which eliminated the need for matches or lighting devices. Various company flyers and catalogs described the Springfield Incandescent Gas Burner as providing excellent light quality with an extremely efficient use of gas. For example, the incandescent burner used 2.5 cubic feet of gas per hour to produce a light of forty candles (or sixteen candles per foot), while an ordinary gas burner consumed 4 cubic feet of gas per hour to produce the light of sixteen candles (or four candles per foot).[61] The company's lightbulb-shaped No. 111 gas burner, its "little wonder," consumed only 1.25 cubic feet of gas per hour to produce the light of 25 candles.[62]

The company produced a variety of models and styles of the Springfield Incandescent Gas Burner. The No. 95 had a simple cylindrical chimney and shade, and the No. 95½ sported a variety of interchangeable glass globes, while the No.

99 required no chimney and worked with standard globes with four to five inch openings at the bottom. The No. 111 featured a small burner and mantle with a globe shaped remarkably, and not coincidentally, like an electric lightbulb. The company's catalogs advertised a wide selection of shades, mantles, and chimneys (mica and glass), and a range of globes from plain to elaborately decorated and

Figure 3.13. Globes for No. 95½ Springfield Incandescent Burner. From Gilbert and Barker Manufacturing, *Springfield Incandescent Burner*, ca. 1900. Shelburne Farms Collections, Shelburne, Vermont.

etched, for each style burner.[63] Chimneys, used to protect the mantle from breezes and to enhance the draw of fresh air into the low pressure burner, and mantles required frequent replacement as they regularly broke or shattered from extremes of heat or cold. Delicate mantles could disintegrate from simply being bumped, from vibration (a particular problem in factories), or when a chimney shattered.

For industrial applications, the company also manufactured and sold an "anti-vibration fixture" that made the use of incandescent mantle burners "practical and economical" in factories. This fixture consisted of a No. 95½ burner contained within a wire holder that hung from the ceiling with a spring; gas reached the burner via a flexible rubber tube.[64] Both the spring and tube dampened vibration from machinery that could destroy the fragile mantle and often the chimney.

Gilbert and Barker employees constructed, shipped, and often installed the Springfield units, providing quality assurance from production through installation. Based on the 1882 price list, the company sold all of the components and fittings necessary to install a Springfield Gas Machine system, including wrought-iron pipe and pipe fittings, brass goods, check valves, tanks for storing gasoline, burners and brass fittings, and the generators and air pumps. The company's price list indicates that the cost of the unit included "machines, cased ready for shipment, and delivered on board cars at our factory in Springfield. The connecting air and gas pipe with their fittings are not comprehended in this price."[65]

Customers could choose to have the system installed by Gilbert and Barker employees, by local contractors, or, in at least one extraordinary case, by themselves. Customer letters to Gilbert and Barker frequently refer to the company's employees who installed the machines. For example, a representative of the firm of H. O. Bernard and Company of Westboro, Massachusetts, wrote in 1873 that "we report that the machine you put in for us nearly four years ago, is giving us perfect satisfaction." Residential customer Byron Weston of Dalton, Massachusetts, noted in 1878 that "the men you sent were good workmen and good men." Local gas or plumbing contractors could also install the machines, as was the case at Alexander Stephens's house in Crawfordville, Georgia, and several of the Ames sites in North Easton, Massachusetts. Intrepid homeowner Warren Fuller of Brunswick, Georgia, stands out as an early "do-it-yourselfer," writing in 1869, "I found no difficulty in properly placing and putting it into operation by the directions given, although I had no previous experience in setting up any kind of gas apparatus, and there was no one in this section who had."[66] Whether completed on the whole by Gilbert and Barker, a local plumber or gas fitter, or even the homeowner, certain parts of the installation likely fell to local contractors, notably construction of the brick vault.

The firm offered several types of installation for the system's generator unit; each had benefits and drawbacks in terms of the initial labor and cost and later maintenance. The original design for the system used a vaulted brick chamber that completely enclosed the generator.[67] Gasoline flowed into the generator from an aboveground container, usually wooden or steel storage barrels. A wooden trap door situated directly over the chamber allowed access to the generator. Both

Figure 3.14. Vault installation of Springfield Gas Machine generator. From Gilbert and Barker Manufacturing, *Springfield Gas Machine*, 1874. Author's collection.

the pipe leading from the air pump and the pipe leading from the generator to the house exited and entered through the side of the vault and remained underground for their entire length. By 1874, the company offered a second version of the brick vault installation. The newer version varied slightly from the original in that the brick vault was set into a hill or slope and entered through a cellar-like, bulkhead door. Like the previous installation, the generator stood on a brick ledge above the floor.[68]

By 1878, the firm had introduced a "flat" gas generator in two new configurations that either simplified or eliminated the brick vault arrangement: the Partially Buried Gas Generator and the Buried Gas Generator, respectively. The new "flat" gas generator design, described in the firm's 1878 catalog, "consists in making [the generator] of greater diameter, and fewer but larger evaporating chambers . . . and so arranging the connections for the pipes that they may be substantially buried in the earth—the only part exposed being the arc to which the gauge, valve, etc., are attached, which projects into the little brick pit."[69] Ashland, in Lexington, Kentucky, had a "Partially Buried" installation in which the new flat generator was buried directly in the ground with the valves and

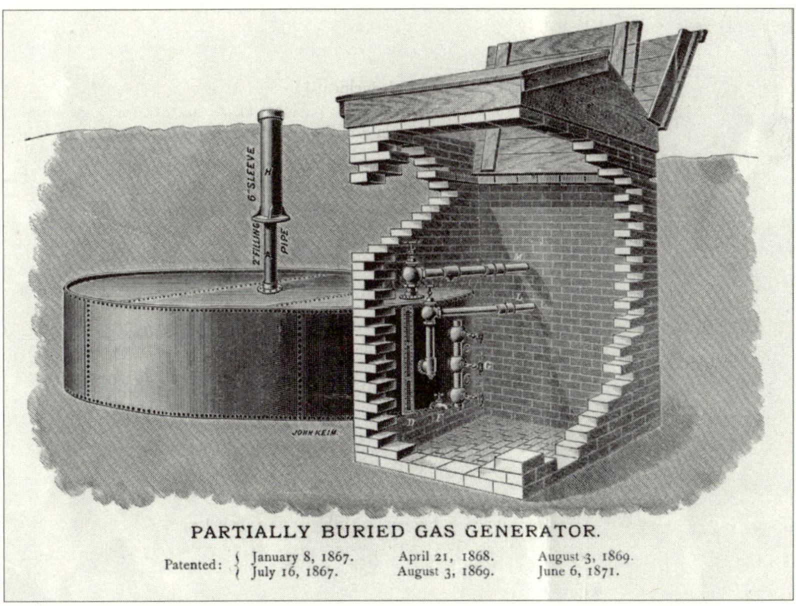

Figure 3.15. Partially buried gas generator installation. From Gilbert and Barker Manufacturing, *Directions for the Care of the Springfield Gas Machine*, May 15, 1897. Author's collection.

gauges accessed in a brick-lined shaft to facilitate maintaining and adjusting the machine.[70]

The "flat" gas generators also were designed to be "buried," so that they sat completely underground, with access pipes for venting, for measuring the gas-

Figure 3.16. Fully buried gas generator installation. From Gilbert and Barker Manufacturing, *Directions for the Care of the Springfield Gas Machine*, May 15, 1897. Author's collection.

oline level in the float, and for filling the unit, running to the surface in small wood-lined and, later, metal-lined shafts. The company compared its "Buried Gas Generator" to "our regular generators, except that the valve-tube and gauge are omitted, and in their place we have the independent float and rod to indicate the amount of fluid they contain." The company's 1885 *Directions for the Care of the Springfield Gas Machine* indicates that shipping and storage barrels supplied gasoline to the buried gas generator through the fill pipe into the upper evaporator pan; when this pan filled, the gas ran down a tube and filled the lower chamber(s). In a demonstration of the company's focus on selling gasoline, the instructions for the Springfield Gas Machine reminded owners that "your machine will give you the best satisfaction if you will use a good quality of gasolene and keep the generator well filled."[71]

The buried Springfield unit eliminated "the expense of the brickwork of a vault." The company cautioned, however, that it recommended this style of generator "only in cases where a moderate consumption of gas is contemplated, as in dwellings of the smaller class. . . . We build Nos. 1, 2, 3, and 4 generators 'to bury'; when larger are required, we advise our regular pattern [vaulted]."[72] Thus Gilbert and Barker continued to build larger machines intended for commercial and industrial buildings and elaborate residences in the original style that required brick vaults. Nevertheless, a footnote in the firm's 1878 catalog recommends "the use of 'Flat' generators [partially buried] over any other we make—this method of manufacture and setting being a mean between the two extremes of either wholly burying in the earth, which has its advantages in certain cases, or placing the generator in a vault."[73]

In 1892, Gilbert and Barker developed another type of gas generator configuration known as the New Feed Springfield Gas Machine, based on "experience and observation of a quarter of a century."[74] Like the "Buried Gas Generator" described above, this New Feed system employed a fully buried (flat) generator, with vent and pump pipes running to the ground surface within a wood or metal shaft or sleeve. The innovative idea of the system was a large storage tank that held several hundred gallons of gasoline. A pipe connected the tank to the generator, controlled by a "self closing feed valve" that was operated from the building being lighted. The New Feed design allowed it to work without using a mixing regulator to adjust the air-gas ratio. As the company boasted in 1893, "Instead of making a gas of 80 candle power and then diluting it to the more practicable 20 candle power, this Machine makes the gas of the latter desirable quality once for all in the gas generator itself, and so avoids the need of dilution by the more or less complicated mechanical devices, inside the house, common to all other machines." The system accomplished this by charging the generator with a base fluid, a "hydrocarbon of gravity so low that it will not evaporate" under the typical working temperatures of the machine. The storage tank contained gasoline of a gravity of 80 degrees Beaumé, which "at a temperature of 40°, [will] mingle with air in a nearly uniform proportion of fifteen parts of vapor to eighty-five parts of air." Thus "the supply of fresh gasoline [to the machine] is promptly diffused through the whole body of permanent liquid absorbent [base fluid], whence its particles are given off slowly and evenly, as needed to produce an approximately uniform result.[75] When the size of the flame in the test burner dropped below a mark etched on the chimney, it was time to replenish the generator with fresh gasoline. As the operating instructions explain, this simply necessitated "holding the valve [between the storage tank and generator] wide open for thirty seconds. Repeat this operation at intervals of ten minutes until the flame of the test burner rises again to the line of the chimney as above."[76]

Gilbert and Barker clearly intended this system as another solution to the problem of maintaining a stable air-gas ratio. It eliminated the need for the Springfield Mixing Regulator and made the overall operation of the system much easier. However, it appears the company quickly abandoned this system with the

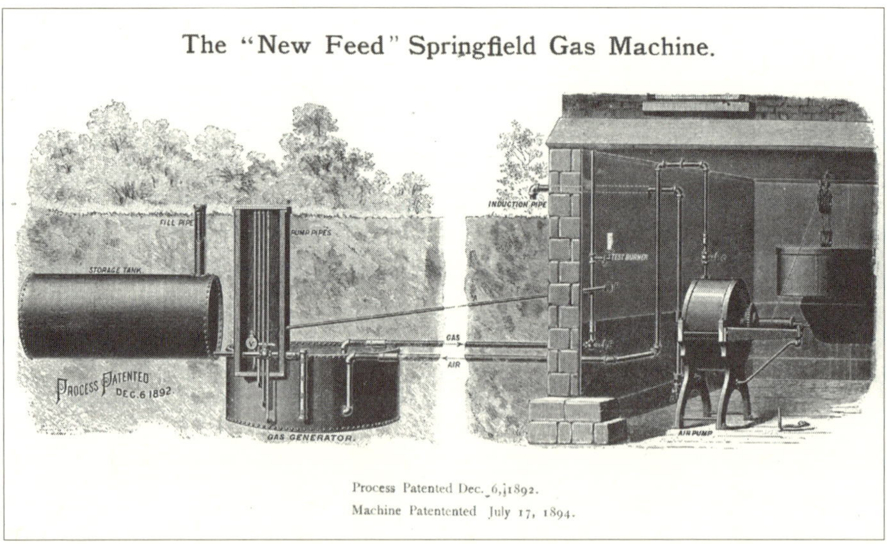

Figure 3.17. The New Feed Springfield Gas Machine. From Gilbert and Barker Manufacturing, *Directions for the Care of the Springfield Gas Machine*, May 15, 1897. Author's collection.

success of the Combination Mixing Regulator in the late 1890s, which provided a simple, low-maintenance way to regulate the changing air-gas mixture of the generator. The addition of the large storage tank also further simplified the maintenance and operation of the system, making the refilling of the generator a very simple operation and reducing the need to handle the gasoline and gasoline barrels to a minimum, in some cases just once a year.

Gilbert and Barker also manufactured and sold a wide variety of gas-powered appliances for both residential and commercial applications. These products expanded the application of the Springfield system, making it more competitive with municipal coal-gas systems and initially with electricity. The residential devices included water heaters, such as the Automatic Instantaneous Water Heater, which provided hot water on demand; laundry iron heaters; kitchen broilers; kitchen stoves; and furnaces/room heaters, such as the Astral, "the biggest little stove ever made." Commercial devices included a variety of specialized burners for brazing and other heating operations; the Springfield Furnace, for heating soldering coppers and light forgings; and the Springfield Bunsen Burner, designed for laboratories and used in many university labs and even by Thomas A. Edison in his laboratory.[77]

Gilbert and Barker consistently claimed that the quality of the light produced by the Springfield Gas Machine rivaled that of coal gas. Despite these claims, the company continually experimented with ways to maintain a constant air-gas ratio

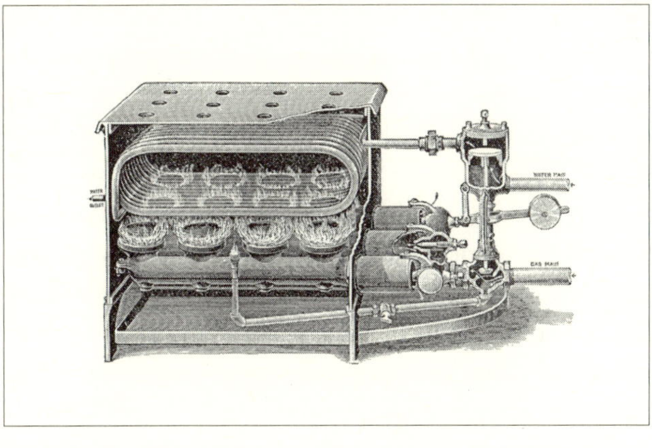

Figure 3.18. "Our Automatic Instantaneous Water Heater." From Gilbert and Barker Manufacturing, *Our Automatic Instantaneous Water Heater*, n.d. Shelburne Farms Collections, Shelburne, Vermont.

Figure 3.19. "The Argand Smoothing-iron Heater." From Gilbert and Barker Manufacturing, untitled advertising flyer, n.d. Shelburne Farms Collections, Shelburne, Vermont.

and with it a stable, high-quality light. "When properly manufactured," as the company noted in 1870–71, air-gas "is found to possess an illuminating power of from 18–20 candles . . . about one-third better than the ordinary gas supplied by our public works." The catalog goes on to report that "the steadiness of the Flame depends upon the working of the machine supplying it; if the machine be perfect and works properly, it burns with equal steadiness to coal gas."[78] In fact, the Gilbert and Barker catalogs touted the light produced by the Springfield Machine as "not only strong and rich, but also exceedingly mellow and soft to the eyes. It has none of that peculiar sparkle and glare found to be so hurtful in coal gas."[79]

Not surprisingly, testimonials in the company's catalogs celebrated the light quality of the machine, often declaring it better than light from coal gas. For example, C. Laflin, a customer in Westfield, Massachusetts, wrote that "I can read by it thrice the length of time I could by coal gas, without tiring the eye." Similarly, L. F. Thayer, also of Westfield, Massachusetts, wrote in 1870 that "for purposes of light I know of nothing as good, for while I do not think it as brilliant as coal-gas, the light is softer and steadier. I can read through an entire evening without tiring my eyes, while it is difficult for me to read an hour by coal-gas."[80] In August 1868, customer Nathaniel Lamson of Franklin County, Massachusetts, wrote to the company that he used the Springfield system to light his office and home and that "the light is certainly very strong and brilliant. I do not believe there is a house in the country lighted better than mine, and I have never, either in New York or Boston, seen a billiard table so perfectly lighted." In 1870, Gilbert and Barker also published a letter from P. T. Barnum in which the infamous showman wrote that "the light is much whiter than that supplied by the Bridgeport Gas Co.; and the whole thing seems thus far to be near perfection."[81]

Comparing the Springfield's light quality to that of kerosene lamps, customer A. H. Graves of Hatfield, Massachusetts, wrote in 1872 that the system "worked to my entire satisfaction, giving a soft, steady, yet brilliant light, one burner giving as much light as two kerosene lamps in their best trim." In an 1873 letter to the company, L. F. Beeler, an agent for the Baltimore and Ohio Railroad, noted that the machine "furnishes us with a brilliant light, that burns steadily without flickering. . . . The light afforded is ample, is bright, beautiful, very steady, and a great addition to the comfort and convenience of our offices." An 1878 testimonial from E. F. Baldwin, architect of the B&O, noted that the Springfield machine produced "a pure, soft, white light that is less trying to the eyes than coal gas." In addition to producing a light of high quality, the company celebrated the gas its system produced as "a remarkably pure gas; contains no sulphurous compounds or impurities of any kind; with proper burners combustion is perfect without smoke or odor."[82] This claim was very important, as the public was wary of how gas systems, particularly municipal companies, might introduce "foreign elements" into their homes.

The company's Incandescent Gas Burner, introduced in the 1890s, paired with the Combination Mixing Regulator, resulted in a significant improvement in the quality and dependability of the Springfield system's light. As the company noted in a brochure, an "incandescent gas burner gives more light for less money than any other artificial illuminant." Customers responded quickly to both the Combination Mixing Regulator and later the incandescent gas burner. In 1904, the proprietor of Sea Shore House in Old Orchard, Maine, declared himself "pleased to state that the 300-light mixer which you put in ten years ago has given perfect satisfaction. We have had a very fine light in every part of the house and free from smoke. In our public rooms we have used incandescent burners. The light has been more brilliant than any we have seen used with the common gas." About the same time, a customer in Ardmore, Pennsylvania, reported that he had bought his Springfield system more than thirty years earlier and had just added a Combination Mixing Regulator: "I can now use the incandescent burners, which give a steady light of great power and brilliancy, superior to the electric light." Fred

Dreer compared the Springfield gas light in his Chester County, Pennsylvania, country house with city gas at his home in Philadelphia, concluding, "I am using your small mantles consuming one and one-fourth feet of gas per hour and have a much brighter light than we get at our city residence with the U. G. I. gas and large mantles with Welsbach burners."[83]

As late as 1905, customers still pronounced the Springfield's light as superior to both the older coal gas and the increasingly popular electric light.[84] In fact, some customers even reported switching from electric to Springfield gas light. Philadelphia customer S. John Lea wrote in 1903, "I had electric lights in my house and I cut them out, as the Gas Machine gave me much better light and cost only about half." Lea added that he switched to the municipal gas company when they ran a pipe by his house but quickly reverted to his Springfield system "on account of the poor light and heavy gas bills."[85]

Business as well as residential customers exclaimed about the Springfield system's light quality. A store owner in Hinsdale, New Hampshire, wrote that "our store is 42 feet wide and 80 feet deep, and to light this store we have twenty Incandescent Burners. The store is so well lighted as to be a source of satisfaction to us, and so that it attracts attention from the street outside." Likewise, Haynes and Company of Springfield, Massachusetts, testified that "we have now lighted our Clothing Store for twenty-two years with . . . the Springfield Gas Machine. . . . Our store was never so brilliantly lighted before."[86] Retailers valued improvements in lighting as a way to attract customers. As marketing and display became central for stores along Main Street, lighting helped demonstrate that a firm was a modern establishment and created a bright and inviting atmosphere. Historian Alan Trachtenberg explains that "the glittering façade, the ceremonial entrances, the bright interiors, all [suggested] a kind of magical appearance of goods as if from nowhere."[87] These changes drove new levels of consumer expectation and helped to establish the economic and social power of consumption from the largest cities to the smallest villages.

Unsurprisingly, almost all of the testimonials reported that they received excellent light from the Springfield system. At least one of the testimonials, however, hinted at dissatisfaction with the system prior to the introduction of the mixing regulator and incandescent burner. A. G. Moody, founder of the Northfield Seminary in Northfield, Massachusetts, offered a left-handed compliment: "We wish to give you our testimony for the efficiency of the Mixing Regulator which goes with your Springfield Gas Machine. The first machine which you put in for any of the Northfield schools was without that adjunct, but since then you have put in for us seven of your Gas Machines, all with the Mixing Regulator attached. These seven machines have not given us so much trouble as the old one without the Regulator."[88]

Two other pieces of evidence also reveal the less than perfect operation of the system and of its light quality. Oakes Ames, who probably lit his home Borderland in Boston with a Springfield system, compared "the household of 'the nineties'" with a modern household in his 1950 autobiography. With the benefit of long hindsight, he recalled that Borderland was "illuminated by means of wheezy gas jets and Welsbach burners." He continued, noting that "our houses were dimly lighted in the 'gay nineties' and after sunset reading would have been an eye-straining pastime had it not been for kerosene oil lamps and the soft yellow glow of student lamps. Until electricity became the general source of illumination we tried every new invention which intensified the weak light of city gas or home made gas."[89] This passage is particularly interesting in that Ames actually preferred kerosene lamps to gas lighting. It also confirms that users of systems like the Springfield kept trying new products, like mixing regulators or improved burners, which increased the quality of light.

Henry B. Hyde, the superintendent of the Jekyll Island Club, offered more specific and forceful critique. In 1895, Hyde, known as the "czar" of Jekyll, complained to the club's New York office that the "gasoline lamps and fixtures in use throughout the club house [are] dangerous to health as well as disagreeable." Not only that, they provided poor light. Hyde's wife apparently complained continually about the smell and "he agreed that it was 'positively sickening.'"[90] Hyde urged the club's board to act on a proposed plan to install an electric plant on the island to improve the lighting situation.

While customers generally found the Springfield's quality of light much higher than kerosene or city gas illumination, they also considered the level of care and maintenance required by the machine much lower compared to previous sources of lighting and other types of contemporary portable gas lighting systems. One had to refill the Springfield generator two to four times a year and wind the air pump weight once or twice a week. In addition, the user had to oil the bearings every month or so in order to keep the system working smoothly. These efforts to maintain the Springfield compared favorably to the onerous daily routine of maintaining kerosene lamps—filling bases, washing chimneys, trimming wicks. W. P. Townsend of New Brighton, Pennsylvania, reported in 1873 that his machine had been running well for two years and that "we have as good a light as can be produced with gas in all the rooms of my house, and *with much less labor than is necessary to keep in order one oil lamp.*" "Ten minutes time each day gives the machine all the care required," commented A. J. Stanton, agent and treasurer of the Hampshire Manufacturing Company of Huntington, Massachusetts, in 1872. "We consider it to be a cheap, safe, and reliable apparatus for lighting manufactories, and should be unwilling to exchange it for any other process of lighting with which we are acquainted."[91]

Customer Mathew Baird of Darby, Pennsylvania, observed in 1873 that "my gardener informs me that ten minutes per day is all the time required to attend to the machine, while with the old rosin works they were obliged to make gas at least twice a week. I am much pleased with the change." Lumber company owner John Knapp of Menomonie, Wisconsin, reported that the machine in his house "requires no other attention than winding the weight, and supplying the gas generator once in three or four months with good gasoline."[92] Steadfastly unimpressed by the system, Oakes Ames remembered the occasional attention it required from family servants in the 1890s: "The gas machine had to be wound up daily and if the man whose duty it was to attend to our comfort forgot to wind it up, a time would come early in the evening when the gas went low during our dinner until somebody rushed to the cellar to correct the oversight."[93] Ames's recollection indicates that his household used a weight-driven air pump, not the more automatic water-driven pump.

Business clients also seemed to find the system low maintenance. Describing their six-month-old Springfield, the owners of J. S. and W. H. Hyde, General Merchandise, of Ridgway, Pennsylvania, wrote in 1878, "Ours is, as you know, a 100-light machine, with a water-wheel pump, so we have nothing to do except keep the tank supplied with gasoline." Another commercial customer, merchant M. Birdsall of Greene, New York, who had used the system for six years, asserted in 1878, "I have never expended one cent on it, as it has never required any repairs, has kept in perfect order and given me perfect satisfaction."[94]

Because of the Springfield's light quality and ease of operation, customers purchased the system to replace just about every type of lighting system then available. Gilbert and Barker boasted that "of the popularity and excellence of our apparatus it is enough for us to say, that though we have replaced with our machine those of almost every kind manufactured in the Eastern States, up to this time no one machine of ours has given place to that of any other whatsoever."[95] The rosin works discussed by residential customer Mathew Baird, for example, was likely one of the earliest types of self-contained gas systems (see chapter 1).[96] It seems clear from testimonials like Baird's and from extant evidence and documents related to several sites, including Alexander Stephen's Liberty Hall in Crawfordville, Georgia, and Mount Holyoke College in South Hadley, Massachusetts, that the Springfield Gas Machine often replaced an earlier system from another manufacturer.[97]

J. M. Seeley, a manager for Monument Mills in Berkshire County, Massachusetts, informed Gilbert and Barker in 1873 that "we had previously lighted these mills with oil gas works, but concluded to adopt your apparatus in place of them." In 1878, Baltimore residential customer W. N. Weyth volunteered that "before buying your machine I used for several years a Maryland Portable [rosin-oil system], and I consider yours superior in every respect." Likewise, D. B. Smith of Pine Meadow, Connecticut, explained in 1878 that the Springfield system "put in place of the Pittman Coal Gas Machine gives full satisfaction." Remarkably, store owner J. L. Sheldon of Wakefield, Rhode Island (1905), found the Springfield system "far superior to the electric light service previously used."[98] By 1906, customers reported that they had installed Springfield systems to replace acetylene systems and, like Sheldon, even electric lights.

Customers also appreciated the economics of owning and operating the Springfield Gas Machine. The actual cost of a unit depended on several factors, including the initial cost of the machine, the cost of installation, the price of gasoline, and the overall efficiency of the machine. The purchase price varied according to the model needed, as determined by the size of the building and the number of burners included (see tables 2.1–2.2 and 2.4–2.5). In 1874, Springfield machines ranged in cost from $300 to $700 for home models. Given that the average annual income of wage laborers in Massachusetts in 1874 was about $750.00, it follows that members of the new middle/upper middle class and upper class purchased the vast majority of systems.[99]

While the company may have initially looked toward a more economically diverse customer base, it appears from Gilbert and Barker's advertising that it eventually focused its sights on a niche market that combined commercial and business interests with the expanding upper-middle-class and wealthy cottage or country house owners. The firm did add a lower cost iron version of the machine beginning in the early 1880s. While this new machine cost 25 to 30 percent less than the copper versions, its purchase price still ran from $250 to $1,700, putting it out of reach for the average working-class family and making it a stretch for even the middle-class homeowner.

The cost of installation depended on site-specific items, such as the type of vault needed, if any, the effort required to excavate the pit, and the cost of constructing the brick vault. Of course, the later "Buried Gas Generator" systems did not require a brick vault at all, which made them significantly easier and cheaper to install.

The price of gasoline had the greatest impact on the cost of operating the Springfield unit. In 1868, naphtha and gasoline had relatively few uses and demand was almost exclusively related to the use of gas machines for illumination.

With the cost of gasoline at 25 to 35 cents per gallon, the light provided by the machine would have cost about $1.50 to $1.75 per thousand feet of air-gas in 1870, or less than 50 percent of the average price paid by coal-gas customers.[100] Typical of many Springfield users, C. Laflin of Westfield, Massachusetts, enthused in 1873 that "the gas is less than half that of coal gas." Likewise, C. O. Morris, a New York banker, reported of "the Gas Machine you put in my house in February, 1871, [that] the cost I should think is nearly two-thirds less than city gas, with a purer and whiter light." The owners of the National Drovers Hotel in New York City congratulated themselves in 1871 for "having purchased the machine. Our gas bill, which formerly, when we used the coal gas, was nearly $1,400.00 per year, will now be about $550.00."[101] Miles Morse of Thomaston, Connecticut, reported in 1878 that he used about one hundred gallons of gasoline per year to light his home, which would have cost about twenty-five to thirty-five dollars. Harrison Robbins explained to Gilbert and Barker in 1878 that he used "some 5 barrels per year" to power his Springfield system to light his home and stable with about sixty lights. At forty-two gallons per barrel, Robbins consumed about two hundred gallons, spending approximately seventy dollars per year.[102]

This theme of economic savings appeared repeatedly in all of Gilbert and Barker's catalogs and clearly served as an important marketing claim. In fact, numerous customers reported that they retained their Springfield systems even when municipal coal-gas service became available due to the lower cost of the gas produced by the Springfield. A comparative study of the cost of naphtha and coal gas prepared by Charles Bailey, the foreman of the Finishing Department at the U.S. Armory in Springfield, confirmed the accuracy of customer's claims of efficiency. In his August 19, 1865, report to the armory's commandant, Bailey wrote that his study over a thirty-six-day period had demonstrated that naphtha gas (air-gas) cost $1.55 per thousand feet while coal gas was running $4.50 per thousand feet.[103]

Gilbert and Barker's 1906 publication, *How Best to Light our Country Homes and Resorts*, asserted that "acetylene light costs three and one-half times as much as gasoline" light. Of his Springfield machine, customer W. E. Greenwood commented that "the net cost to me is extremely small and I am getting better results in this machine at about one-fourth the cost than were received by me through the use of a modern acetylene machine which was removed to make room for this one."[104] With increased production and demand, gasoline fell to only sixteen cents per gallon by 1906, inducing the company to exclaim that "the Springfield Incandescent Burner with this gas affords the *cheapest artificial light in the world.*"[105]

The Springfield Gas Machine's remarkable safety record also appealed to consumers. Locating the generator that contained the gasoline underground at a safe distance from the house radically reduced the chance of fire or explosion. In its 1870–71 catalog, Gilbert and Barker assured potential buyers that "the greater security and safety from fire in the use of gas delivered through permanent pipes and fixtures over that attained in the use of kerosene can not be questioned. Not a day passes without record of losses in property, and often of life itself, caused by fires resulting from the breakage or explosion of kerosene lamps; and when we consider the extent to which kerosene is now frequently adulterated, and the dishonesty of manufacturers and dealers, we are not surprised at such losses, as the risk of using it can hardly be overestimated." As of 1874, the manufacturer reported, no buildings had burned because of accidents caused by the Springfield machine. The firm claimed that the gas machine was safer than kerosene lamps, particularly those using kerosene "enhanced" with gasoline or other volatile liquids, because it used permanent pipes and fixtures. Attaching fixtures to walls and ceilings eliminated the risk of knocking over kerosene lamps, a major cause of fires. Customer J. P. Chamberlin of Connecticut ended his 1878 testimonial letter by pointing out that "we also think it gives greater security from accident than kerosene oil." Remarkably, the proprietors of the Crescent Paper Mills at Russell, Massachusetts, explained that "our mill was burned in December, 1870, during a night in which a high wind prevailed, yet there was no explosion of gas, nor any fear of it. We are now using the same [Springfield] generator in our new mill, and are as well pleased with its operation and the light it furnishes as before the fire."[106]

Gilbert and Barker also consistently reminded its customers that insurance premiums cost about the same for owners of an approved gas machine as for those using municipal coal gas and that both cost less than premiums for kerosene lamp users. E. B. Fellows, president of the Rutgers Fire Insurance Company of New York and chairman of the "Committee on Gas Machines, &c." of the New York Board of Fire Underwriters reported in 1872 that "your Company have uniformly put up your Machines in a safe manner, and in compliance with the regulations adopted by the Board of Fire Underwriters of this city. If care is taken to make all the connections air-tight, I unhesitatingly say that, when so put up, there is little or no danger of an accident and it can be operated or used with safety."[107]

Both the New York Board of Fire Underwriters and the National Board of Fire Underwriters regularly published and updated rules and regulations for the proper installation and use of gas lighting machines like the Springfield. Obtaining insurance for buildings using gas machines required adherence to these specifications for installation and operation. Companies could refuse to write policies or could charge higher premiums if a lighting system violated the Underwriters' rules.

The rules and regulations of both the New York and National Board of Fire Underwriters established a general list of ten "recommendations" that needed to "be substantially complied with to ensure safety" of gas machines.[108] The National Board of Fire Underwriters' 1876 report on gas machines states:

1st. The Vault or Gas-house to be removed from the premises insured the distance required by this Report.

2nd. The machine and all the apparatus to be made of good materials and in a substantial and workmanlike manner.

3rd. All the apparatus containing Gasoline and other inflammable fluid, or any Gas-holder, is to be placed in a vault outside the premises to be insured.

4th. Stop-cocks must be placed on both the gas and air-pipes near the machine in the vault, and also on the gas pipe within the building, and on the air-pipe near the air pump when the pump is in the cellar or building.

5th. The vent-pipe and filling-pipe area to be so arranged that one cannot be opened without opening both.

6th. All the main gas-pipes leading to the premises lighted, must have an inclination towards the Gas Machine, so as to return all the condensation that may take place in the pipes.

7th. The premises to be lighted must be securely piped, the pipes thoroughly tested by competent persons before the gas is let on, and the pipes so put up as to avoid, as much as possible, all accumulation of any condensation that may occur inside the building lighted.

8th. Particular inquires are to be made as to the competency of the persons who are to have charge of filling the machine, and to ascertain if they are informed as to the importance of having the vent open and the air-pump shut off whilst the machine is being filled with fluid, or the great danger of using a light in or near the gas-house or vault.

9th. No barrels containing Gasoline or other like fluid, or from which Gasoline has recently been emptied (yet full of vapor), are to be allowed to be kept in any cellar, barn, shed or outbuilding where other property is kept, or where there is a liability to use a fire or light.

10th. In case the air pump of the machine is placed in the cellar of the building, an automatic check-valve is to be inserted in the air-pipe to prevent backward flow, and [the] escape of gas through the pump into the cellar, or the pump is to be so constructed and set that the air supplying it is drawn through an induction pipe leading from without the building.

The Springfield Gas Machine complied with all of the requirements and, as discussed in chapter 2, Gilbert and Barker's system seems to have been the model on which the rules and regulations were based. Insurance companies that covered properties using gas machines clearly viewed these requirements as very important. The insurance policy of John S. Ames for his home, Langwater, located in North Easton, Massachusetts, actually includes much of the language of the "Rules and Regulations" in the text of the policy.[109]

By the turn of the twentieth century, the National Board of Fire Underwriters was regularly updating its "Rules and Regulations . . . for the Construction, Installation and use of Gasolene Vapor Gas Lighting Machines, Lamps and Systems."[110] This booklet established five classes of gas machine systems: "Class A—Machine having Outside Carburetors, Class B—Machines having Inside Carburators; Class C—Systems having outside Tanks and Inside Flame Heated Generators; Class D—Gasolene Vapor Lamps; Class E—Systems having Inside Tanks and Inside Flame Heated Generators." Class A machines, like the Springfield, according to the board's rules and regulations, were "regarded from an insurance viewpoint as constituting the least dangerous type of gasoline gas machine"; the board considered each successive class more dangerous than the preceding class.[111]

In order to improve the safety of gas lighting machines, the rules and regulations of the Board of Fire Underwriters focused on both specific construction details and installation configurations as well as operation of the systems. The extremely detailed rules "should be," the board warned, "rigidly observed." They address each component of Class A "cold process" systems, including the carburetor, blower, mixer, pressure regulator, outside piping, inside piping, and care and attendance. For example, the carburetor needed to be a minimum of thirty feet from the building to be lit, located below the frost line, constructed in a substantial and rigid manner, and painted on the outside with a heavy coat of tar to prevent corrosion. Furthermore, the carburetor had to be constructed of sheet copper or galvanized sheet steel (not less than No. 19 UDS gauge in thickness), with joints of the outer shell double seamed or riveted. It also had to have sufficient evaporating surface to ensure the production of saturated non-explosive vapor, and its fill and vent pipe had to be designed so that the fill pipe could not be opened without also opening the vent.[112]

Many Springfield customers wrote to Gilbert and Barker to express their comfort at using the Springfield Gas Machine over other types of lighting systems. Randolph Ladd, "Springfield's most prominent Underwriter," wrote in 1866 that "as an underwriter, I heartily approve of it [Springfield Gas Machine], and shall cheerfully recommend its introduction to any one who wishes to light their residence or manufactory with something better and safer than kerosene lamps." Customer Edward Hulburt of Portland, Connecticut, wrote to Gilbert and Barker in 1868 to confirm the Springfield system's approval by his insurance company: "I regard it as perfectly safe, and it is so admitted by the companies in which my insurance is made." Similarly, the owner's of Gambrill Sons and Company of Baltimore, who lit their Druid Mills at Woodbury, Maryland, with the Springfield system, explained in 1870 that the "machine was set in accordance with the regulations adopted by the National Board of Fire Underwriters to govern in such cases, and, after a personal inspection by the agents of the various companies in which our Insurance is made, our arrangements were approved, and their consent to the use of the machine indorsed upon our policies."[113]

The Springfield system's excellent safety record continued into the twentieth century: "We now have many thousands of these machines in use, covering a practice of thirty-six years," the company declared in 1906, "and we don't know that any insurance company has ever paid a dollar of loss caused by one of them."[114] Both the company's catalog statements and customer testimonials make clear the importance of the safety issue for both domestic and commercial customers and emphasize that much of the system's continued success was wrapped up in its excellent safety record and subsequent endorsement by insurance underwriters.

CHAPTER 4

A Bright Light for the Home: Domestic Lighting with the Springfield Gas Machine

The Springfield Gas Machine, among the higher priced gas machine systems available to the public, nevertheless gained in popularity with the upper middle and upper classes, particularly for rural and suburban homes. In many ways, the company's catalog listing of customers reads like a late-nineteenth- and early twentieth-century social register. According to the firm's catalogs, Gilbert and Barker installed upward of twenty thousand Springfield systems in businesses and homes nationwide along the East Coast and across the county by the end of the nineteenth century; domestic installations represented close to 50 percent of the total units sold.

By the last quarter of the nineteenth century, the Springfield system predominated in country houses and cottages, both of which became popular as part of the rural, romantic revival of the period.[1] Gilbert and Barker produced a catalog titled *How Best to Light Our County Homes and Resorts* specifically targeted at this market. Published in 1888 and 1906, this catalog contained illustrations of some of the best known country estates in America.[2]

The country house or "cottage" movement had its origins in Europe. Alexander Jackson Davis's *Rural Residences* (1837) and Andrew Jackson Downing's *Cottage Residences* (1842) and *The Architecture of Country Houses* (1850) introduced the movement in the United States and became influential in shaping the nation's picturesque country estate and suburban retreats. The country house phenomenon tapped into a growing interest in the rural and suburban retreat as an escape from the stress and pressure of an increasingly urban and industrial lifestyle. Businessmen and industrialists, note historians Gary Cross and Rick Szostak, were attempting "to regain the 'garden'—that is, traditional pastoral values—not by returning to the farm and village, but by a train or car ride from the city to the country home or leafy suburb."[3] Ironically, technological improvements in transportation, particularly the expansion of the railroads, allowed for easier movement and access to these more remote and tranquil locations. Cheaper, quicker, and easier travel, combined with building costs lowered by mass production, made the country house craze available to upper-middle-class as well as upper-class consumers.

Significantly, these cottages and estate homes typically employed the latest in modern conveniences at the same time they offered retreat and refuge in an idyllic rural setting. "In the late-nineteenth-century suburb," explains Cross and Szostak, "affluent Americans combined their longing for the countryside with an equal passion for urban technology and the wealth and comfort that it created." "Late nineteenth-century technological utopians," argues Cowan, "envisioned a perfect world in which mechanical ingenuity would subdue nature in order to bring prosperity and leisure, which would help ever larger numbers of men to reach the Romantic goals of creativity and free expression." Critics of the rise of technology, like early neurologist Dr. George M. Beard, felt that "technological society" and its products like the Springfield "exacted a stiff price for its comforts and conveniences in the historically unprecedented strain it placed upon the nervous system."[4]

Members of the expanding business class who enjoyed the economic means to build second homes purchased new technological innovations like the Springfield Gas Machine that flooded the marketplace in the second half of the nineteenth century. As customer Joel Baily of Darby, Pennsylvania, outside Philadelphia, wrote of the Springfield Gas Machine, "I consider it one of the chief modern improvements in house furnishing, and that no country house is complete without it." Similarly, H. S. Haywood, an executive with the Manhattan Savings Institution, expressed surprise that "so many of our beautiful suburban homes are still wanting in this almost indispensable luxury, good light."[5]

As early as the 1850s, advertisements for gas machines targeted at owners of country homes and estates began to appear. For example, an 1858 ad in the *New York Times* touted "Coates' Improved Portable Gas Works for the Country."[6] Gas machine manufacturers began to advertise in architectural design and plan books focused on country and suburban houses by the late 1860s. *Woodward's National Architect,* published in 1869, contained a full-page ad for the Empire State Gas Machine, one of the many systems appearing about this time in competition with the Springfield Gas Machine. Interestingly, Woodward's book also contained a full-page ad for "Ives' Patent Lamps" that announced the sale of "kerosene lights perfected"; kerosene lamps still played a role lighting country places and remained the principal lighting option available to more modest pocketbooks.[7]

Gas machines for lighting began to appear in real estate advertisements for houses for rent and sale in the 1870s. An 1875 advertisement for a "large, handsomely-situated double house" in Brooklyn noted that it "has all improvements; is heated by Boynton furnace, and lighted by improved underground gas machine." By the 1880s, the *New York Times* had classified sections for "Country Real Estate" and "Country Houses to Let" containing notices of country homes on

Long Island, along the Hudson, and in New Jersey, all of which touted "modern improvements" and luxuries. A notice for a private residence in Plainfield, New Jersey, with "all modern improvements; windmill, gas machine, &c," typified the marketing of gas lighting as a necessary modern convenience and marketable attribute.[8]

By the turn of the century, notices for country houses for sale and rent regularly included language boasting of modern improvements such as furnaces, indoor plumbing, and gas lighting. A 1903 ad for "A Gentleman's Country Place" on Long Island noted that the house "is modern and up to date in all appointments," highlighting that it was "lighted by a gas machine." While most ads simply provide a generic reference to a "gas machine" or "private gas machine," some did specify particular brands. An 1887 ad for an "elegant country seat known as 'Cherrywood'" in fashionable Elberon, New Jersey, noted that the property was "provided with the Imperial Gas Machine, and supplied with hot and cold water by a caloric pump."[9] An 1892 classified ad for a "large square country house" located outside Hartford specifies that the "house [is] lit from [a] Springfield gas machine." Companies' literature and testimonials also stressed gas light as a modern improvement, particularly in suburbs and on country estates. For example, Springfield Gas Machine customer George A. Burr wrote to Gilbert and Barker to laud its machine as "the most practical luxury among the many modern improvements that can be introduced into a county home."[10]

Modern improvements, like gas lighting systems, brought about significant changes in the social realm, giving domestic users a public image of prosperity and currency with prevailing styles and innovations. "Living by lamplight and candlelight," notes lighting historian Brian Bowers, "was not always idyllic. It affected people's outlook and way of life." "Lighting up the night with gas," writes Schivelbusch, "stirred people's feelings because it represented a triumph over the natural order. . . . Gaslight offered life, warmth, and closeness."[11] Gas's bright and steady light literally lengthened the day, permitting people to work in homes at night as they did by day, and wrought changes in the rhythm of daily life. "Gaslight," explains Bowers, "made it easier to read and write, or do other things at home in the evening." Irish author Alice Taylor recollected: "The pace of life was slow for those who grew up with oil lamps and candles. It was not advisable to go rushing around in candle-light as the way forward was not brightly lit."[12] Good quality lighting, first gas and then electric, enabled a faster pace of life in the home and on the factory floor.

"In addition to improving on the illuminating power of candles and lamps," writes historian Merritt Ierley, "gaslight allowed for new possibilities in lighting the average house." Chandeliers using candles and oil lamps, Ierley notes, were "too expensive to be practical for most people. The gasolier . . . on the other hand, made room lighting practical for the first time for the average home." The introduction of gas light into the home also altered domestic labor by eliminating at least some of the work involved in maintaining older lighting devices like kerosene lamps. "The spread of artificial illumination on streets and in homes, by means of gas, kerosene, and electricity," writes Trachtenberg, also "increased the practice, after 1870, of publishing evening newspapers in the afternoon, especially to catch homebound shoppers and workers and downtown evening crowds."[13]

Both within and outside the home, gas light enabled new levels of social interaction; large rooms such as ballrooms and parlors could be brightly lit at a reasonable cost, so that parties, dances, and other social engagements could take place at night in festive surroundings. Robert Louis Stevenson noted in 1881 that "when gas first spread along a city . . . a new age had begun for sociality and corporate pleasure-seeking . . . ; mankind and its supper parties were no longer at the mercy of a few miles of sea-fog; sundown no longer emptied the promenade; and the day was lengthened out to every man's fancy." Writing in the *Gas-Consumer's Guide*, Richards proclaimed that gas lighting "has afforded great facilities for persons to assemble at nightly scientific, literary, musical, and other meetings; it has thus facilitated the reunion of society of every grade." Gas lighting, explains historian Christopher Castaneda, "allowed people to watch indoor and nighttime theatre performances; oil lamps were comparably ineffective."[14]

Most likely, the general populace first encountered the wonders of gas lighting through the introduction of street lights. Robert Louis Stevenson captured the magical effect of street lights and immortalized those who lit them in his poem "The Lamplighter":

> My tea is nearly ready and the sun has left the sky;
> It's time to take the window to see Leerie going by;
> For every night at teatime and before you take your seat,
> With lantern and with ladder he comes posting up the street.
>
> Now Tom would be a driver and Maria go to sea,
> And my papa's a banker and as rich as he can be,
> But I, when I am stronger and can choose what I'm to do,
> O Leerie, I'll go round at night and light the lamps with you!
>
> For we are very lucky, with a lamp before the door,
> And Leerie stops to light it as he lights so many more;
> And O! before you hurry by with ladder and with light;
> O Leerie, see a little child and nod to him to-night![15]

Gas lighting radically changed the city streetscape, "making it safer and easier to go out at night." Before street lights, going out of doors after dark could be a dangerous venture.[16] Watchmen were necessary to help ensure public safety, and citizens were often called upon to help light the streets. Brightly illuminated streets made it safer and more pleasant for people to move about the city at night and eliminated the need for the night watchman. Gas lighting, Richards enthused, "tended to prevent crime" and "rendered life and property more secure."[17]

Gas lighting also altered social experience by changing the way things looked. In low light, everything seems to be colored in shades of gray, black, and white. "Colours seen by candle light, will not seem the same by day," as poet Elizabeth Barrett Browning noted. Light from candles and oil lamps cast a flickering, tentative light that affected the way people experienced the world around them after dark. Candlelight, remembers Irish author Alice Taylor, "was kind . . . to ageing faces, cobwebs and bad housekeeping, its soft flickering glow casting gentle shadows over many a blemish, human and otherwise."[18]

Artists faced technical and aesthetic challenges in painting scenes that took place at night and in correctly representing the colors and tones revealed by weak candle and lamplight. "The technical qualities of gas lighting and its impact on perceptions," explains Schivelbusch, "can be summed up in a single word: distance. The candle and the oil-lamp were extremely intimate forms of light, as they put out only enough light to illuminate a small area." People looked different in brighter gas light from candle or lamp light which may have played a role in changes to clothing color and styles and makeup. For example, the lower levels of light provided by pre-gas lighting sources encouraged the use of brighter colors and sharper contrasts. Ironically, the intensity of gas light was so powerful, however, that many felt is should be "covered with shades made of material such as frosted glass, which dissolved the concentrated core of the light."[19]

As with all new inventions, however, not everyone was overly impressed with the new ways of lighting. For example, Edgar Allen Poe complained that "we are violently enamored of gas and of glass. The former is totally inadmissible within doors. Its harsh and unsteady light offends. . . . The huge and unmeaning glass chandeliers, prism-cut, gas-lighted and without shade, which dangle in our most fashionable drawing-rooms, may be cited as the quintessence of all that is false in taste or preposterous in folly." Likewise, socialite Frederika Bremer complained in 1850 that "evening parties do not agree with me; the heat produced by the gaslights of the drawing-rooms makes me feverish."[20]

A common complaint and fear about gas lighting in general was its effect on the quality of air in a room and the effects of the combustion process on human health. The burning of gas lights produced byproducts, such as carbon monoxide and carbonic acid, along with heat and moisture vapor. The production of carbonic acid was of particular concern for municipal gas customers, causing the proponents of gas light to caution users about proper ventilation of all buildings using gas and the need for an adequate supply of fresh air in all structures. As gas industry proponent Richards points out, however, byproducts are produced whether one burns candles, oil lamps, or gas for light.[21]

The introduction of gas lighting offered many changes in household chores and activities related to household lighting. The demands of maintaining lamps were quite onerous and frequently commented on by nineteenth-century writers such as Laura Ingalls Wilder and Catherine E. Beecher. Discussing his lighting options in the rural village of Shelburne Falls, Massachusetts, Springfield customer J. W. Gardner commented that "kerosene I dreaded, not alone on account of the trouble and care of the lamps, but more especially because of their liability to breakage and explosion." Historian Susan Strasser enumerates the "tasks associated with lamps" as "daily chimney wiping and wick trimming, weekly washing of chimneys and shades, and periodic rewicking and dismantling for thorough cleaning." Rachel Haskell, a Nevada housewife and diarist, continues Strasser, "complained of these operations, calling lamp cleaning 'a hard operation always.'"[22]

The self-contained gas machine had another very important social implication: It eliminated the difference between lighting in urban and rural homes for those who wanted and could afford it. People in country homes could thus entertain guests in grand style just as their city cousins could. In fact, the use of gas lighting in remote and rural areas was a demonstration of power, both the technological and economic power of the homeowner to purchase and use such a system. Control over light was increasingly important, explains Schivelbusch. The period, he notes, witnessed the "evolution of homogeneous light into two increasingly different species, outside light and inside light." This transformation, he continues, was "related to the wider social process by which the public and the private were increasingly separated in bourgeois life." On the one hand, these systems eliminated some of the urban/rural social disparity, bringing people in rural areas up to date with those in urban settings. On the other hand, writes Ierley, "gaslight by home manufacture was clearly for the well-to-do, and even then, the gasworks in the cellar was not necessarily in use all the time. Sometimes it was turned on only in advance of special occasions, so as to have the gas lamps and gasoliers all aglow for guests. And when the duly impressed visitors had left, the family perhaps went back to using kerosene lamps."[23]

By pitching the Springfield as ideally suited for lighting cottage or country homes, Gilbert and Barker positioned themselves to take advantage of several trends: the developing romantic revival in architecture and landscape architecture, the rapidly expanding suburban landscape and its culture of acquisition and display, and architectural reforms of the period related to improvements in plumbing, lighting, and heating systems. The company's self-contained system also addressed an increasing suspicion of light produced from a source outside the home. "Ostensibly burning in the middle of the room in the lamp," argues Schivelbusch, "its real origin was in the gas-works . . . that is in 'big industry,' from which the bourgeois psyche tried to separate itself as it did from the public sphere."[24]

By 1906 gas machine companies like Gilbert and Barker were also promoting the use of their machines to fuel new lines of "labor saving" products, including ranges, broilers, laundry iron heaters, stoves, gas logs, water heaters, and radiators. The introduction of these very types of appliances, notes historian Ruth Schwartz Cowan in her study of the "Industrial Revolution" in the home, also brought changes in household structure and the work life of middle-class women. Cowan argues that while these products markedly improved the average American's standard of living, they "did not substantially decrease the amount of work that women had to do in their homes." Cowan writes that "even when they had access to cast-iron stoves and tin washtubs, piped water and gas lamps, carpet sweepers and treadle sewing machines, women found that there was a great deal of backbreaking labor still left to be done at home. . . . A woman's work, it seemed, would never be done, industrialization notwithstanding."[25]

The significant technological and social aspects of gas lighting, specifically gas produced in self-contained gas machines like the Springfield, are explored in detail below by examining several surviving residential systems. For example, the Alexander H. Stephens and Robert Toombs houses in central Georgia serve as wonderful examples of early (mid-1870s) domestic Springfield installations; the system at the Stephens House has survived extraordinarily complete. A slightly later version of the Springfield system, a Partially Buried Gas Generator from the early 1880s, exists at Ashland, the home of the McDowell family, relatives of the estate's original owner, politician Henry Clay. Post-1900 examples of Springfield installations in rural country houses or cottages include Sagamore, the Adirondack great camp of industrialist William West Durant and later Alfred Gwynne Vanderbilt, and Hill-Stead, the country home of iron magnate Alfred A. Pope. The two Springfield machines installed at Sagamore stand out as the most intact discovered to date, probably due to the site's remote location in the backwoods of central New York. Finally, documentation of installations at W. Seward Webb's model Shelburne Farms in Shelburne, Vermont, and several homes of the Ames family in North Easton, Massachusetts, provides additional information for discussion.

Countless installations have not survived. Air pumps and mixing regulators, accessible in the basements of homes, often ended up as scrap during World War I and II. For example, Oakes Ames of Massachusetts wrote that he "had the pleasure of throwing [our acetylene system] on the scrap pile when the Government called for scrap metal in the early days of World War II."[26] At many locations, only the buried generator, piping, and remnant fixtures survive. At still others, the only evidence of the system lies in correspondence, invoices, fire insurance maps, and other documents of the period.

Liberty Hall, the Alexander H. Stephens House

Liberty Hall, the home of Confederate vice-president Alexander Hamilton Stephens in Crawfordville, Georgia, received a No. 2 Springfield Gas Machine in 1875. The Springfield system at Stephens's Liberty Hall remains largely intact, although it appears to have been moved and later reassembled.[27]

Alexander Hamilton Stephens was a lawyer who worked in Georgia's northern circuit, and there met and became friends with fellow lawyer and Springfield Gas Machine owner Robert Toombs of nearby Washington, Georgia. Stephens practiced law in Crawfordville until his 1836 election to the Georgia House of Representatives. His friend Robert Toombs joined him in the legislature in 1837. Stephens ran for and won a seat in the state senate in 1842.[28] Small in stature and with a "frail physical constitution" from childhood, Stephens suffered from various ailments, including "melancholy" or severe depression. On top of his other health problems, Stephens was permanently disabled in an accident in 1869 and remained in a wheelchair for the rest of his life.[29]

Following his service in state office, Stephens ran successfully for the U.S. House of Representatives in 1843. A large slaveholder, Stephens emerged as a firm defender of slavery and states' rights.[30] In 1859, after fifteen years of service, Stephens retired to Liberty Hall, as he began to call his plantation.[31] Soon, however, southern succession would draw Stephens out of retirement.[32] He served as a delegate in the Confederate Provisional Congress and was elected vice-president of the Confederate States of America in February 1861; his longtime friend Robert Toombs accepted an appointment as the Confederacy's secretary of state. Following the war, Stephens found himself imprisoned in Boston's Fort Warren. With his pardon and release from prison in October 1865, he returned to his Georgia home.[33]

Back at Liberty Hall, Stephens wrote a monumental two-volume history of the war, *A Constitutional View of the Late War Between the States.*[34] In 1873, Stephens again won election to the U.S. House of Representatives, serving in Congress for the next nine years. In October 1882, the weak and aging Stephens was elected governor of Georgia, but he died on March 4, 1883, after just a few months in office.[35]

The *Savannah Morning News* (1875) described Stephens's original Liberty Hall as "a comfortable one, but constructed in the primitive style. . . . In 1845, all this was changed, a new roof was put on and other improvements made." In *Liberty Hall*, a history of the property and its owners, architectural historian Martha Norwood describes the house as a "two-story, four-bay, gable-roofed, frame structure with a four-over-four central-hall plan, four exterior end chimneys, and a central, one-bay, one-story, temple-form portico with simple Tuscan columns." In 1858, an article in the *Augusta Constitutionalist* reported that "the house is without any pretension to modern architectural style, but is built after the fashion and in conformity with the plans of the country residences of wealthy Georgia planters thirty years ago."[36]

Stephens seems to have installed his first portable gas lighting machine between the mid-1860s and 1874. A pamphlet for a gas machine sold by the Canster Gas Works and dated 1869 may give some clue to this earlier machine.[37] Letters found with the pamphlet were dated 1869, 1870, and 1871, and a handwritten note on the pamphlet stated that "Aunt Dora Stephens told Mr. Nash that A. H. Stephens installed gas lights in 1872." As architectural historian Martha Norwood writes, this original "gas system exemplified Stephens's standing in the county as one of its most eminent and prosperous citizens."[38]

It is probably not a coincidence that the pamphlet is dated 1869, the year Stephens had his disabling accident. It seems likely that the newly disabled and wheelchair-bound Stephens sought out an improved lighting source for the house in which he was increasingly confined. Stephens maintained regular work hours and had a structured regimen in the afternoon and evening that demanded good light. He would typically work "on his prodigious correspondence" from three until seven in the evening and then read until his bedtime at midnight.[39]

Stephens, who suffered from severe depression most of his life, reportedly felt quite cheerful during the rebuilding of Liberty Hall in 1875, with few bouts of "melancholy." His total focus on a major writing project probably ameliorated his depression, since he typically treated his melancholy with hard work. In addition to writing the second volume of his history of the Civil War, which he finished in 1870, Stephens began writing a school history of the United States in 1871. This focused, demanding work, however, may have negatively affected his already bad eyesight, further necessitating improved lighting in his home.[40]

Like the later Springfield system, the Canster Gas Works machine was a cold process gasoline system that created an enriched air-gas for lighting fixtures.[41] In order to combat the inherent problems of uneven evaporation rates, which resulted in variation in light quality, the Canster system injected small amounts of gasoline into a carburetor that evaporated the fuel rather than run air over large quantities of gasoline in a larger gas generator.

The early gas works at Liberty Hall probably stood in a gas works building, no doubt the same one used for the subsequent Springfield machine. The Canster brochure recommended that the "most desirable method of construction is to make an excavation, say six or ten feet square, according to the size of the machine. This space will be sufficient to contain the machines, tank etc.; but in order to get an ample fall for the weight, it is desirable to cover this . . . with pitched roof [or as a tool or storage shed]. . . . It would be well to erect a one story building."[42] This description very closely approximates the current gas house configuration, with its six-foot-square brick-lined pit and one-story, gable-roof structure.

Expanded in 1875, Stephens's newly configured central passage dwelling had four rooms on each of the two floors. He intended this mansion, journalist Sidney Herbert reported, "simply for the comfort and accommodation of the guests," and by all reports Stephens entertained a continuous stream of visitors. The upstairs rooms, according to a newspaper account, were "set apart for the use of the guests at 'Liberty Hall.'" "Every room," Herbert continues, "is provided with a grate for burning coal, is handsomely papered and lighted with gas, which is manufactured on the premises."[43]

It is unclear how Stephens discovered the Springfield system, although by the mid-1870s the company had become well known, particularly among the country's elites. Stephens did make several trips to the New England and the mid-Atlantic states, which had the majority of houses with Springfields at that time, and may have actually seen one in operation or heard good reports about the system. His close friend Robert Toombs installed a Springfield Gas Machine in late 1874, suggesting that Toombs may have recommended the system to Stephens. Likewise, Stephens's sister-in-law Mary Salter Stephens installed a Springfield machine in her home in 1875.[44]

The final bill from C. A. Robbe, "Practical Plumber, Gas and Steam Fitter" of Augusta, dated November 1, 1875, indicates that he removed an older gas system and installed a new No. 2 Springfield Gas Machine. Robbe charged $591.64 for the new machine, fixtures, two barrels of gasoline, and labor to install the entire system.[45] Robbe's invoice indicates that he probably removed the old gas pipes and system on May 31 and spent August 14 and September 8 piping the new or newly remodeled house. A letter from Stephens to Robbe dated September 21 indicates that Robbe had not yet installed the machine and fixtures: "I am becoming exceedingly impatient about my gas fixtures. It is now two weeks past the time you said you would have the new machine here and put up forthwith. . . . If you are going to put the works up I want it done at once." Robbe still hadn't completed the final work by October 13, when newsman Herbert commented that "just now every portion of the house is in a confused state, so far as 'putting the finishing touches' is concerned."[46]

The Springfield Gas Machine system at Liberty Hall is among the best preserved and most complete examples of the early design and configuration of the Springfield technology. The system as restored by Civilian Conservation Corps workers in the 1930s is housed in a reconstructed gas house that sits over the original brick-lined pit. The pit contains the copper gas generator, while the ground level portion of the gas house holds the copper air pump, weight, and mixing regulator. A single buried pipe delivered gas from the generator to the main house, indicating that the air pump was originally located in the gas house and not in the basement of the main house (as was typical during this period).[47]

Liberty Hall's No. 2 Springfield system had four evaporator pans or chambers, almost identical to the unit illustrated in the company's 1874 catalog.[48] The machine could power approximately twenty-five burners and held one hundred

Figure 4.1. Springfield Gas Machine generator, Liberty Hall. Photograph by author.

Figure 4.2. Springfield Gas Machine air pump, Liberty Hall. Photograph by author.

gallons of gasoline (see table 3.2).[49] The generator is constructed entirely of sheet copper and the valves and gauge are all brass; the windows in the gauge are isinglass. The gas generator, housed in a brick-lined pit, sits on a brick ledge above the floor so the unit could be drained. The air intake, located on the side of the twenty-eight-inch-high machine, delivered air to the bottom chamber. The enriched air-gas vapor moved up through the generator and exited through a distribution pipe at the top of the machine. This pipe ran underground for about sixty-two feet to the northeast corner of the main house. A shut-off valve is located just below the point where the pipe entered the house, running through the sill and into the wall.

The air pump is a weight-driven model manufactured by Gilbert and Barker and virtually identical to the one illustrated in the 1874 catalog.[50] Constructed of sheet copper with brass fittings, the pump provided pressurized air to the gas generator. The large concrete weight used to drive the air pump, suspended from the rafters of the building, conforms exactly to the weight typically used to power a No. 2 system.[51]

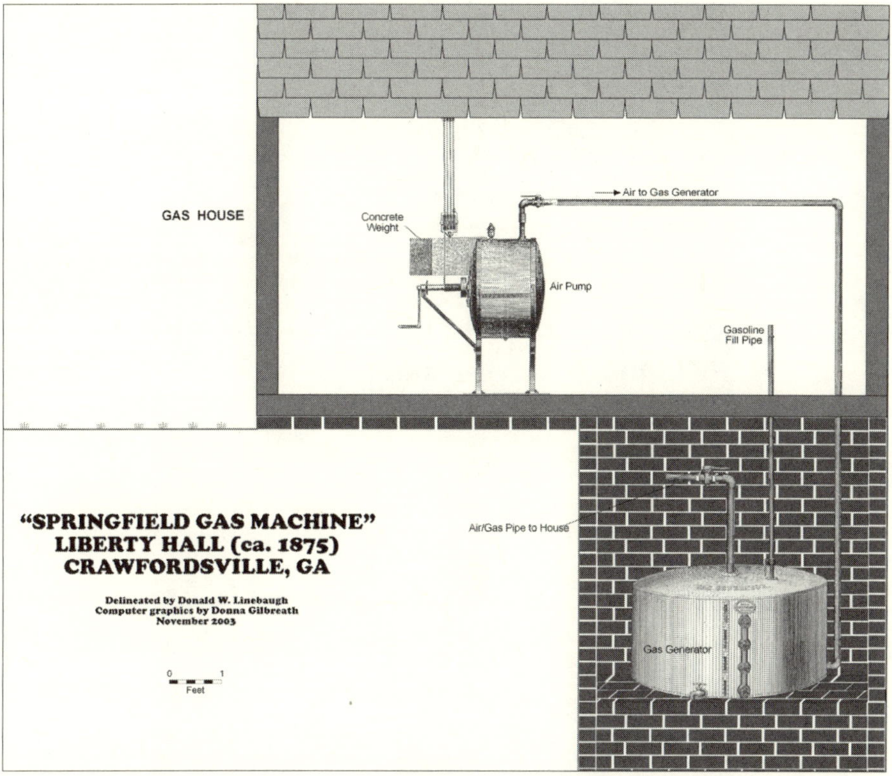

Figure 4.3. Conjectural drawing of Springfield Gas Machine, Liberty Hall. Delineated by Donald W. Linebaugh, computer graphics by Donna Gilbreath.

The mixing regulator for the Liberty Hall system was added after the initial installation, perhaps as late as the 1930s restoration. Gilbert and Barker did not develop and begin to sell these components until the early 1880s and it is not mentioned in Stephens's 1875 invoice. The unmarked mixing regulator differs from the type sold by Gilbert and Barker for use with their system.[52]

Most of the original gas fixtures and burners in Liberty Hall have been replaced with reproduction lighting fixtures, but the structure's original piping remains intact.[53] A two-light pendant located in Stephens's bedroom/office in the southeast corner of the house is the only original fixture. One of the burners has been fitted with a brass connection for a flexible tube, likely the "drop light ferrule" listed in the 1875 invoice, which ran to a portable lamp.[54] The portable lamp has an Argand-style burner identical to a burner in the collection at the nearby

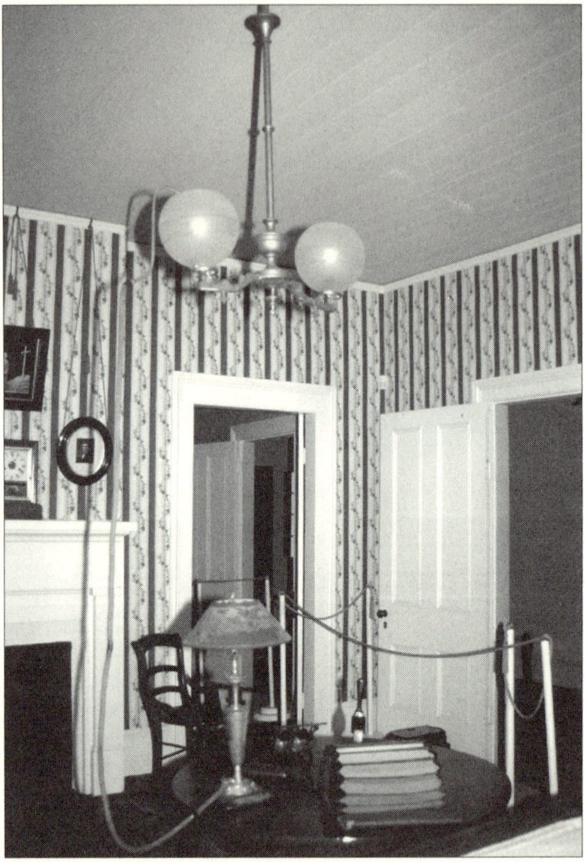

Figure 4.4. Hanging pendant fixture and portable desk lamp in Alexander Stephens's bedroom/office, first floor, Liberty Hall. Photograph by author.

Toombs House. The decorative portable lamp, with its Egyptian Revival–style, urn-shaped base and patterned mica shade, reportedly is original to the house and was used by Stephens on his desk.[55]

The first floor of Liberty Hall had hanging pendants in all four main rooms and in the central passage, as did the two rooms in the rear ell of the house. The four second-floor rooms and central hall each had one wall mounted fixture. The physical evidence at Liberty Hall indicates that it had a minimum of twelve fixtures probably with double burners. The twenty-four estimated burners also correspond to the 1875 invoice's list of one two-light chandelier, ten Argand burners, and fifteen adjustable burners, for a total of twenty-five burners.

Figure 4.5. Detail of portable desk lamp in Alexander Stephens's bedroom/office, first floor, Liberty Hall. Photograph by author.

The No. 2 Springfield Gas Machine system at Liberty Hall furnished light for the entire main house and appears to have remained in use until Stephens's death in 1883. In addition to providing a wealth of technical information on the installation and use of the early Springfield Gas Machine units, Liberty Hall presents an intriguing and socially important use of gas light by a renowned public figure. As noted above, Stephens entertained a steady stream of visitors and overnight guests at his home and remained in the public eye throughout his career.

Newspaper articles of the period suggest that Stephens expanded the structure for the comfort of his guests. A bachelor, Stephens had many permanent and semi-permanent residents, including both family and friends, and regularly hosted scores of travelers. A newspaper interviewer noted in 1875 that the new house existed "simply for the comfort and accommodation of the guests." Stephens also housed numerous young lawyers and clerks who worked for or studied with him.[56] The "new apartments" constructed in 1858 provided Stephens with a quiet

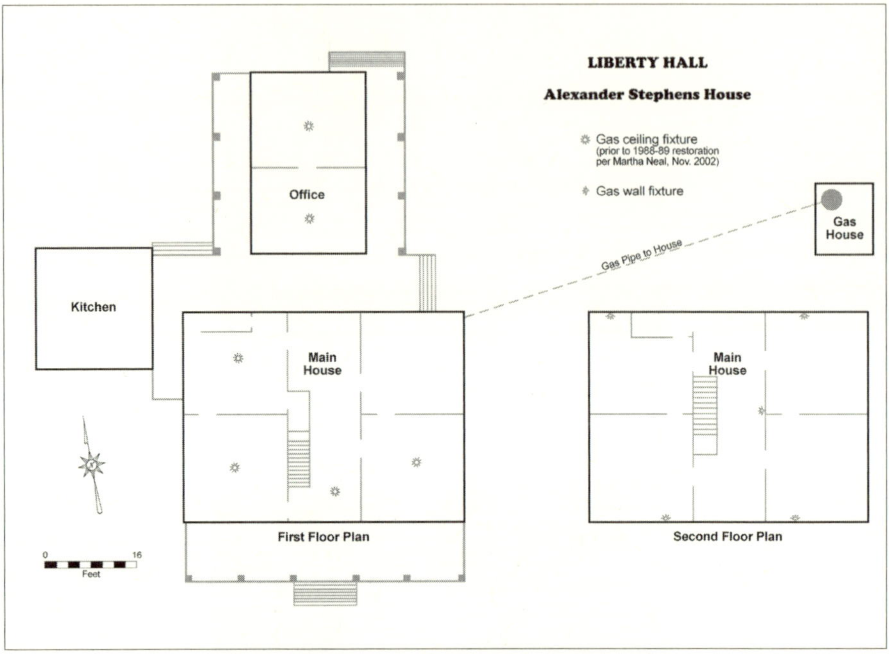

Figure 4.6. Plan of Liberty Hall showing locations of gas fixtures. Delineated by Donald W. Linebaugh, computer graphics by Donna Gilbreath.

bedroom and office space entirely separated from the regular comings and goings of guests in the main house. He found it difficult to turn anyone away, and clearly many took advantage of his generous spirit.

In rebuilding Liberty Hall in 1875, Stephens did not stint on amenities like fancy gilt wallpaper, window shades, floor cloths, a large and comfortable parlor and dining room for relaxation and conversation, and gas lighting in all the rooms.[57] Fresh wallpaper and floor cloths would have enhanced the gas lighting, adding to the house's bright quality. These amenities also enabled Stephens to deliver his idea of proper hospitality and highlighted his affluence and social position. The house, on a slight rise above and on the edge of Crawfordville, became a symbol of political and economic power in the county as well as a clear sign of prosperity. With its bright gas lights glowing each evening until midnight, Liberty Hall literally reflected the knowledge and power that shone in the life and career of Alexander H. Stephens.

The Robert Toombs House

The Robert Toombs House, located relatively close to Stephens's Liberty Hall, in nearby Washington, Georgia, was built in several stages during the late eighteenth and early nineteenth centuries and remodeled in the late nineteenth century. Like his close friend Stephens, Robert Toombs was a successful planter, lawyer, and politician. Toombs purchased his Springfield Gas Machine in 1874, about the same time as Stephens purchased his system.[58] While much less physical evidence of a Springfield machine remains at Toombs's home, several documentary clues and artifacts indicate the configuration of this lighting system. A group of extremely ornate gasoliers that still hang in the first-floor rooms of the house provide the most notable evidence of a gas machine. The use of these remarkable fixtures demonstrates the ways in which gas lighting could become an important aspect of a home's decoration and furnishings, reflecting the more up-to-date Victorian interior elegance and lavishness of a house otherwise referred to as "old fashioned."

Robert Toombs came from a privileged background. Expelled from the University of Georgia, he graduated from Union College in Schenectady, New York. He entered the University of Virginia's Law school in 1828, but lasted only

Figure 4.7. Elaborate gasolier fixture in living room of the Toombs House. Photograph by author.

one year there. Toombs was admitted to the bar in 1830 and married Julia Ann DuBose several months later. Settling in Washington, Georgia, he began to practice in the county courts; during these early years at the bar he met his lifelong friend Alexander Stephens. Between the Panic of 1837 and the beginning of his service in Congress, Toombs amassed a fortune from his legal work.[59]

In 1836, Toombs joined his colleague Stephens in the Georgia legislature. A Whig with "strong pro-slavery beliefs and a fervent states' rights advocate," Toombs served the state legislature until 1843. In 1844 he won election to the U.S. House of Representatives, where he again joined Stephens. Toombs was elected to the U.S. Senate in March 1853. The new junior senator owned property in five Georgia counties, large-scale speculative landholdings in Texas, and a home in Washington, D.C., where Julia reportedly entertained with unrestrained enthusiasm.[60]

Toombs resigned from the U.S. Senate and in January 1861 returned to Georgia, where he accepted the post of secretary of state in the new Confederate government. Increasingly opposed to Jefferson Davis's handling of the war, Toombs resigned his post and accepted a commission as a brigadier general in 1862. He resigned his commission in 1863 and returned to his home in Washington, Georgia. Following the arrest of his close friend Stephens in 1864, Toombs fled to Europe to avoid imprisonment.[61]

Returning from his self-imposed exile in 1867, Toombs resumed his law practice in Georgia and his land speculation in Texas. He regained a prominent place in the Georgia bar and once again became active in Georgia politics. By the end of the decade, however, Toombs's health declined; he reported that "troublesome cataracts were progressing slowly on his eyes." Following the deaths of Stephens and Julia in 1883, Toombs's health failed rapidly and he died in his home in December 1885.[62]

Biographer Pleasant A. Stovall described Toombs's Washington, Georgia, dwelling as "a three-story frame house . . . , built after the manner of the olden time, with the spacious piazza, heavy columns, the wide door, and the large rooms." Toombs purchased the house, reportedly built about 1797, in 1837. Beginning in 1850, Toombs made numerous improvements, including reworking the front facade in the popular Greek Revival style with huge Doric columns. When he returned from Europe and resumed his lucrative law practice, Toombs began a major building campaign, adding two Victorian wings to the main central block of the house. Documentary evidence suggests Toombs had a Springfield Gas Machine installed in late 1874, probably as part of the expansion project. Toombs continued to live and work in the house until his death in 1885.[63]

Relatively little documentary and physical evidence regarding the Springfield system at the Toombs House survives. Chronological evidence for the machine exists in the form of various company catalogs between 1875 and 1879. In a testimonial letter to Gilbert and Barker dated September 26, 1876, Toombs indicates that "the machine has done admirably well up to this time (about 2 years)."[64] Given the close relationship between Toombs and Stephens and the timing of the installations, it seems reasonable to speculate that Stephens first saw the Springfield Gas Machine system at Toombs's house.

While the machine itself no longer exists at the Toombs House, the museum's collection contains a name plate from the generator unit, identical to the one on the gas generator at Liberty Hall, suggesting a machine of similar vintage. Available evidence indicates Toombs purchased a No. 4 Springfield Gas Machine that could power a maximum of seventy-five burners.[65]

In addition to the name plate, a large concrete weight used to drive the air pump also survives at the Toombs House. The weight most closely resembles the twenty-by-thirty-four-inch weight designed to drive a No. 4 Springfield Gas Machine air pump. A photograph of one of the back, basement rooms shows the weight at the time the restoration was begun in 1976. This location represents the original placement of the air pump, as two supply pipes run from the corner of this room toward an outbuilding located some fifty feet behind the main house.[66]

The site's collection also contains several fixtures, gas pipes, and burner parts. Numerous gas fixtures survive in the house along with evidence of mounting plates and capped gas pipes where fixtures once existed. This evidence, compared with photographs of the house during the restoration, helped estimate the total number of fixtures powered by the Springfield system. Based on physical examination and restoration photographs, the basement level of the house had four wall and two ceiling fixtures, the first floor included three large gasoliers in the front hall and main rooms, along with two wall and two ceiling fixtures, and the second floor had eight separate wall fixtures in the hall and three main rooms. In total, there is evidence of fourteen wall and four ceiling fixtures, along with the three gasoliers. Oral tradition suggests that Toombs moved these large, ornate gasoliers from his Washington, D.C., residence to the Georgia house in 1861.[67] The gasoliers are highly ornamented with sculptural figures and decorative flourishes and hang from stylized ropes; etched glass globes protect the fixtures' single, flat-flame burners. The gasolier fixtures in the two main first-floor rooms each contain six separate burners, while the hall gasolier has three separate burners, for a total of fifteen. Wall and ceiling fixtures typically had one or two burners each. Thus the Springfield system at the Toombs House would have powered approximately seventy-one burners, very close to the rated capacity of the No. 4 Springfield system at seventy-five burners.[68]

In addition to the gasolier fixtures, two wall fixtures with movable arms survive, one in a second-floor bedroom and one in the hallway. These fixtures have Argand-type burners, similar to another burner in the collection stamped "PAT. 1867 REISS 1875."[69] Another burner part in the collection comes from an adjustable, single flat-flame burner produced by Gilbert and Barker in the 1870s.[70] This burner allowed for the adjustment of the air-gas mixture at the burner head which would control the richness of the flame and light.

Like Stephens, Toombs may have had another brand of lighting system prior to the installation of the Springfield system, given the tradition that he brought the large gasoliers from Washington, D.C., in the 1860s.[71] Alternatively, he may not have had any gas lighting system at the Georgia house and simply stored the gasoliers until he installed a Springfield in the mid-1870s.

In any event, the machine would have produced the extensive lighting suitable for the Toombs's large-scale, lavish parties. Toombs regained his financial health and statewide political power quickly following the war and remained a well-known lawyer, politician, and figure of national standing until his death. The machine may also have been important for his work, as he used one of the newly constructed wings of the house as his law office. He would have also required high-quality illumination when cataracts caused his eyesight to deteriorate in the mid-1870s.

The Springfield system at the Toombs House, unlike Stephens's more subdued Liberty Hall, also clearly reflects the family's embrace of proper Victorian decorating; the installation brought the interior of the house, "built after the manner of the olden time," up to date.[72] By the 1870s, gas light set the gold standard for good lighting and upper-class households considered it indispensable. The Toombs family certainly shared the notion that a properly furnished Victorian home required gas light. The light provided by the Springfield Gas Machine brought a brilliance and intensity to the Toombs House unique in the small county seat of Washington, Georgia.

Ashland

Ashland, the Henry Clay Estate, is best known as the home of the prominent Kentucky politician who served in the U.S. Senate for much of the first half of the nineteenth century.[73] Senator Clay, an associate of both Alexander Stephens and Robert Toombs, operated the then rural property as a large farm worked by slaves. By the early 1880s, Major Henry Clay McDowell had purchased the property. McDowell, a well-heeled Lexington lawyer and businessman, and his heirs continued to own Ashland into the twentieth century.

Henry Clay died in 1852, and his son, James, purchased the rundown house and farm at public auction in 1853. James subsequently razed the structure and hired noted architect Thomas Lewinski to rebuild the house in the same style as the original. The family lived there until the Civil War, when James, a southern sympathizer, fled the country. James died in 1864 and his widow, Susannah, sold the 324-acre estate to Kentucky University in January 1866.[74] Major Henry Clay McDowell and his wife Anne purchased Ashland in 1882.[75]

The McDowell family almost certainly installed the gas lighting system as part of renovations begun in 1882. While McDowell focused his energy on transforming the estate into a horse breeding farm, known as Ashland Stud, Anne directed her time and talent to completely redecorating the "old-fashioned" house. She reportedly "oversaw the redesigning of the grand staircase from its original elliptical design to the more fashionable Eastlake style. She also added the back stairway, Butler's pantry, the first indoor plumbing and introduced gas lighting into the house." A 1904 article in *Country Life in America* described the house as "a spacious two-story brick structure with one-story wings at both ends, which

Figure 4.8. Front facade of Ashland, ca. 1900. Courtesy of Ashland, The Henry Clay Estate, Lexington, Kentucky.

project out beyond the main portion of the house, making, as it were, three sides of a long, narrow quadrangle. The house stands on a slight eminence and faces toward Lexington."[76]

Ashland became a Lexington showplace in the late nineteenth century. The estate's associations with Henry Clay made it an important tourist "sight" for distinguished foreign visitors. Major McDowell exercised a kind of public hospitality, "his charm and graciousness in those days were proverbial as was his kindness."[77] The extensive renovation and addition of gas lighting in 1882 both resulted from and added materially to the McDowells' generous hospitality. A successful lawyer, businessman, and horse breeder, McDowell became a prominent member of Lexington and Kentucky society. With his wife, he served as the host at one of the region's and state's most notable homes. The socially conscience McDowells would have valued the Springfield Gas Machine's reputation as the best lighting system available to them in 1882.

Archaeological investigations of a small brick feature revealed both the buried Springfield gas generator and the pipes connecting it to the house.[78] The "Partially Buried" model of the generator rested on a brick ledge that stood above the floor of the rectangular brick lined shaft and extended into the chamber.[79] Made of riveted iron or steel, the cylindrical generator contained four evaporating chambers or pans, as indicated by the valves and fuel gauge on the side of the generator.[80] Although the dimensions of the unit do not correspond to any of the ten sizes listed in the company's 1874 catalog, a calculation of the unit's fuel capacity suggests that it was a No. 5 generator. This size held 350 gallons of gasoline and could power one hundred burners.[81] The installation is almost identical to the "Partially Buried Gas Generator" illustrated in Gilbert and Barker's 1885 pamphlet.[82]

The pipes conducting air to the machine and returning gas vapor to the house were identified running in a northeast-southwest direction.[83] Subsequent archaeological investigations revealed two other buried gas pipes, suggesting that the Springfield system lit several of Ashland's outbuildings; a historic photograph also shows gas lights along several of the walks.[84] The two-inch-diameter pipes between the generator and house further indicated that the McDowells bought a No. 5 Springfield. According to the 1906 catalog, Gilbert and Barker recommended this size pipe for the No. 5 system, which powered between 70 and 140 burners.[85] Examination of the concrete weight, marked "Springfield Gas Machine," in the basement of Ashland also suggests a No. 5 generator. While the weight remains, the air pump and mixing regulator are no longer present.

Extant gas fixtures in the house include chandeliers and pendants mounted in the ceiling, bracket sconces mounted on walls near windows, fireplaces, or doors,

Figure 4.9. Partially buried Springfield Gas Machine at Ashland. Photograph by Donald W. Linebaugh. Courtesy of the William S. Webb Museum of Anthropology, Lexington, Kentucky.

and a lamp mounted on the main stair newel post. This evidence suggests the house had a minimum of thirty-eight gas fixtures, the majority located on the first floor. The prevalence of fixtures on the first floor makes sense, since this is where entertaining, business, and other public activities took place. The second floor has very little evidence of gas lights, although later restoration may have altered or obscured this evidence. Two fixtures connected to supply pipes were found in the third-floor attic rooms, a space likely used to house servants in the 1880s.

Historic interior photographs include five images of rooms containing gas fixtures and one image showing an electric fixture. A photograph of the home's front entrance hall includes a five-globe lamp mounted on the newel post of the Eastlake-style staircase. The dining room photograph shows a six-globe chandelier and a pair of two-globe bracket fixtures over the fireplace; these fixtures, now electrified, still hang in the room. A photograph of the High Victorian–style drawing room shows an eight-globe chandelier, also electrified. The pair of two-globe bracket fixtures shown over the fireplace are no longer present in the house.

Figure 4.10. Parlor showing gas light fixtures at Ashland. Courtesy of Ashland, The Henry Clay Estate, Lexington, Kentucky.

The billiard room contains a four-globe chandelier with early electric fixtures attached to it and a single-globe bracket fixture at the fireplace. Finally, the library contains an unusual ceiling fixture with no burners or globes. This chandelier may have had detachable or "portable" fixtures connected to the gas source by flexible pipe, so that one could place them on desks or tables for more focused lighting, as seen at the Alexander H. Stephens House.[86]

The extant fixtures, early photographs, and mounting bracket evidence suggest a minimum of sixty-four burners at Ashland. Twenty-six of these are Argand-style burners with a cylindrical glass chimney that enclosed the flame; ornamental globes or shades rested on the frame of the fixture.[87]

Interestingly, McDowell purchased gasoline from various suppliers but apparently not from Gilbert and Barker.[88] In 1889, McDowell purchased two barrels of gasoline from the Detroit Heating and Lighting Company and must have asked them for information about maintenance of the machine. The company wrote: "We enclose [for] you our circular for re-plumbing a Detroit Combination Gas Machine and as we do not know whether you have a Springfield buried Carburetter [*sic*], or a vault Carburetter, we hardly know what to say."[89]

The Springfield Gas Machine system represents a significant step in the modernization of Ashland. Installation of gas lighting brought the rural estate up to the urban standard of interior lighting for upper-class residences; urban Lexington had established a gas company in the 1850s. The large number of fixtures on the first floor and substantially fewer fixtures on the upper floors suggests that the McDowells cared as much about social status and public image as increased comfort and convenience, since they would have used the first-floor rooms for entertaining and interaction with friends and business associates. The installation of the Springfield system as part of the McDowells' major remodeling and redecoration of the house in the Eastlake style, also points to the social importance and role of gas lighting. Ashland's gas lights thus served the dual purpose of improving the interior lighting and demonstrating the family's stylishness and wealth. The system remained in use in the house until approximately 1907, when Mrs. McDowell negotiated a contract for the installation of electric service. While the two systems may have operated concurrently for a time, electricity eventually made the gas works obsolete and it was abandoned.

Newspaper coverage of Nannette McDowell's wedding on the Ashland grounds in 1892 indicates the family had electric lights temporarily "introduced into the mansion. The extemporized banqueting hall running the width of the house at the back will be thus illuminated. This newly constructed supper apartment embraces a long verandah, the conservatory and a bit of freshly covered-in space—the whole presenting one uninterrupted length of brilliancy. . . . Electric lights will blossom from pink rosettes draping the ceiling."[90]

Several letters exchanged between Anne Clay McDowell and Henry Jr. in 1906 discussed the conversion of the house's lighting system. In a letter dated December 23, 1906, Henry comments, "While there is no 'money to burn' out of your income, I do think that you have put up with the very poor lights at Ashland long enough." Henry went on to suggest that she contact an agent of Gilbert and Barker Manufacturing for an assessment and cost estimate to overhaul the Springfield Gas Machine system: "It is quite probable that some rather inexpensive repairs to the evaporating machine (tank) and a new pipe from the tank to the house would give as good light as formerly. Neither electricity nor natural gas strikes me favorably. If necessary a new plant of the present type would be best."[91] Interestingly, while Henry was not favorably impressed with electric lighting, a May 29, 1907, letter acknowledged that electricity would soon light the rooms of Ashland.[92]

Sagamore, an Adirondack Great Camp

The Adirondack great camp known as Sagamore, located near Raquette Lake, New York, houses two virtually identical Springfield units installed between 1896 and 1902. Both systems employed "Buried" 150-light generators, water-powered air pumps, and mixing regulators to power gas lights in multiple buildings in the remote Sagamore complex. Although Sagamore's owners switched to hydroelectric power in 1914, they continued to use the gas machines for light along with electricity until the camp was connected to the power grid in the early 1930s.

Industrialist William West Durant had Sagamore built between 1895 and 1897. Durant's father, Thomas Clark Durant, was the general manager of the Union Pacific Railroad and speculated widely on land in New York's Adirondack Mountains to secure the railroad rights. Working with his son William, Durant eventually began to develop the land that he had acquired in central New York. William Durant located his first camp, Pine Knot, on Raquette Lake. He assembled a crew of more than two hundred to build and maintain this and subsequent camps and developed the infrastructure to house and supply the families of the workers.[93]

William Durant eventually sold Pine Knot to his father's colleague Collis P. Huntington, a founding partner of the rival Central Pacific Railroad. William built another camp, known as Uncas, on nearby Mohegan Lake between 1893 and 1895. Like Pine Knot, Uncas had a large assemblage of structures with totally separate guest and worker areas. Durant used Uncas as collateral for a loan with J. P. Morgan and lost the camp when he defaulted on the loan. He seemed to recover quickly, however, and wasted little time moving his crew to Sagamore Lake and beginning construction of his third camp between 1895 and 1897.[94]

Durant sited the Sagamore complex on a piece of land that jutted into the lake, providing a beautiful, idyllic setting. He spared no expense on the fifteen-hundred-acre property, installing hot and cold running water, indoor plumbing, a septic system, and, probably, the gas lighting system. Durant also had the lake stocked with fish to ensure the success of the "sports" (wealthy sportsman) who visited the camp.[95] The Springfield Gas Machine system was installed in the basement of the picturesque Main Lodge, a three-story, rustic-styled structure modeled after a Swiss chalet. This system provided light to both the lodge and the Dining Hall. In addition to the Main Lodge, the complex included the Dining Hall/Kitchen building, several cottages, and a series of support buildings for the workers carefully hidden from guests' view over a slight rise. The workers' complex lacked the amenities of the main camp structures; instead of gas lighting, workers relied on less effective and more dangerous kerosene lamps. The entire project reportedly cost Durant a then-staggering $250,000.

Due to several unfortunate events, including divorce, the death of his financial backer Collis Huntington, and a lawsuit accusing him of mismanaging the family fortune, Durant had to sell Sagamore to raise cash. He rather quickly interested Alfred G. Vanderbilt, then the wealthiest young man in America, in the property, and after several visits, Vanderbilt purchased Sagamore in 1901.[96]

Vanderbilt married Elsie French in 1901 and reportedly intended Sagamore as their honeymoon "cottage." A "sport" himself, Vanderbilt spent most summers at

Figure 4.11. Main Lodge, Sagamore. Courtesy of Sagamore, Institute of the Adirondacks, Raquette Lake, New York.

the camp. In 1901 and 1902, he had the Recreation Hall, or Play House, as it came to be known, built and installed a second Springfield Gas Machine in its basement. Probably about the same time, Vanderbilt expanded the Wigwam, a guesthouse, and added gas lights powered by the machine in the Play House. In February 1903, Vanderbilt wrote a testimonial letter to the Gilbert and Barker: "In response to your inquiry, I would say that the two gas machines and automatic mixers which you installed at Sagamore Lodge, have so far proved very satisfactory."[97]

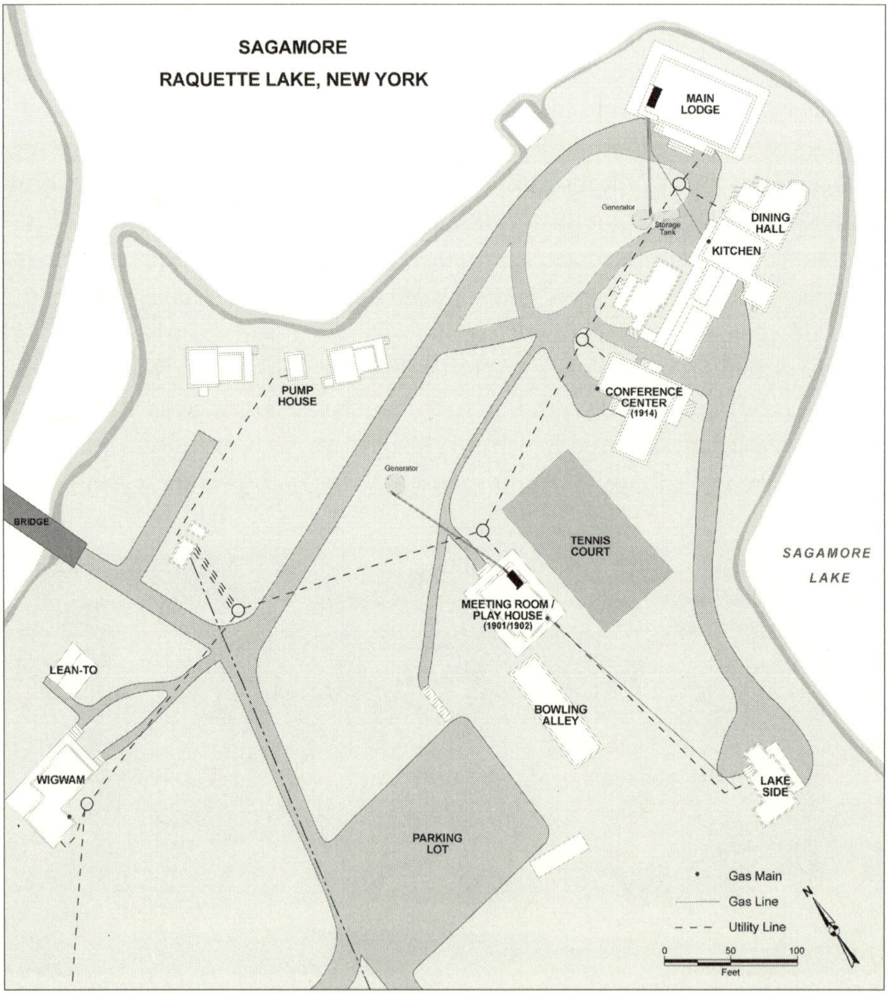

Figure 4.12. Plan of Sagamore showing the two Springfield Gas Machines. Delineated by Donald W. Linebaugh, computer graphics by Donna Gilbreath.

After a messy divorce from Elsie in 1908, Vanderbilt quietly married Margaret Emerson McKim, the daughter of Issac Emerson, the inventor of Bromo-Seltzer, in 1911. The new couple and their growing family further developed the property, completing another addition to the Wigwam, adding a hydroelectric plant in 1914, building a tennis court and bowling alley, and constructing a new laundry in 1915.[98]

The two Springfield Gas Machine systems at Sagamore are the most complete examples known to exist. The "Buried" Springfield Gas Machine in the Main Lodge, which also includes a bulk gasoline storage tank that supplied the gas generator, provided gas for light fixtures in the Main Lodge and Dining Hall/Kitchen; later, it may have also powered the Conference Building or the New Laundry. It appears that the gas generator for the Main Lodge system is a three–evaporator pan model similar to the ones illustrated in the company's 1897 *Directions for the Care of the Springfield Gas Machine.*[99] A fill pipe, with shut-off valve and vent pipe, and three "pump pipes" lie just below the surface, approximately eighty-five to ninety feet west and south of the northwest corner of the Main Lodge. This configuration of surface pipes is illustrated on a 1904 Gilbert and Barker letterhead.[100]

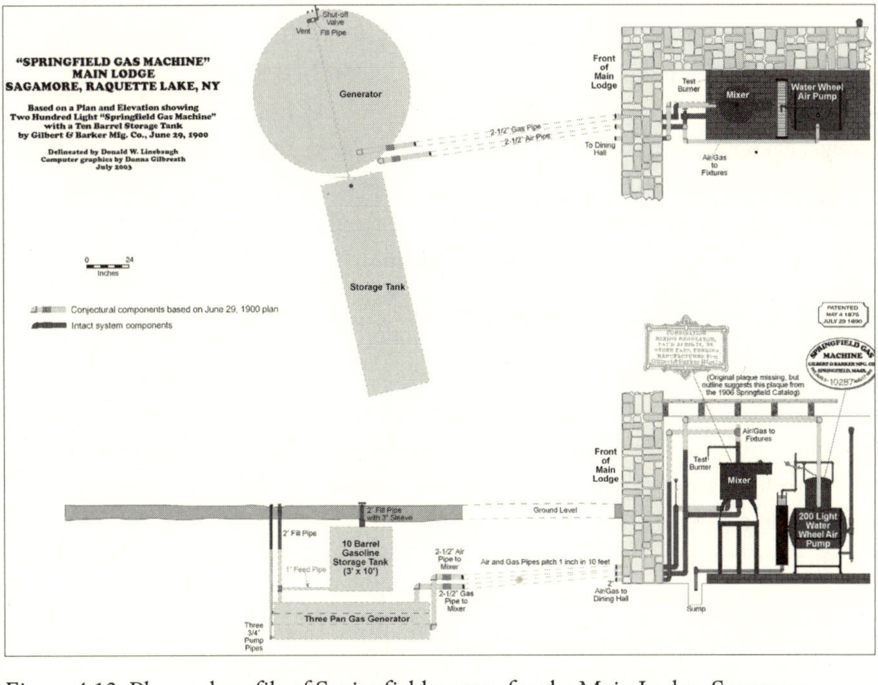

Figure 4.13. Plan and profile of Springfield system for the Main Lodge, Sagamore. Delineated by Donald W. Linebaugh, computer graphics by Donna Gilbreath.

Approximately seven and a half feet south of the valve and pump pipes lies a larger two-and-a-half-inch capped pipe, the fill pipe for the bulk gasoline storage tank. The 1906 Springfield catalog indicates that the system's two two-and-a-half-inch pipes would deliver sufficient gas to power a 150-light system.[101] The entire Main Lodge system is estimated to have held over 950 gallons of liquid gasoline. At approximately 1,000 cubic feet of air-gas from 6 gallons of liquid gasoline, the system could provide almost 160,000 cubic feet of air-gas between refills.[102]

Sagamore's system used a water-driven air pump manufactured by Gilbert and Barker, identical to the ones illustrated in the company's 1885 and 1897 operating instructions for the Springfield system.[103] The pump, located in the basement of the Main Lodge, provided air under pressure to the gas generator; Sagamore's reservoir supplied water to power the water wheel that drove the pump.[104] This automatic feature was important at Sagamore, guaranteeing steady, dependable gas light production whenever the Vanderbilts were in residence.

The mixing regulator for the Main Lodge system joins the gas generator and the air pump via two-and-a-half-inch pipes.[105] Air-gas flowed from the generator into the mixing regulator, which then adjusted the air to gas ratio. The Main Lodge system has test burners on the supply pipe from the generator and on the exit pipe from the mixer; these were used to test the flame quality of the air-gas before and after mixing to adjust the system for the best flame quality.[106]

While most of the gas fixtures and burners have disappeared from the Main Lodge, evidence of fixture locations in the form of capped pipes exists on all three principal floors and in the basement, suggesting that the Main Lodge had a minimum of sixty-five fixtures with either single or multiple burners.[107] Although several of the chandelier fixtures in the dining room have been converted to electricity, they are clearly gas fixtures and appear as such in photographs taken prior to the electrification of the complex. The Vanderbilts expanded the Dining Hall/Kitchen in 1901, doubling the size of the structure, and the new section also had gas lighting. The building has evidence of two four-burner chandeliers or gasoliers and nine wall fixture locations, including one intact single-burner fixture, in the form of capped pipes. The bedrooms on the second floor of the building contained an additional six fixtures. The Main Lodge system, therefore, powered a minimum of eighty-eight fixtures. Later, it may also have run the lights within the Conference or New Laundry building, another thirty-seven burners.[108] All of the fixtures in this later building are "combination" fixtures that utilized both gas and electric light using one or two burners and one lightbulb.[109]

The second Springfield system, located in the Recreation Hall or Play House, also appears to be a 150-light "Buried" Springfield Gas Machine with a water-driven air pump and mixing regulator. The system provided gas for light fixtures in the Play House, Wigwam, and Lakeside Cottage.

This gas generator also appears to have three evaporator pans, similar to the one in the Main Lodge; no evidence for an auxiliary tank was located. A fill pipe with shut-off valve and vent pipe and three pump pipes lie just below the surface, approximately 102 feet northwest of the northeast corner of the Play House. The generator connects to the water-driven air pump and mixing apparatus via two two-and-a-half-inch pipes that could accommodate a 150-light system. This configuration matches to the Main Lodge system exactly. The mixing regulator for the Play House system differs slightly from the one in the Main Lodge system, suggesting that the two systems were installed at different times.[110]

The Play House had approximately twenty-three gas fixtures, most likely with single burners.[111] The Play House system also lit the Wigwam, which had at least fifty-three single-burner fixtures.[112] The surviving fixtures in the Wigwam are marked "Springfield Incandescent Burner." These "latest and most improved" incandescent burners did not become available until about 1900, which corresponds with the construction of the Play House and the installation of the second

Figure 4.14. Springfield Gas Machine, water-powered air pump (right), and Combination Mixing Regulator (left), Play House, Sagamore. Photograph by author.

Springfield system in 1901 or 1902.[113] The Play House system also served the tiny Lakeside Cottage, which had approximately twelve single-burner fixtures.[114] Altogether, the Springfield Gas Machine system located in the Play House powered a minimum of eighty-eight burners and could have also run the thirty-seven burners within the Conference or New Laundry building.

The survival and preservation of the two Springfield systems at Sagamore no doubt have resulted from Sagamore's remote location. In addition to providing important technical information on the installation and use of the Springfield machines, Sagamore presents an intriguing example of use of gas light in a largely uninhabited and inaccessible place. Set in the type of sublime and romantic wilderness increasingly valued as an important antidote to industrial and consumer society, Sagamore served as a stage for expressions of power, influence, and social clout in Victorian America. The use of gas light in the remote woods of the central Adirondacks reflects the prevailing ideas of the period regarding the power of culture over nature and civilizing the frontier. Ironically, it also reflects the desire on the part of families like the Vanderbilts and their visitors to retain every possible luxury of modern society in the midst of a wilderness retreat.

Figure 4.15. Alfred Vanderbilt and "sports" in the kitchen building showing gas light fixtures. Courtesy of Sagamore, Institute of the Adirondacks, Raquette Lake, New York.

Just getting the Springfield system components to Sagamore was a major demonstration of power and resources. The initial gas generator and steel storage tank probably traveled by train to North Creek, New York, and then covered some 70 miles by steamboat and on dirt roads in horse drawn wagons. By 1901, when the Play House was built, the railroad reached all the way to Raquette Lake, but the trip to the camp itself still required a four-mile trek on dirt roads via a horse drawn wagon.

The Durants and Vanderbilts used Sagamore at a time when many Americans sought to return to a simpler, natural setting like the Adirondacks to recover from the stresses of modern urban living. Energized by the notion of the rapidly disappearing frontier in America, people began to view "wilderness" as heaven on earth rather than a place of desolation and terror. As expressed by nineteenth-century writers such as Henry David Thoreau and John Muir and painters such as Thomas Cole and Thomas Ekins, America's expansive lands had become sublime and sacred refuges.[115] Paradoxically, the camps built by wealthy, powerful men ameliorated much of the simple wilderness experience that they purportedly offered. Sagamore and camps like it provided the most current amenities of modern life, including hot and cold running water, septic/sewage systems, indoor plumbing, and gas and eventually hydroelectric lighting. Owners employed a full complement of servants to build, maintain, and operate the camps. These modern amenities were neatly packaged within Sagamore's rustic-styled buildings in a way that allowed visiting "sports" and family members to immerse themselves in the "wilderness" experience with a minimum of discomfort and inconvenience. Sagamore also shows the divide that occurred in the classes in terms of conditions such as lighting. The estate's servants and laborers lived over a rise and out of sight of the Main Lodge complex; their modest cabins were lit with kerosene lamps and lanterns.

Hill-Stead

Hill-Stead, the country home of retired iron industrialist Alfred Atmore Pope of Cleveland, Ohio, also featured many late Victorian luxuries. Built between 1898 and 1901, the house was a collaboration among Mr. and Mrs. Pope, their daughter Theodate, and the architectural firm of McKim, Mead, and White. The Popes purchased a Springfield Gas Machine system about 1900 and had it installed during the construction of the house. While little physical evidence of the system remains, the Hill-Stead Museum archives retains an installation drawing for the house's two-hundred-light New Feed Springfield Gas Machine.[116]

As a young man, Alfred A. Pope, born in New England in 1842, worked in his father's woolen business. He married Ada Brooks, also the child of an old New England family, in 1866, and the couple's only child, Effie Brooks Pope, was born in 1867; Effie changed her name to Theodate (meaning "gift of God") in 1886.[117] Alfred Pope left the family firm in 1869 to join the Cleveland Malleable Iron Company. He rose rapidly through the ranks and became president of the company in 1879. Under his leadership the firm became a national leader in the production and sale of malleable iron. The success of the firm provided the Popes with immense wealth. The family built a large brownstone house in the Romanesque style in a fashionable Cleveland neighborhood, just down the street from Standard Oil executive and fellow Springfield Gas Machine owner Henry Flagler and from Flagler's business partner John D. Rockefeller.

The family embarked on a year-long European tour in 1881, during which Alfred Pope emerged as a serious art collector. Over the next several decades, he amassed an amazing collection of Impressionist art and an extensive collection of decorative arts from around the world. In 1886, the Popes sent Theodate to Miss Porter's School for Girls in Farmington, Connecticut, where she became enamored of small-town life.[118]

Following an allegedly disastrous coming out party in 1889, Theodate convinced her parents to let her rent a small house in Farmington after completing school. Her move to Farmington not only distanced her from the marriage market of her parents' world but also allowed her to enjoy a simple and unpretentious lifestyle. During the family's travels in Europe, she had become very interested in architecture and dreamed of studying the subject. While in Farmington, Theodate joined other women who taught arts and crafts classes to the poor and eventually opened a tea shop where she sold students' work.[119]

In August 1898, Theodate convinced her parents to purchase 150 acres of land in Farmington, on which to build their country estate and retirement home.[120] Alfred Pope selected the firm of McKim, Mead and White, then the largest architectural firm in New York, to prepare the drawings for the house. Pope well knew the firm's reputation and design work on large country homes and townhouses, notably the Colonial Revival–style country home of friend and business associate John Howard Whittemore in nearby Naugatuck, Connecticut. Whittemore installed a Springfield Gas Machine in his house by about 1895.[121] Pope gave Theodate a free hand to "make the new estate according to her own ideas" and made her responsible for dealing with the architects. She became the project manager and conduit between the firm and her father and played a major part in determining the eventual design. The Popes and McKim, Mead and White reportedly had a stormy relationship as a result.[122]

This uneasy collaboration produced Hill-Stead, a vernacular Colonial Revival house designed in the style of the traditional New England extended farmstead. The Popes had the finished house, which reflected the romantic and historic sensibilities of the period, decorated specifically to "show off the works of art in sympathetic surroundings." They prized natural light and spared no expense in their demand for numerous large windows; many rooms had windows in three compass directions. When author Henry James visited, he commented on the structure's air and light quality and called the house "an exquisite palace of peace, and light and harmony."[123]

Like many rural country houses built at this time, Hill-Stead had all of the modern conveniences, including insulation, central heating, a cooling system for summer, indoor plumbing, and gas lighting provided by a Springfield Gas Machine. The Springfield system, in particular, would have interested the Popes greatly, given their focus on proper lighting for their art collection. Writing to her father in June 1899 or 1900, Theodate reported that she "went to the architect yesterday and had a satisfactory talk with Mr. Swartwout. . . . Had a representative of Springfield Gas Machine down the other day. Advises no separate machine for that house—might have to put in 300 light machine at [your] house instead of 200 to supply both needs though 300 much too large & nothing between them—they run 100, 200, 300, etc., etc. Yet it might draw on our source too much to pipe from

Figure 4.16. Hill-Stead. Photograph by author.

ours and use from our 200 light machine. He did not urge the matter and thought with the use we make of that house and the smallness of it we would better just use lamps as everyone else does here." Theodate apparently considered lighting another home on the property from the system planned for Hill-Stead, until dis-

Figure 4.17. "Chess Game" (1921) by G. K. Peterson. John Wallace Riddle (*left*), Theodate Pope Riddle, foster sons Donald Carson and Paul Martin, and the boys' tutor (*standing*) in the second library, Hill-Stead. Archives, Hill-Stead Museum, Farmington, Connecticut.

couraged by the Springfield salesman. The Popes realized that the Springfield would be a novelty in Farmington, as indicated by Theodate's comment about the use of kerosene lamps "as everyone else does here."[124] The Popes ordered the two-hundred-light Springfield system for Hill-Stead by June 1900, and the system was operational by the house's completion in the spring of 1901. The Popes "gave their first dinner, by gaslight, as new residents" on May 30, 1901.[125]

Mr. Pope died in 1913, and shortly thereafter Theodate opened an architectural office in New York.[126] The same year, she married diplomat John Wallace Riddle and the couple moved to Hill-Stead. Riddle traveled frequently and held several overseas assignments, but Theodate remained at Hill-Stead, where she maintained an office for her architectural practice.[127] A 1924 Sanborn map of Hill-Stead does not mention the Springfield machine, suggesting that the property had been electrified by this time.[128]

The Springfield Gas Machine installed at Hill-Stead was a two-hundred-light system consisting of a buried gas generator, a water-driven air pump, a mixing regulator, and a gasoline storage tank that fed the gas generator. The basement of the house still contains two two-and-a-half-inch pipes that connected the generator and air pump.[129] In addition to the larger pipes, a small pipe to supply water to the air pump was also located in the basement.

Based on Gilbert and Barker's June 29, 1900, drawing, the Popes purchased a four–evaporator pan buried generator. Notes on the drawing, possibly made by Theodate, indicate that the machine held approximately 697 gallons of gasoline when full.[130] In addition to the generator, the system included a ten-barrel storage tank that attached to the generator with a one-inch feed pipe; this pipe had a valve operated by a control rod for refilling the generator. The storage tank could hold approximately 420 gallons, making the entire system about 1,100 gallons of gasoline.[131]

The air pump, identical to the pumps at Sagamore, was a water-driven model manufactured by Gilbert and Barker that provided "automatic" operation with little regular maintenance. The mixing regulator illustrated is very similar to the Combination Mixing Regulator shown in the 1906 Springfield catalog. Handwritten notes on the drawing report that "this mixer is entirely automatic, it needs no care, whatsoever."[132]

Although the Springfield system components are no longer extant, most of the gas fixtures and burners in Hill-Stead have survived. These include a range of gasoliers and wall scones that held both single and multiple burners. While the installations at most of the other houses investigated have relatively plain, factory-produced fixtures, the Popes spared no expense and had their "bronze wall gas

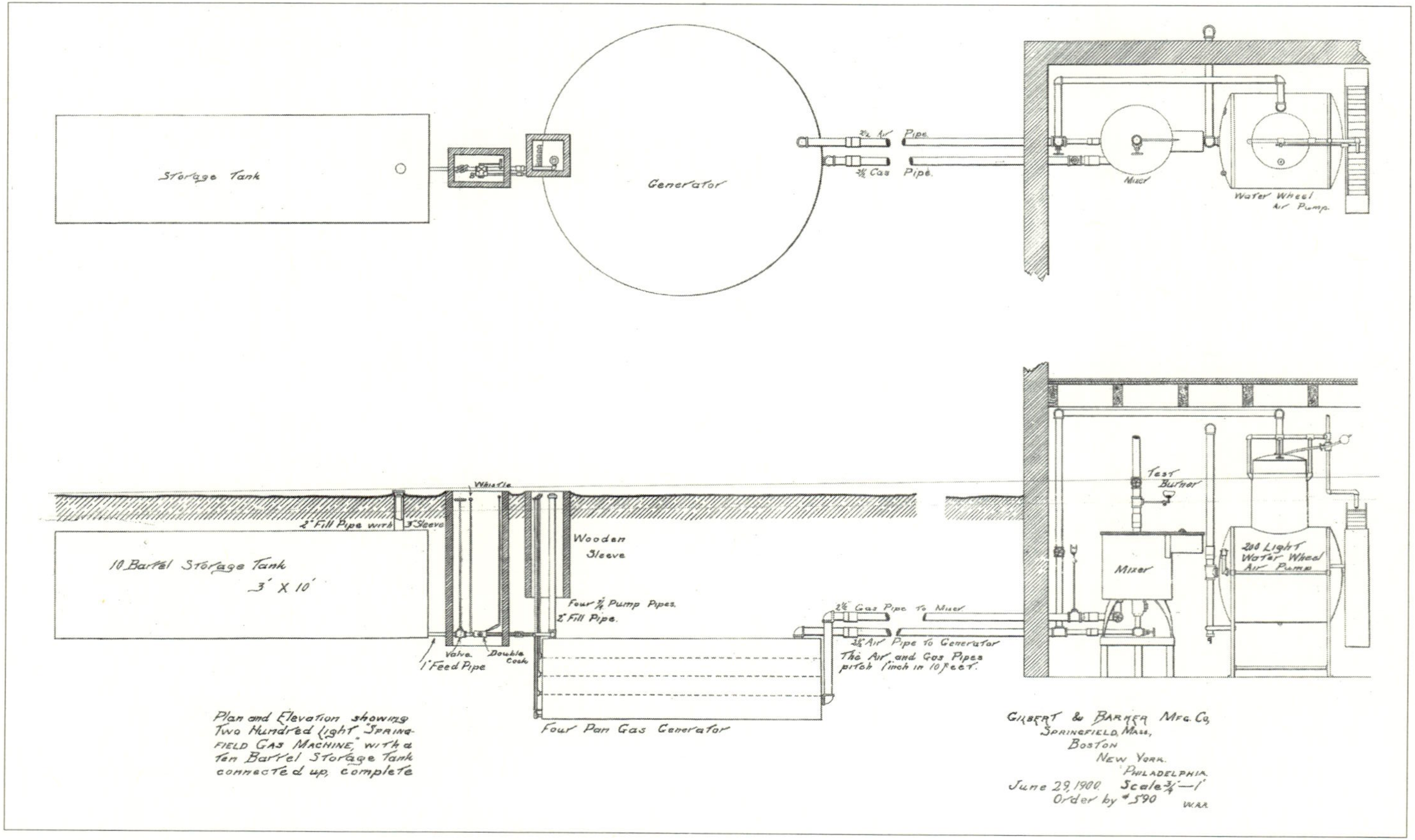

Figure 4.18. Plan of Springfield Gas Machine for Hill-Stead, June 29, 1900. Archives, Hill-Stead Museum, Farmington, Connecticut.

fixtures" produced by the Boston jewelry firm of Shreve, Crump and Lowe. These stylish fixtures, with their etched glass shades and glass pendants, were clearly integral elements of the interior design. "An overlay of decoration," write Hazen and Hazen, "transformed the overpowering rawness of a fire into genteel 'warmth' and 'radiant beauty.'"[133]

The installation of the Springfield Gas Machine system at Hill-Stead suggests the use of gas lighting for both practical and social purposes. Given the location of the house near the still rural village of Farmington, the system provided the only lighting option available other than kerosene lamps. In fact, Theodate Pope wrote that all town residents used kerosene lamps and that the Popes could choose to use them as well.[134] Due to her work with the architectural firm of McKim, Mead, and White, however, Theodate would have known about the latest systems for central heating, cooling, indoor plumbing, and light. Ironically, properly outfitting a country house as a retreat from the stresses of business and industry required the use of the most up-to-date and technically sophisticated inventions.

Figure 4.19. Hill-Stead dining room, showing Shreve, Crump, Lowe gasolier fixtures. From *Architectural Record*, August 1906. Archives, Hill-Stead Museum, Farmington, Connecticut.

Like many other country home owners, the Popes probably discovered the Springfield system through connections with satisfied Gilbert and Barker customers. At Hill-Stead, it seems that the Popes' friendship with the Whittemore family, whose house was also designed by McKim, Mead, and White, may have influenced their decision to purchase a Springfield system.[135]

Shelburne Farms

The Springfield system at Shelburne Farms, the country estate and model farm built and owned by William S. Webb in Shelburne, Vermont, also exemplifies the gas machine's place in providing lighting to country homes and estates. The Springfield Gas Machine at the Webbs' property, a quintessential country estate of the late nineteenth and early twentieth century, is well documented with maps, catalogs, and installation drawings. As with many other Springfield system owners, the Webbs occupied a top rung on the social ladder and participated fully in the revolutionary industrial changes taking place at the time. William Webb's interest in breeding work horses illustrates the many ironies and dichotomies of the period and the seeming anti-industrial sentiment of the county house movement. While serving as the president of a firm that built the latest in passenger rail cars, Webb created a model farm to improve the breeding of work horses to pull carriages and farm machinery.

Webb graduated from the College of Physicians and Surgeons of New York in 1875 but abandoned the medical profession for Wall Street in 1879. He went to work for railroad magnate William Henry Vanderbilt, son of "Commodore" Cornelius Vanderbilt and uncle of Alfred Vanderbilt of Sagamore, and quickly rescued "a floundering Wagner Palace Car Company" from economic ruin. In 1881, Webb became president of the company and married the boss's daughter, Lila Vanderbilt. Lila had grown up in the family's palatial home on Fifth Avenue in New York, and like Theodate Pope of Hill-Stead and Madeline McDowell of Ashland, she had attended Miss Porter's School in Farmington, Connecticut. Webb had first traveled to Vermont in 1880, when Vanderbilt asked him to investigate the Rutland Railroad as a possible takeover candidate. In 1882, the Webbs began construction of a "modest country estate," Oakledge, in Burlington.[136]

The couple regularly summered in Burlington, and Webb soon began to purchase property along the Lake Champlain shoreline near Shelburne. He planned to acquire some of the thirty small farms that made up his grandfather's post–Revolutionary War holdings in this area. By 1886, Webb had acquired several thousand acres and engaged the elderly Frederick Law Olmsted as an advisor to his project.[137] Although Webb did not follow Olmsted's recommendations precisely, he did craft an estate influenced by Olmsted's "Natural Style." Ironically, writes Shelburne historian Joe Sherman, "men [like Webb] whose wealth came from the dynamics of change . . . desired retreats reminiscent of their pre-industrial past, on the land of their grandfathers."[138] These fathers of industry spent lavishly to fashion their own pastoral oases. The resulting landscape depended heavily on the very industrial and commercial world that their builders sought to escape, particularly in terms of the equipment and technology used to create and then operate these large country estates. The Webbs, for example, employed the latest technologies to furnish household services—gas and electric lighting, central heating, indoor plumbing—which matched or exceeded anything available in a major city.

While Shelburne Farms exemplifies the ideal country estate of the period in many ways, it also functioned as a very large working farm. Webb envisioned, Sherman notes, a "farm of the new age, vast in dimension, functioning outside typical economic restraints, embracing both the pastoral landscape of yesterday and the scientific farm of tomorrow." For example, Webb built a breeding barn, the largest of its kind in America, that could stable over three hundred horses. He was, as Sherman states, "convinced that Vermonters had crossbred too many trotters and mares, saddling farmers with a skinny horse that could carry them quickly to the station but hardly pull a plow. He intended to change this situation by breeding a refined all-purposed hackney, a strong stocky horse that could both pull the plow and take the farmer to town in style."[139]

In 1887 and 1888, the Webbs built a Shingle-style "cottage" on their Shelburne Farms property. Designed by architect Robert H. Robertson, the dwelling was two stories tall, 128 feet long, and 50 feet wide. By 1892, the estate had grown to over three thousand acres and its buildings included the cottage, the breeding barn (ca. 1892), a massive five-story farm barn complex (ca. 1890), three acres of greenhouses, and a host of other smaller, specialized farm buildings. In 1895, the Webbs had their "cottage" converted into a new principal residence three times the size of the original. Robertson's plans called for raising the roof one story, adding several new wings, and decorating the entire complex in the Queen Anne style.[140]

One of the several model farm buildings erected by Webb was the Farm Barn, largely completed by 1890. This massive structure, lit with a Springfield system, was "built on three sides of a square, the fourth side consisting of a massive stone wall, with a gateway in the middle, enclosing a large yard over 400 feet long and 260 feet wide. The main or central portion of the building is about 416 feet in length, has a depth of about 50 feet, and is no less than five stories high ." The central building, according to Webb, stabled "seventy or eighty head of well-fed and grandly conditioned mules." Other parts of the building housed farm implements and stored hay and corn; the hayloft could hold fifteen hundred tons of hay. The building also contained a series of offices and paint, carpenter, and blacksmith shops.[141]

The entire complex of domestic and farm buildings featured the latest technology, particularly in terms of fire protection. Webb painstakingly described the elaborate water supply system for the property, which pumped water from the lake to a reservoir, "the entire force capable of being directed on to any particular building, reducing the possibility of fire to a minimum." He boasted in 1893 that "the electric light is in use everywhere on the property, a large and powerful plant having been constructed in a building especially erected for the purpose," adding that "there is also an elaborate gas plant on the place, so that in the matter of illumination every department is fully provided for."[142]

Although the farm operation seemed to do well financially for several years, by 1920 the "agricultural divisions of the Farms generated little if any profit." As farmers increasingly bought cars and tractors, the venture into breeding hackneys failed. In declining health, Webb saw his railroad interests and investments largely disappear due to a patent lawsuit that resulted in the loss of the Wagner Palace Car Company to George Pullman.[143] Webb died in 1926, followed by Lila in 1936, and the farm was inherited by Vanderbilt Webb, who used it as a summer home and continued to run the agricultural operations into the mid-1950s.

The "elaborate gas plant" that Webb described in 1893 referred to a Springfield Gas Machine located in the Farm Barn complex. Sanborn Fire Insurance maps confirm that this was the only gas machine on the huge Shelburne Farms estate. An inspection of the Farm Barn complex did not reveal any aboveground evidence of the system, so research into the Springfield system focused exclusively on archival materials. Maps, photographs, company catalogs, and correspondence provide a reasonably clear idea of the Shelburne Farms system.

Sanborn maps from 1894 to 1942 and Gilbert and Barker catalogs in the Shelburne Farms Archives provide the best documentation of the use of the Springfield Gas Machine at the farm. The 1894 Sanborn map records that "gas & elec." provided the lighting, and shows a "Gasoline Tank, Underground," located in the south-central portion of the courtyard of the Farm Barn complex, just north of a short wall.[144] It is possible that the Springfield machine lit other buildings, but these may also have been lit exclusively with electricity or oil lamps. It seems likely that the main house had electric lighting from its initial construction or expansion, given the extant fixtures and lack of evidence for gas piping.

An undated plan drawing in the Shelburne Farms archives illustrates a Springfield Gas Machine consisting of a partially buried generator, measuring

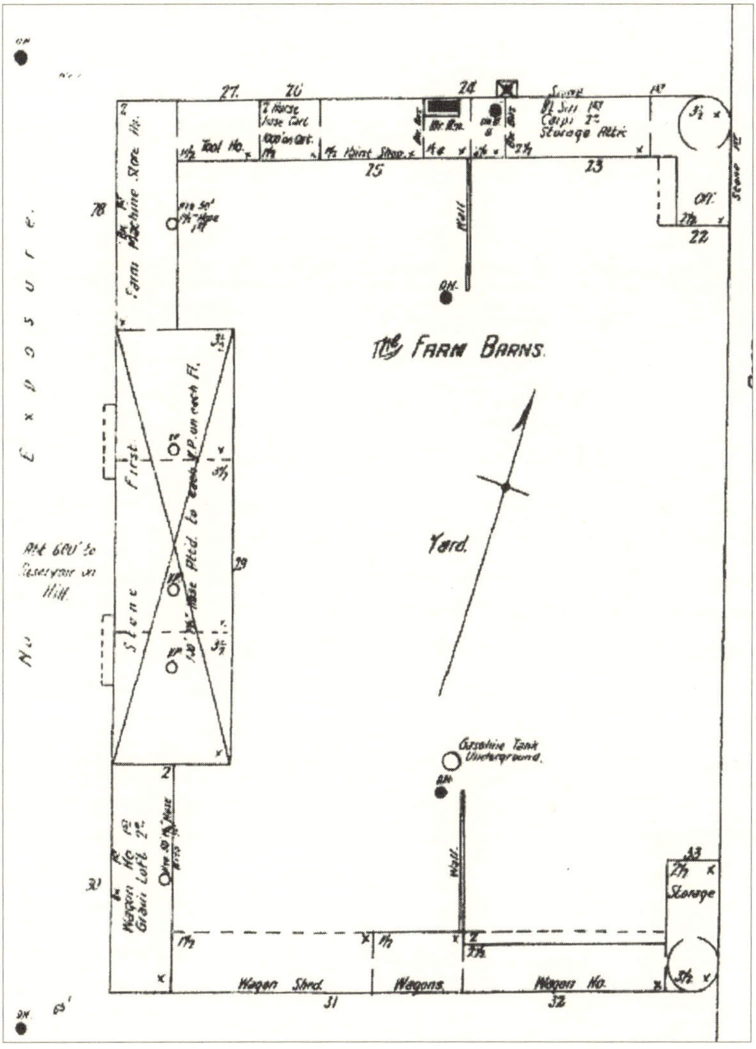

Figure 4.20. Sanborn Insurance map of Farm Barns, 1894. Sanborn Library, Sanborn Fire Insurance Company, Burlington, Vermont, 1894.

30 inches high by 108 inches in diameter, accessed from a four-foot-wide brick shaft in a configuration exactly like the circa 1882 Ashland installation. Based on its measurements, the generator would have held approximately one thousand gallons of gasoline and could supply up to five hundred burners.[145] Unlike the Ashland unit, the Shelburne Farms system included a large 36-by-240-inch storage tank that would have held another thousand gallons of gasoline. Invoices from 1913 to 1915 show that the Webbs purchased "10 Steel Barrels" or 536 gallons of

"86 gasolene" at least once a year.[146] Each yearly purchase would have produced approximately 89,333 cubic feet of gas for lighting. The incandescent burners used by the Webbs each consumed a maximum of 2.5 cubic feet of gas per hour to produce a light of forty candles. Thus the Webbs' system could run one burner for 35,733 hours or all five hundred burners for approximately seventy-one hours. More likely, only a fraction of the burners would typically be used at any one time (e.g., about 290 hours with 125 burners operational). According to surviving documents, the system closely resembled a "flat" gas generator, as pictured in the brochure "Directions for the Care of the Springfield Gas Machine" in the Shelburne Farms collection.[147]

The plan of the Springfield Gas Machine system at Shelburne Farms indicates that the generator and storage tank were installed fifteen feet from a wall, probably the inner courtyard wall shown on the 1894 Sanborn map just south of the notation for "Gasoline tank, underground." Given this location, it seems that the air pump was installed in one of the first-floor rooms of the Farm Barn, perhaps the wing due south of the generator. Although no extant evidence of the air pump exists, the archive at Shelburne Farms contains a brochure dated May 1, 1890, for the company's "Water Wheel Air Pump."[148] The property's excellent water supply system would have facilitated the use of this type of air pump. The system at Shelburne Farms may also have included the Combination Mixing Regulator, as the archive at Shelburne Farms contains two brochures for this component.[149]

The use of the automatic mixing regulator, the large gasoline storage tank and generator, and the water-powered air pump would have made the Webbs' system both efficient and easy to operate and maintain. The brochures specifically discuss the use of the mixing regulator in conjunction with the No. 95 or 95½ Springfield Incandescent Burner. Several invoices in the archive document the Webbs' use of this particular burner, as they frequently ordered No. 95 mantles as well as the six-inch mica chimneys that the burner required.[150] Two photographs of the Farm Barn, taken about 1900–1903, show wall-mounted, lantern-like fixtures that appear to contain incandescent burners.[151]

Shelburne Farms' catalog collection also includes marketing publications illustrating numerous appliances run by the Springfield Gas Machine, including "Our Automatic Instantaneous Water Heater."[152] Other appliances advertised in the various catalogs include hot water boilers, gas ranges, irons, and room heaters. While it is not known if the Webbs actually purchased any of these, it does seem very possible that they had a water heater in the barn complex, as well as room heaters for the upstairs apartments that housed staff members. It appears that the Springfield system at Shelburne Farms operated until at least 1912. The

Figure 4.21. Farm Barn interior, ca. 1895. Shelburne Farms Collections, Shelburne, Vermont.

1900 Sanborn map notes that lights were provided by "gas, electric, and oil." The Webbs may have used combination gas and electric fixtures at some point, or at least considered using them, to mitigate the unreliable nature of early electric power systems. The archive contains several catalogs from companies that sold combination fixtures.[153]

Like the 1894 map, the 1900 version also illustrates a "gasoline tank underground," in the southern yard of the Farm Barn. The 1906 Sanborn map still shows the "gasoline tank underground," although it doesn't reference the various types of lighting used. By 1906, the Webbs had expanded the Pumping Station, or private water works, which by then also included a new "electric light station." The 1912 Sanborn map includes the "gasoline tank underground" in the Farm Barn yard; it also shows a water line and hydrant locations. The power station remains unchanged. By 1919, the Sanborn map no longer shows the "gasoline tank underground," suggesting that the Webbs had abandoned the system. Correspondence in the property's archive supports this conclusion.[154] In 1916, when farm manager E. F. Gebhardt wrote to Gilbert and Barker to order mantles and chimneys, the company replied that they "recently retired from the gas machine and gas appliance business, disposing of our interests to the Detroit Heating and Lighting Company of Detroit, Michigan."[155] The Webbs may have discontinued using their gas machine at about this time, rather than deal with a new supplier.

The use of the Springfield Gas Machine at Shelburne Farms indicates the continuing utility of the system as a backup or supplement to electric lighting. Webb boasted of the cutting edge nature of his electric lighting infrastructure but added that "there is also an elaborate gas plant on the place."[156] At Shelburne Farms, the Springfield system seems to have been used principally for agricultural buildings. It may also have provided light and powered equipment for support trades, such as painting, carpentry, and blacksmithing, for the offices of various farm managers, and possibly for the apartments of farm staff. Webb prided himself on employing the latest technologies to provide household services—gas and electric lighting, central heating, central plumbing—that matched or exceeded anything available in the city. While seeking to reinvigorate the hackney workhorse of old and create a traditional pastoral setting, he spent lavishly on the most modern and technically sophisticated equipment to operate his "model" farm.

The Ames Houses

Like the Popes and Webbs, several members of the Ames family used Springfield Gas Machine systems to light their large country properties. The Ames, owners of the Ames Shovel and Tool Company in North Easton, Massachusetts, also installed Springfield systems in municipal buildings they supported, including Ames Memorial Hall, the Ames Free Library, and Unity Church (see chapter 5). Like Hill-Stead and Shelburne Farms, most of our knowledge of the Springfield systems used in the Ames houses, including Langwater, Spring Hill, and Sheep Pasture, comes from documentary sources including architectural drawings, correspondence, and invoices and receipts for gasoline and replacement parts for the machines.

Langwater

Langwater was the estate of Frederick L. Ames, the son of Oliver Ames Jr. Ames, a Harvard graduate, inherited his father's business interests in the Union Pacific Railroad and joined the firm of Oliver Ames and Sons in 1863. In addition to his work as treasurer for the family firm (as of 1876), Ames reportedly served on the boards of at least fifty-two railroads at the time of his death in 1893. In addition, he served as an officer in banks and as a trustee of Harvard University. He owned large amounts of Boston real estate, including the Ames Building, then the tallest building in Boston. Interestingly, he also helped found General Electric Company, having invested heavily in Thompson-Houston, the company that became the nucleus of General Electric. Considered "a power in industrial circles" and one of the most knowledgeable men regarding railroads in the country, Ames was a colleague and confidant of railroad men such as Durant, Vanderbilt, and Webb. Ames also had an intense interest in horticulture; his friendships with architect H. H. Richardson and landscape architect Frederick Law Olmsted led them to take on many projects in Easton.[157]

In 1859, Ames had Langwater designed and built by Snell and Gregerson in the French Mansard style. The main house subsequently underwent four renovations, becoming increasingly chateauesque. The estate's property contained numerous other buildings, including the Gardener's Cottage and Gate Lodge, designed by H. H. Richardson; Olmsted provided the estate's landscape design.

A visual inspection of the main house at Langwater did not reveal evidence of a gas machine. Documentary evidence, however, clearly indicates that Langwater

Figure 4.22. Langwater, North Easton, Massachusetts. Langwater Album, used with permission, digital copy by Greg Galer.

had a Springfield system. A set of undated architectural drawings by the Boston firm Shepley, Rutan and Coolidge Architects for an "alteration to residence" of F. L. Ames shows a gas machine located in the basement of the home. In addition, the firm, successor to the office of H. H. Richardson in Boston, provided a detailed lighting plan for each of the house's four floors.[158] The lighting plan identifies the positions of both gas and electric lights as well as combination gas-electric fixtures, indicating that the drawings postdate the house's electrification. The combination gas-electric fixtures likely mark original gas fixture locations.

According to the lighting plan, the first floor of Langwater had a total of eighteen fixtures located throughout the front two-thirds of the house. The dining room and library addition from about 1900 has neither gas nor electric fixtures; many homes of the time eschewed gas light for the dining room in favor of the ambience created by candlelight. The hall, or entry, had two gasolier fixtures, while the drawing room had eight wall fixtures, with two fixtures positioned on each side of the fireplaces and one large window. The second floor at Langwater, made up primarily of five bedrooms and the hall, had twenty-three fixtures, while the third floor had twenty-four fixtures, including four in a bedroom in the dining room and library addition. The billiard room, located over the library, did not have any fixtures according to the light plan.

In all, the lighting plan indicates that Langwater had a minimum of seventy-seven fixtures. The basement probably had several more. The house would have needed a Springfield Gas Machine system with a one-hundred-light generator at least and a larger one if it also lit any of the estate's ancillary buildings. All types of rooms and spaces, including service areas and bathrooms, used gas lighting. The plan called for placing lights on either side of windows and fireplaces and on long, unbroken stretches of wall. An architect's drawing for a mantel in the third-floor guest room, for example, shows a two-burner fixture on either side of the mantel. The placement of fixtures on either side of windows was a typical design element of the period intended to simulate natural light from windows; this technique also provided consistency of lighting, allowing seating arrangements to remain the same during the day and evening.

Gasoliers principally lit hallways and stairways, which needed well-distributed light. The first-floor drawing room, the main space for entertaining and receiving visitors, had eight wall fixtures, the most extensive lighting in the house. The bedrooms in the house had anywhere from two to five fixtures per room with the largest number in the principal second-floor bedrooms and the fewest in the small attic bedrooms of those in service.

Figure 4.23. Interior, drawing room, Langwater, North Easton, Massachusetts. Langwater Album, used with permission, digital copy by Greg Galer.

Although the dining room and library addition clearly was built after electricity had become available, the addition's third- and fourth-floor bedrooms have gas fixtures. Equally curious, neither the two large second-floor bedrooms nor the third-floor billiard room had either gas or electric lighting according to the lighting plan.

The gas light at Langwater provided the massive house with good quality light well into the age of electricity. It is significant that Frederick Ames continued to use the Springfield system even after the introduction of electricity, especially given his role as a founding director of General Electric. This continued use of gas lighting reflects the slow acceptance and geographic spread of electricity in the late nineteenth and early twentieth centuries as well as its technical limitations. Early electric lighting often required the installation of a private dynamo system, not unlike the self-contained Springfield system.[159] Similarly, the perceived "harshness" of electric lights was viewed by country house owners like Ames as offensive and annoying when compared to the "softness" of gas light; gas was still the sine qua non of country house lighting.

Spring Hill

Spring Hill, a large country house designed in the French chateau style by architect Carl Fehmer, belonged to William H. Ames and his wife. William's father, Governor Oliver Ames, presented the house as a gift in 1894. William Ames, an MIT-trained engineer, also sat on the board of Ames Shovel and Tool and several other businesses. Sited on a large property on Elm Street in North Easton, Spring Hill featured landscaping designed by Frederick Law Olmstead. The estate became the home of John S. Ames, son of Frederick L. Ames, about 1918. John, a Harvard graduate, served on the board of directors of numerous firms, including the First National Bank of Boston and Boston Edison.

The story of gas lighting at Spring Hill is also based largely on documentary sources because the house could not be physically inspected. A 1919 letter from George Burris of Gilbert and Barker to J. McKion clearly describes the system and references a repairman's visit to the home of Mr. John S. Ames in North Easton. John Ames, who had just taken possession of the house the previous year, apparently wanted to use the gas machine again. The letter describes the system as consisting of "a water wheel pump and a Detroit mixer with a four pan generator." Although the machine had not been used for about three years, D. J. Mahoney, the Gilbert and Barker repairman, found it to be "in first class condition" after receiving minor maintenance. The most critical issue, he notes, was that "either an electrician or other workman had taken out piping for the gas machine and had failed to plug up elbows, etc. . . . It will be necessary to find the open pipes in the walls before using the machine."[160]

In addition to this description of the system, the archives at Stonehill College contain a series of receipts from Gilbert and Barker to W. H. Ames for the purchase of gasoline. For example, in 1902, Ames purchased both "Pratt's Patent Prepared" gasoline and "Liquid Fuel" gasoline from the company. The volume of these purchases—eight barrels (approximately four hundred gallons) of "Pratt's Patent Prepared" rated at either 90 or 86 gravity and eighteen barrels (nine hundred gallons) of the "Liquid Fuel" rated at 76 gravity—seems quite large. However, as the typical four-pan machine, such as that at Hill-Stead, held approximately 697 gallons of gasoline, Ames would have needed such quantities to keep the machine filled, particularly if he used the system to fuel other appliances like stoves and hot water heaters.[161]

John's desire to utilize the machine after it had lain dormant for three years and after the introduction of electricity to the building makes his use of the Springfield Gas Machine at Spring Hill particularly interesting. Although William and his household may have abandoned the machine in favor of electric light, it seems that John found the machine an entirely suitable lighting system, at least as a backup to the electric lights. Since he was a director of Boston Editon, John's selection of gas over electricity is quite remarkable. While it might simply reflect his desire for a backup system at his semi-rural estate and speak to the unreliability of electricity, it may also reveal his desire to maintain gas lighting as the socially correct usage for fashionable country estates.

Sheep Pasture

The property known as Sheep Pasture was the home of Oliver Ames, the son of Frederick L. Ames. Following Oliver's graduation from Harvard in 1886, he worked at Ames Shovel and Tool. With the sudden death of his father in 1893, Oliver assumed many of Frederick's business pursuits, including serving as treasurer and director of Ames Shovel and Tool and as president of the First National Bank of Easton. In addition, he became active in the private water company founded by his father and held directorships in General Electric Company, Edison Company of Boston, Western Union Telegraph Company, and four or five railroads.[162]

Oliver and his wife, Elise West Ames, hired Rotch and Tilden to design the half-timbered Tudor Revival–style house at Sheep Pasture and had the house built between 1893 and 1894 on land given to them by Frederick as a wedding gift. As with the other Ames family homes, the landscaping for the property was completed by Frederick Law Olmsted between about 1891 to 1893. In addition to Sheep Pasture, Oliver and his family had two other properties: a house in Boston on Commonwealth Avenue and a three-story, French Renaissance–style summer home designed by Ogden Codman in 1904, on Boston's North Shore. The family regularly used all three properties but considered Sheep Pasture their home. With Oliver's death in 1929, Elise and their children continued to use all three homes until her death in 1945. Sheep Pasture was demolished in 1945, thus no physical evidence of the actual structure exists.[163]

Two Gilbert and Barker catalog references indicate that the Ames family had a Springfield Gas Machine in their Easton, Massachusetts, home and one in their summer house on the fashionable North Shore. Oliver's testimonials regarding the systems provide what little information exists about them. In a letter published in the 1906 catalog, *How to Light Our Country Homes and Resorts*, Oliver notes, "I would say that in the last fourteen years I have had five of them put in, and up to date they have given most perfect satisfaction in every way. We have had practically no repairs on them and they have caused not the slightest trouble or anxiety." Interestingly, in addition to printing Oliver's testimonial, Gilbert and

Barker chose to illustrate the main house at Sheep Pasture, one of just sixteen dwellings featured in the catalog.[164] Oliver answered a query from the company in about 1910, which published his reply in a brochure titled *Are You Aware? Some Suggestions Relating to Gas for Light and Fuel.* "In reply to yours asking me how I found the working of your Springfield Gas Machine and Mixing Regulator," Oliver wrote, "I would say that I put it into my house in the country six years ago, and since that time it has given perfect satisfaction."[165]

It is unclear which three other units he references, although he might mean systems installed by his father at Unity Church (1875) and Langwater or at his brother's house, Spring Hill. Given Oliver's directorships with General Electric Company and Edison Company of Boston, his enthusiastic embrace of the Springfield Gas Machine seems both perplexing and revealing. His testimonials may say as much about the unreliability of electric light at the time as they do about the reliability and performance of the Springfield system. As with other Ames family members, Oliver's embrace of the Springfield and gas lighting might also relate to his understanding of gas lighting as the socially correct usage for fashionable country estates.

Domestic Lighting with the Springfield System

The examples described above are just a few of the literally thousands of domestic Springfield systems documented in the company's late-nineteenth- and early twentieth-century catalogs. The geographic distribution of the machine in the United States mirrors that of the country's centers of power and wealth, particularly along the East Coast. The system's overwhelming representation in the northeastern United States is not surprising given the location of the company headquarters in Massachusetts and its principal sales offices in New York City and Boston. More important, the Springfield machine's sales distribution pattern corresponds closely with the great concentration of social and economic power in the region from Philadelphia to Boston. Approximately 47 percent (n = 919) of the domestic Springfield installations occurred in the mid-Atlantic region, with New England a close second at 35.8 percent (n = 700) (table 4.1).[166] Thus while Springfield systems could be found in homes across the country, over 82 percent were located in the northeastern United States from roughly Philadelphia to Boston. The Midwest, principally Chicago, had the third greatest concentration with 7.8 percent (n = 152) of residential systems. The South had 5.8 percent (n = 114) of residential systems, while the West had just 3.7 percent (n = 72).

Across the country, the domestic use of the Springfield system existed mostly in country and suburban homes. For example, in the mid-Atlantic region, New York state had the largest concentration of Springfield Gas Machines (n = 492,

Table 4.1
Residential Installations of Springfield Gas Machines by Region and State

Region	State	No. of residences	Percentage of region	Percentage of total
Mid-Atlantic	Delaware	2	0.20	
	District of Columbia	2	0.20	
	Maryland	17	1.80	
	New Jersey	220	23.90	
	New York	492	53.50	
	Pennsylvania	186	20.20	
Subtotal		**919**		47.0
New England	Connecticut	242	34.60	
	Rhode Island	42	6.00	
	Maine	37	5.30	
	Massachusetts	297	42.40	
	New Hampshire	31	4.40	
	Vermont	51	7.30	
Subtotal		**700**		35.8
Midwest	Iowa	7	4.60	
	Illinois	46	30.30	
	Indiana	5	3.30	
	Kansas	10	6.60	
	Michigan	14	9.20	
	Minnesota	6	3.90	
	Missouri	17	11.20	
	Ohio	29	19.10	
	Wisconsin	18	11.80	
Subtotal		**152**		7.8
South	Florida	24	21.00	
	Georgia	19	16.70	
	Kentucky	19	16.70	
	Louisiana	1	0.90	
	North Carolina	27	23.70	
	South Carolina	5	4.40	
	Virginia	17	14.90	
	West Virginia	2	1.75	
Subtotal		**114**		5.8
West	California	55	76.00	
	Hawaii	17	24.00	
Subtotal		72		3.7
Total		**1957**		

SOURCES: Gilbert and Barker Manufacturing, *Springfield Gas Machine*, 1870–79; *How Best to Light Our Country Homes and Resorts*, 1888–1906.

or 53.5 percent), with the majority on Long Island or north of New York City in Westchester County and up along the Hudson River. These places were all popular "country house" or "cottage" locations for wealthy, upper-class New Yorkers who sought to escape the city. New Jersey had the next largest concentration in the mid-Atlantic region, again mostly suburban and summer homes in the vicinity of New York City or along the Jersey shore.

As Gilbert and Barker grew, it increasingly targeted the rapidly growing country estate and suburban markets. This approach proved profitable as industrialists and businessmen with residential systems often purchased them for their workplaces. Conversely, business owners who purchased Springfield systems for their workplaces regularly bought them for summer and country homes. For example, the firm of Poole and Hunt in Baltimore, Maryland, notes in a testimonial letter that "the success of the seventy-five light Machine our Mr. Poole has in use to light his dwelling, induced us to introduce it in our works."[167]

Architectural historian W. Barksdale Maynard has written "appropriate to its size, [New York] was ringed by the greatest number of suburban homes of any city on the continent, all in a diverse and attractive setting." While an 1817 visitor to Long Island noted that "country seats are rather numerous," most outlying Long Island locations were not intensely developed until rail transportation made commuting feasible. Writing in 1847, architect William H. Ranlett commented that "probably no portion of the globe, offers a greater variety of beautiful country seats than the vicinity of New-York." Maynard notes that the "coming of the steamboat and later the railroad rapidly suburbanized the Hudson Valley north of New York."[168] From New York City to Hyde Park and beyond, the country estates and summer cottages of city dwellers lined the Hudson's banks.

At the same time that late-nineteenth-century New Yorkers sought to escape its technological and commercial whirlwind, they outfitted their rural retreats with the modern conveniences and luxuries that they enjoyed and depended on in the city. Sales ads highlighted good lighting, heating, and plumbing to attract buyers for country properties along the Hudson. For example, a classified ad for a "gentleman's residence" in Irvington-on-Hudson touted "exposed plumbing, gas machine . . . and every necessary convenience." Another ad, for "one of the finest estates on the Hudson River," a "modern mansion of brownstone and brick" located near Poughkeepsie, listed a gas machine as the principal amenity.[169]

Long Island also had a large concentration of country and suburban houses that utilized the Springfield system and had many of the largest estates lit by the Springfield Gas Machine in the entire country. For example, W. Bayard Cutting, a lawyer and railroad investor, built a huge home named Westbrook in Great

River in 1884–85. Designed by prominent architect Charles C. Haight, the huge "brick, shingle-and-stucco house in the Queen Anne style" sat "amid grounds laid out by Frederick Law Olmsted in the English naturalistic style." Both the 1888 and 1906 Gilbert and Barker catalogs included illustrations of the property. Focusing on the technical aspects of the Springfield system on his Long Island estate, Cutting wrote to Gilbert and Barker in 1898: "First, that the Automatic Regulator is capable of reducing the gas to a substantially uniform quality which can be burned through plain burners. Second, the Air Pump of my machine is operated by a Water-wheel which has given entire satisfaction."[170] Cutting subsequently confirmed that he had had the machine for fourteen years, indicating that he had it installed when the house was built in 1884–85.

Springfields also lit well-known country estates in Glen Cove, Bay Shore, Islip, Sayville, Shelter Island, and Westbury, designed for the country's captains of industry, commerce, and finance by architects such as McKim, Mead, and White, Carrère and Hastings, Charles A. Platt, and Charles C. Haight. Glen Cove, located on what would become known as the Gold Coast of Long Island, had many residences that used the Springfield system, including a compound belonging to family members of Charles Pratt Sr. Pratt was the founder of Charles Pratt and Company, a petroleum refining firm, and a director of Standard Oil. His sons, Herbert L., Frederick B. and George D. Pratt, all built huge country homes on the families' eleven-hundred-acre property in Glen Cove. In each case these homes relied on Springfield Gas Machines until electricity became available. Of course, the Pratts' use of the Springfield is not entirely surprising given that Pratt Manufacturing Company, a subsidiary of Standard Oil of New York, acquired Gilbert and Barker Manufacturing in 1884.

Businessman Bradish Johnson also lit his Long Island country estate at Bayshore with a Springfield Gas Machine. Johnson, a major investor in New York real estate and owner of B. Johnson and Sons, a firm that owned distilleries and sugar refineries, also owned a house in New Orleans and several major sugar plantations along the Louisiana coast.[171] Likely built about 1860, Johnson's Bayshore residence, a Gothic Revival cottage, would seem small in comparison to the later nineteenth-century estate homes of the Cutting and Pratt families. Johnson's testimonial in the 1870 Springfield Gas Machine catalog indicates that in 1869 he also installed a system in his Woodland Plantation in Plaquemines Parish, Louisiana, which worked well.[172] Johnson's use of the Springfield in at least two of his residences reflects his confidence in the system.

Many New Yorker's built their country estates across the river in New Jersey. For example, Gilbert and Barker's 1906 catalog featured Bernardsville, New

Jersey, located west of New York City (near Morristown), for having twelve country estates that used Springfield Gas Machine systems. Almost all of the owners of the estates worked in or around New York City, including F. P. Olcott, president of the Central Trust Company (known as the "bank of the sugar crowd"); Charles Pfizer Jr., president of Pfizer Pharmaceuticals; Judge John F. Dillon, a New York jurist; Percy Pyne, president of the National City Bank of New York and vice-president of the Delaware, Lackawanna, and Western Railroad; and Julius Stursberg, treasurer of the Brooklyn Sugar Refining Company, which later became part of the American Sugar Refineries Trust.

This aggregation of country houses in Bernardsville, all lit with Springfield Gas Machines, demonstrates the close business linkages and friendships among the country house class. For example, Olcott, as president of the Central Trust, would almost certainly have known Julius Stursberg, given his status as banker to "the sugar crowd." Percy Pyne, a director of the Central Trust Company of New York, would have known both Olcott and likewise Stursberg. These owners of Bernardsville country houses represent the type of social and business web that Gilbert and Barker successfully exploited in both their publications and marketing materials for the Springfield Gas Machine and its components.

After New York City, the largest concentration of Springfield systems was in the vicinity of Philadelphia, Pennsylvania (n = 186, or 20.2 percent), primarily in the suburbs and rolling landscape west of the city in what became known as the "Main Line" communities.[173] Finished in the 1830s, the eastern section of the Main Line was purchased by the Pennsylvania Railroad in 1857. Later in the century, the railroad promoted the development of the picturesque property it owned along the line. To attract speculators and thus future riders, the railroad built residential enclaves "along the portion of its track closest to Philadelphia. The construction of sprawling estates attracted Philadelphia elite, many of whom had one house in the city and another larger 'country home' on the Main Line." The villages that grew up along the Main Line, such as Overbrook, Merion, Ardmore, Haverford, Bryn Mawr, Rosemont, Villanova, Radnor, and Paoli, represent the locations of many of the Springfield residential units in the Philadelphia area. Railroads, notes Maynard, "did their part to spread a Picturesque ideal, even as they spoiled the earliest suburban neighborhoods close to towns."[174]

A list of Springfield customers from suburban Philadelphia—Brown, Drexel, Biddle, Godey, Lippincott, Shoemaker, Wharton, Ingersoll, Clothier, Pullman, Wannamaker, Van Rensselear, and Whitney—includes some of the most powerful and influential Philadelphians of the period. Customer Joel J. Bailey of Darby, a suburban Philadelphia community, wrote of the Springfield system, "I consider it one of the chief modern improvements in house furnishing, and that no country house is complete without it." Likewise, investment banker F. A. Drexel of Drexel and Company reported in 1878, "I have had the Springfield Gas Machine in use at my house near Torrisdale for seven years, and have found it to be very satisfactory in every respect."[175]

New England also had a large proportion of country homes and estates that utilized the Springfield system (n = 700). The greatest number of these houses were in Connecticut (n = 242, or 34.6 percent) and Massachusetts (n = 297, or 42.4 percent), but they also were along the Maine coast and even into Vermont and New Hampshire (see table 4.1). In Massachusetts, the western suburbs and north and south shores of Boston and Cape Cod had a great many customers with country estates and suburban retreats, as did western Massachusetts, particularly the Berkshires in and around Stockbridge and Lenox (then known as the "Inland Newport"). Of course, Springfield, Massachusetts, the home of the Gilbert and Barker factory, also provided a major customer base for the firm.

By the late 1880s, Gilbert and Barker regularly issued catalogs specifically tailored to the country home set. For example, their 1906 catalog, *How to Best Light our Country Homes and Resorts*, pictured several estates in western Massachusetts, including the residence of Mrs. John Swann in Stockbridge and the homes of H. H. Cook and John Sloane in Lenox. In the photographs highlighting these palatial "cottages," the firm clearly conveys the message that the grandest country homes used the Springfield system, a fine example of the marketing of conspicuous consumption. Sloane co-owned W. and J. Sloane, one of New York City's oldest and largest home furnishing stores, while H. H. Cook had amassed tremendous wealth as a "railroad man and banker."[176]

The illustration of Sloane's Wyndhurst demonstrates an important advertising strategy that Gilbert and Barker used to cultivate country and estate home owners as customers. By the turn of the twentieth century, W. and J. Sloane "set the national decorating taste of the United States, and over the next sixty years decorated the homes of the most prominent people in the country."[177] As seen at Hill-Stead, proper light and lighting fixtures were very much a part of the increasing focus on interior design and decorating. Firms like W. and J. Sloane established the standard for an appropriately furnished country house; Sloane in particular had a reputation as *the* New York firm for fashionable interiors. By linking itself to a name like Sloane's, Gilbert and Barker placed its products at the apex of consumer culture and left no doubt about the importance of having a similar system in one's country or suburban home.

In Connecticut, many of the homes lit with Springfield systems were located along the coast. Gilbert and Barker's 1906 catalog references the "Eastern Point Gas Co.," a voluntary association of "thirteen seaside summer cottages at Eastern

Point," which sits at the mouth of New London Harbor. Association treasurer George Coit described these cottages as "of fair size," probably averaging "about twelve rooms in each house."[178] Connecticut also provided rural and country home settings within reasonable commuting distance of major cities such as New York and Boston. While businessmen could not commute daily from Connecticut during this period, they could join their summering families for relaxing country weekends.

Places like Connecticut also saw an influx of wealthy entrepreneurs from across the country who, having made their fortunes, sought to return to their "rural" roots in old New England. The Popes moved from Cleveland, Ohio, to rural Farmington, Connecticut, in their retirement in order to be closer to their daughter Theodate and to their family's geographic hearth. Their country home, Hill-Stead, was relatively modest in comparison to many country houses of the period, but its Colonial Revival design suited their need and desire to reestablish the family's colonial roots. The home also became a social and cultural center because of the Popes' outstanding collection of Impressionist art. The family's choice of the Springfield Gas Machine and designer gas fixtures from Shreve, Crump and Lowe indicates both their desire for good quality lighting and for the delivery of this light within a carefully considered and artful interior design scheme. The Popes' selection of a New England extended farmhouse design is a clear nod to their roots, while the home's many modern conveniences, like gas lighting, display their firm embrace of twentieth-century technology.

Another nationally prominent Connecticut customer was P. T. Barnum (1810–1891), whose internationally famous American Museum in New York City served for many years as the "cultural hub of New York." Interestingly, among its many exhibits and curiosities, the museum displayed a host of "new technological devices," perhaps some of the early "portable" gas light systems. By the early 1850s, Barnum, a true marketing genius, had built a reputation as "one of the wealthiest men in the country" and a "prominent social player in New York City." In the 1880s, Barnum joined up with James Bailey, a manager of the Great London Show circus, and they formed the Barnum and London Circus. In 1887, the two established a partnership and launched the Barnum & Bailey Greatest Show on Earth.[179]

Given Barnum's interest in displaying the latest technology alongside his many outlandish freak shows, he may have become familiar with the Springfield system during the final years of his New York museum in the late 1860s. His awareness of the system may have begun or increased circa 1867, when Gilbert and Barker opened a sales office in New York City. Testimonials printed in Gilbert and Barker catalogs, indicate that the infamous showman had a Springfield Gas Machine installed in his Bridgeport, Connecticut, home, known as Waldermere, in 1869.[180] He wrote to Gilbert and Barker in 1870 that "your Gas Machine, placed in my Country house last June, was not touched all summer. We had not occasion to replenish the gasoline, and the vault containing it has not been opened since it was put in. The light is much whiter than that supplied by the Bridgeport Gas Co.; and the whole thing seems thus far to be nearly perfection. We never before had so nice a light, nor one so free from objections. The cost of the light is merely nominal."[181] The choice of the Springfield Gas Machine to light the home of the greatest showman of the era offered another priceless marketing association for Gilbert and Barker, particularly given his previous use of municipal gas for lighting. The name recognition of P. T. Barnum was widespread, and so his endorsement would have generated interest in markets across the country.

The addition of the Springfield Gas Machine system in the Midwest promoted by Gilbert and Barker's Chicago office probably also grew out of the company's ongoing relationship with the Detroit Heating and Lighting Company. The Detroit firm, which manufactured the Combination Gas Machine, originated in Chicago and continued to have a Chicago office. Illinois had approximately 30.3 percent of the systems (n = 46) installed in the Midwest, with Chicago and its immediate suburbs accounting for over half of these residential installations (n = 26) (see table 4.1). Ohio had the next largest concentration of Springfield systems (n = 29), focused largely in Cincinnati and along the southern border of the state in cities such as Portsmouth and Chillicothe; a Gilbert and Barker sales office in Cincinnati likely explains this concentration. States such as Missouri (n = 17), Michigan (n = 14), and Wisconsin (n = 18) had similar numbers, but the systems were disbursed across each state without any significant concentrations. In Michigan, for example, the systems were scattered across the small cities and towns of the state, while very few existed in the Detroit area. Not surprisingly, the Combination Gas Machine, manufactured and sold out of Detroit, seems to have dominated the market for country homes and estates in suburban Detroit communities like Grosse Pointe.

The Springfield systems found in southern U.S. residences represent about 5.8 percent (n = 114) of the total documented residential installations (see table 4.1). While North Carolina had the largest number of domestic systems in the South (n = 27, or 23.7 percent), they were distributed across the state with no significant concentrations, as seen in much of the Midwest. Florida had the next largest group of residential systems (n = 24, or 21.0 percent), Kentucky and Georgia were at 16.7 percent each (n = 19), and Virginia was at 14.9 percent (n = 17) of domestic installations.

Predictably, households in the western United States represented the smallest number of residential customers (n = 72, or 3.7 percent) (see table 4.1). Given Gilbert and Barker's agency relationship with Nye and Acheson in San Francisco, which began in 1875, it makes sense that California had the majority of Springfield sales (n = 55, or 76 percent). More surprising, the only other location represented on the company's lists of western installations is Hawaii (n = 17, or 24 percent). Fifteen of these installations were in Honolulu, while the other two were in Waikiki. Interestingly, the 1888 catalog reports that the Honolulu installations included a three-hundred-light system in the palace of HRH King Kalakaua in Honolulu.[182]

During the late nineteenth century, many of the smaller towns and cities of the nation still lacked organized municipal gas companies that served residential customers. Even in places with such companies, country estate and suburban homes by definition stood on the outskirts of town, beyond the reach of municipal gas service. This limitation created a broad market for systems like the Springfield in a wide range of residential settings, including more modest cottages belonging to merchants and entrepreneurs based in small towns and rural areas across the country.

Architects of the period who designed country houses for middle- and upper-middle-class clients often promoted the use of gas machines as a modern convenience. Many architectural books and periodicals focused on these midrange clients. In *Homes in City and Country* (1893), the authors note that "in viewing the work of this period it is not to the point to consider the larger establishments of Newport, Mount Desert, Lenox, or the great places that have been raised up all through different parts of the country; it is either the permanent home or the summer residence of the man of moderately independent means that interests us."[183] Architect Frank T. Lent's 1895 book, *Sound Sense in Suburban Architecture,* provides "hints, suggestions, and bits of practical information for the building of inexpensive country houses." According to a review in the *Newark Times,* "Mr. Lent goes into the details of light, ventilation, drainage, water, and all those seemingly little things, the absence of which interferes with the comforts of a home." Although Lent concedes that "private gas machines are generally deemed too expensive" for those building homes of this size and cost, he goes on to outline the many arguments "in favor of private gas machines, for instance the saving in labor in taking care of and the wear and tear on the number of lamps required to light the house; diminution of danger; convenience, and the superior lighting, which is secured by the gas. As a matter of fact a gas machine is a good investment, if one has the money for it."[184] Lent clearly articulates the advantages of gas light over kerosene lamps, including greater safety and convenience and most importantly, better quality light.

Lent's book contains a full-page ad for Gilbert and Barker Manufacturing that touts their "Automatic Instantaneous Water Heater" available for use with the Springfield Gas Machine and Combination Mixing Regulator. This ad placement suggests that the company actually sought a wide range of clients, not just the highest echelons of the social register.[185] Customer J. W. Gardner, an agent of Lamson and Goodnow Manufacturing Company in Shelburne Falls, Massachusetts, a middle-class businessman, explained in a letter to the company that "when I decided to make this village my permanent home and to build for myself a residence, I found that I could, without difficulty, introduce into it almost all those little luxuries which are so common in all our best city dwellings." While he found water and heat easily provided, "the only . . . serious question left was—how to light? Kerosene I dreaded, not alone on account of the trouble and care of the lamps, but more especially because of their liability to breakage and explosion."[186] Gardner clearly relished his ability to secure the "little luxuries" of city houses in his relatively modest village home and well understood that portable systems like the Springfield Gas Machine offered one of the few choices for gas light in his rural community.

While most homeowners initially used the Springfield Gas Machine primarily for lighting, by the later decades of the century they also used the system to fuel a wide assortment of appliances that they increasingly saw as necessary modern conveniences. "Late in the nineteenth century," writes Cowan, "gas stoves, ranges, water heaters, and furnaces began to appear on the market, frequently either sold through the gas companies themselves or manufactured by their subsidiaries." Because of this, systems like the Springfield had a broad impact on the operation of rural households. For example, the Gilbert and Barker ad in Lent's *Sound Sense in Suburban Architecture* announces their "Automatic Instantaneous Water Heater," along with gas ranges and gas laundry appliances, suitable for the modern suburban home.[187] In its 1906 catalog, Gilbert and Barker notes that

> one of the most interesting uses to which gas may be put in promoting domestic comfort is undoubtedly that in connection with the preparation of food. It is a matter of surprise that it has not been much more extensively adopted by the general public. Its use for these purposes is of special interest to those having gas machines, because of the great economy of air gas. The first cost of gas ranges is not half that of good coal ranges, while they will last a lifetime. The exact degree of heat required for any special purpose is at once obtained. Since combustion is perfect, there is no smoke or odor, and no flue is required. On the score of economy, it will be at once acknowledged, after trial, that the cost of a coal fire doing the

same amount of work is greater than the cost of gas, while the certainty of its results, its cleanliness, convenience and comfort are obviously in favor of the latter fuel.[188]

This argument was particularly cogent to Gilbert and Barker's middle-class customers, who were inundated with advice on creating the best home environment possible. The "technological revolution" of the period was seized on by entrepreneurs who offered customers a "traditional home" with less work. For the middle class, this apparent lessening of workloads bought about by new innovative products was also correlated with their aspirations to achieve more refined lifestyles like the upper-class owners of estate homes. Yet Strasser explains that "with one major exception (the stove) . . . American food preparation in 1900 looked much as it had in 1800." The bulk of the many inventions of the period, however, "were well beyond most families' budgets."[189]

Many Springfield customers wrote testimonials celebrating the application of the Springfield Gas Machine to power household appliances, "eliminating household drudgery." Customer Leonard Hewett of Saluda, North Carolina, wrote that the Springfield system provided "a splendid light, gas for all kitchen purposes, and an abundance of hot water for bath room and lavatories, through the Columbia Water Heater, which you also put in for me in connection with the gas machine. As an eliminator of household drudgery, and a most economical service of light and fuel, it is invaluable, and should make it indispensable to any country home." Alton B. Parker, the owner of a country home known as Rosemount Hall in Esopus-on-the-Hudson, New York, reported that his Springfield system provided excellent light and powered a wide range of conveniences that even included gas logs for the fireplace. Parker explained that his family "added gas logs in a few rooms, then conveniences for the laundry and butler's pantry, and hot water heater for summer, and finally we added a gas-power engine to be operated at the barn together with a pumping engine. And it does all of these things well."[190]

In addition to providing fuel for all of the lighting, cooking, and laundry at his summer home in Matunuc, Rhode Island, G. K. Webster of North Attleboro, Massachusetts, reported that the system has "heated effectually, all of the chambers that I have connected with the gas radiators." Webster styled himself as a very satisfied Springfield user who depended on such modern conveniences: "I can say that I would not have it taken away at double the expense of the installations, if I could not replace it."[191]

While most testimonials were written by men, it seems likely that women, as the principal household managers of all sorts and styles of houses, also had clear opinions about the conveniences offered by gas-powered appliances. For example, customer Laura A. Fellows of Newburgh, New York, wrote that "applied to a gas log [the Springfield Gas Machine] gives all the warmth and cheerful appearance of a wood fire; and what was at first considered a luxury has become a necessity to the comfort of our country home."[192] For women, however, the availability of "labor-saving" appliances proved a double-edged sword.

Undoubtedly, the use of gas machines reduced the maintenance connected with older lighting sources such as oil lamps and simplified many tasks around the home. Yet as Cowan argues, while these products made a huge difference in the average American's standard of living, they "did not substantially decrease the amount of work that women had to do in their homes." In many cases, labor saving devices did "reduce the drudgery" of housework but in the process often substituted one task for another and even increased the work load for women in many cases. As Cowan writes, "Modern labor-saving devices eliminated drudgery, not labor."[193] For example, while the gas stove reduced the work of building and tending the woodstove's fire, expectations of what the new stove could produce and the speed at which meals could be prepared, simply substituted one type of work for another. Thus the time saved could be used to prepare more complex meals and enable more refined dining, activities that helped satisfy middle-class aspirations to respectability.

The linkage of lighting and housework chores has a long tradition, made explicit by Catherine Beecher and Harriet Beecher Stowe in *The American Woman's Home*. The book introduces readers to the "Principles of Domestic Science" and provides an important and influential "guide to the formation and maintenance of Economical, Healthful, Beautiful and Christian Homes."[194] Of course, the book's publication in 1869 coincided with the very moment at which the Springfield and similar portable gas machines were becoming more widely known and gaining in market share.

In a section called "Fires and Lights," the authors suggest that the most important aspects of choosing good lighting were "the influence on the eyes, and secondly, economy." "It is," they admonish their reader, "poor economy to use a bad light." The authors principally focus on gas and kerosene lighting, declaring "sperm-oil, lard, tallow, etc," superseded by kerosene, while "candles are used only on rare occasions." "Modern houses in cities, and even in large villages," write Beecher and Stowe, "are furnished with gas. . . . Gas is the cheapest (or ought to be) and if properly used, is as good as any."[195]

Beecher and Stowe's application of the principals of domestic science to house hold labor and chores of all types conveys the promise that new technologies like the Springfield held for improved lighting in the second half of the nineteenth century. Gilbert and Barker's marketing of their system clearly paralleled Beecher and Stowe's two primary concerns regarding household lighting: its quality and economy. In fact, the Springfield Gas Machine catalogs typically began with sections such as "Regarding the Quality of the Gas" and "The Cost of the Gas," along with the safety and insurability of the system.[196]

While Beecher and Stowe clearly favored gas for lighting "modern homes," they warn the reader that "there is no more dangerous mode of heating a room than by a gas-stove." They even maintained that gas stoves presented a real danger when used for cooking. With complete condescension, they explain that "in the view of the almost inevitable stupidity and carelessness of servants, who generally have charge of such things, and the frequent thoughtlessness even of intelligent women who manage their own kitchens, the writer believes she is doing a public service by offering her own experience as a guide to simpler, cheaper, and more wholesome means of living and preparing the family food."[197]

Beecher and Stowe's comment about servants is particularly interesting, given the prevalent problem of hiring servants during this period. As industry expanded, job opportunities for women grew and typically offered better pay and more independence than household service. Many manufacturers introduced new "labor-saving" devices in the late nineteenth century, particularly electric appliances, to combat "the servant problem." As an added value, manufacturers noted, appliances didn't talk back or slack off. Edison Lamps Works (later General Electric), in a *McClure's* magazine advertisement marketed its "lamp that lights the way to lighter housework," along with household appliances, explaining that "electric servants can be depended on—to do the muscle part of the washing, ironing, cleaning, and sewing."[198]

Despite a diminishing pool of servant labor, Victorian-era country estate and suburban householders continued to use servants; in fact, many of these houses had more rooms allocated for servants quarters than the average house had in total number of rooms. Even in wealthy families, women served as domestic managers, responsible for organizing and overseeing the daily operation of the household staff. The introduction of new appliances for cooking, heating, lighting, and cleaning often increased the complexity and stress of women's jobs, as these improvements resulted in heightened expectations for delivering a better quality of life to their families.

Residential customers who bought the Springfield Gas Machine to provide light and power appliances made up a large and significant part of Gilbert and

Barker's market. The thousands of Springfield customers across the United States regarded the need for good light quality at a reasonable price a top priority. For those without access to municipal gas, portable, self-contained gas machines offered the only alternative to lighting with kerosene lamps. While the machine was expensive to purchase and install, the gas light produced by the system cost no more or even less than most other types of lighting of the time. Customer comments make clear that the system produced a superior quality and quantity of light when compared with competing systems like kerosene lamps and municipal gas.

The country house market provided, and not by accident, the niche for systems like the Springfield that offered city-quality services in rural locations. Gilbert and Barker's success relied on the owners of these country homes, the nation's growing ranks of upper- and middle-class industrialists and entrepreneurs, who adopted the Springfield system for both their businesses and homes. Country and suburban homes of the period served as both the locus of entertainment and sociability and the central node in many families' culture of consumption. As such they gave companies like Gilbert and Barker the opportunity to sell their systems via intricate social and business networks of customers who embraced the promise of progress through technology. Businesses that used the Springfield system, like the era's grand tourist hotels, provided the setting for extensive public display and social exchange. They thus presented Gilbert and Barker with a rich vein of marketing opportunities, as explored in chapter 5.

CHAPTER 5

Extending the Day: Commercial and Institutional Springfield Systems

The introduction of gas light in the early nineteenth century also dramatically altered commercial and institutional lighting. For the first half of the century, the use of gas light for commercial and institutional structures was largely restricted to urban areas with municipal gas works, which themselves were limited by infrastructure concerns such as the cost of laying gas mains. As early as 1813, inventor David Melville touted gas light for "the growing manufactories of our country, in as much as they are less expensive . . . and free from the inconvenience and danger" of lamps and candles.[1]

This chapter explores the use of portable gas lighting in industrial, commercial, and institutional settings, examining Springfield systems at several locations, including the Springfield Armory, a host of large mills and manufacturing plants, Mount Holyoke College, and a selection of resort hotels in New Hampshire, Florida, and Michigan. The Springfield systems at several of these sites remain largely intact and were carefully documented; systems at other sites are known almost exclusively through the historical record. Taken together, field documentation and archival evidence provide a detailed understanding of large, complex installations from a technological standpoint.

The examples of the commercial and institutional systems discussed in this chapter range from the late 1860s to the early 1900s, a critical and transitional period in the industrial history of the United States. The systems considered here also represent various stages in the business development of Gilbert and Barker Manufacturing and sophistication of gas lighting technology. The second half of the nineteenth century saw the rapid rise of machine production in shops across the country, along with changes in work hours and conditions. Trachtenberg notes that the "unsettled economic conditions made manufacturers obsessed with efficiency, with the breaking of bottlenecks, the logistics of work flow, the standardization of parts."[2] Gas lighting in general and the Springfield system in particular were technologies that supported this obsession.

The superintendent of the Springfield Armory had a system installed there by 1867, providing Gilbert and Barker with the first sale of its new configuration that separated the generator and the air pump and helping to speed the granting of the company's patent application on the machine by citing national military interests. Systems at Mount Holyoke College and the Royal Poinciana Hotel in the 1890s serve as examples of Gilbert and Barker's later systems and largest machines.

In addition to detailing the organization, installation, and operation of these large, complex systems, this chapter reviews the economic, safety, and social implications of gas lighting, particularly for large institutions, factories, and businesses. The introduction of gas lighting in general and the Springfield system in particular had significant implications for the world of work and labor, bringing existing labor issues into an entirely new light. For example, the technology allowed mill owners to light large structures that required proximity to rivers at major falls, usually beyond the boundaries of large municipal gas works. As Schivelbusch writes,

> As long as the work that needed to be lit up was tied to individual craftsmen and only the winter morning and evening hours required extra light, the glow provided by traditional candles and oil lamp was adequate. This changed with the introduction of industrial methods of production. Work processes were no longer regulated by the individual worker; they became integrated, comprehensive operations. The new factories needed new sources of light. Artificial light was needed to illuminate larger spaces for longer periods of time. In the factory, night was turned to day more consistently than anywhere else.[3]

Similarly, architectural historian James Marston Fitch argues that "industrial technology could not develop in a natural luminous environment. The need of industrialism for a luminous environment which could be extended in time and space, and manipulated in any desired direction, sprang from two distinct but closely related facts." These were, Fitch continues, the realization that machines are most economical when run continuously and that a precision machine requires "visual acuity on the part its operators."[4]

Portable gas lighting systems presented both disadvantages and benefits for workers. Artificial light allowed employers to extend the work day and staff multiple shifts, placing greater stress on workers; however, the addition of multiple shifts eventually led to the creation of new jobs. The "continual refinement and rationalization of machinery" and the factory building itself, writes Trachtenberg, "represented to workers a steady erosion of their autonomy, their control, and their crafts."[5] But improved lighting conditions provided a generally safer work environment, particularly in manufacturing settings like mills, which had fast-moving and dangerous machinery. Cowan notes that "before electric motors, every machine was attached by an exposed belt to the engine that powered it, which meant that workers were constantly in danger of being caught in the belting: drawn in and crushed, or thrown across the room." Poor lighting only enhanced this danger.[6]

Businesses like canning factories, which had to be located near the resource to ensure freshness and economy, also used the Springfield system. Springfield machines, for example, powered soldering equipment in canneries in fishing towns along both the East and West Coasts and in many farming communities. The system also ran equipment for brazing rifles at the U.S. Armory in Springfield and at a host of metalworking plants that produced everything from cutlery to silver-plated goods, bicycles and tin toys, and even eyeglasses.

Company catalogs reveal the broad geographic distribution of businesses, factories, schools, and government buildings that had Springfield systems. As seen with the domestic Springfield systems, commercial and institutional installations represent concentrations of extraordinary economic and social capital and power. For example, the magnificent hotels and inns that used the Springfield system to light their establishments catered to those with money and power by providing the best and latest in accommodations and amenities for travelers and tourists.

Likewise, Springfield systems at Mount Holyoke College and Middlebury College, served the children of upper-class families that expected top-notch facilities. As Gilbert and Barker well understood, the various commercial uses of the Springfield encouraged cross-selling among business owners and managers and offered lucrative opportunities to market the system to wealthy patrons of high-end businesses.

Industrial Installations

Numerous manufacturing firms utilized Springfield Gas Machines to fuel lighting systems and various kinds of production equipment. As noted in the company's 1874 and 1878 catalogs, the gas produced by the Springfield system was "used in the Mechanical Arts in the various ways in which coal gas is applied: by manufacturers of Britannia and Silver-plated ware, and by those who make tin and japanned goods, and by workers in metals generally."[7] The company explained in its 1906 catalog that the gas was ideally suited to processes such as "heating light forgings; melting gold, silver, brass, glass, and the like; soldering, brazing, bluing," and so on.[8] In general, gas had several advantages over other more traditional heat sources such as charcoal or wood: Temperature could be more finely regulated for precision operations; time and effort tending fires was greatly reduced, allowing greater focus on production; fuel consumption was reduced by the ability to turn gas on and off quickly; impurities introduced by charcoal, coal, and wood were eliminated; and the risk of devastating fires and poor air quality for workers were reduced.

As with residential Springfield installations, Gilbert and Barker found its greatest business markets in areas not served by municipal gas systems. However, testimonials and catalog copy also portray gasoline air-gas as very competitive with, and often costing less than, coal gas produced by municipal gas companies. As a result, even some companies in urban areas with relatively easy access to municipal gas works used the Springfield system.

Metalworking and "Mechanical Arts"

Gilbert and Barker actually installed its very first Springfield Gas Machine in an industrial setting: the Finishing Department of the U.S. Armory at Springfield, Massachusetts. Writing in November 1866, Armory foreman Charles Bailey explained that "we have used this gas for about two years for brazing purposes, and for melting pots of lead and other metals. We have by its use brazed the sights to as many as eleven hundred musket barrels in a day. I consider it not only more convenient but also more economical for these purposes than is coal or any other fuel."[9] Charles F. Warner, author of *Picturesque Hampden* (1892), reported that a Springfield Gas Machine had been in use for many years "at the Springfield Armory, displacing some 400 tons of coal per annum in making all the forgings used in the manufacture of 40,000 rifles; also for brazing, melting metals, tempering, etc."[10] Unfortunately, little primary documentation on the armory's use of the Springfield system seems to exist. The scant remaining evidence of the armory's system consists of correspondence related to the sale of gasoline. In September 1866, J. W. Wheelock, "Dealer in Gasoleine [*sic*], Suitable for Gas Machines," of New York City, wrote to the commandant of the armory, offering to deliver seventy-five barrels of 80 to 85 gravity gasoline by steamboat for thirty-two cents per gallon, plus two dollars per barrel (refundable if returned in good order).[11] Similarly, several letters exist from New York supplier William Cooke indicating that his firm had shipped gasoline to the armory via its "New Haven boat."[12] At the time, suppliers shipped gasoline in barrels by steamboat and train, making the use of the Springfield system dependent on having either abundant storage capacity or a regular and reliable shipping connection.

Gilbert and Barker sold another early system to the Manning, Bowman and Company's "Britannia Works" located in Meriden, Connecticut. The firm installed a Springfield system in 1866 to use for soldering and for lighting the works:

> We are using it for soldering purposes in about fifteen blow pipes. Our men are very much pleased with it, and they as well as ourselves are of the opinion that it is quite as good if not better for this purpose than Coal-Gas, while the expense is not more than half as much. Formerly we used portable lamps, filled with a prepared fluid which was very objectionable, not only on account of the trouble and care which they required, but also on account of their smoke and odor: and more than that, we considered that so large a number of lamps, each filled with inflammable fluid, endangered our property very considerably, as they were at any time liable to be overturned and the building fired in consequence. We have done away with all these, and also with all movable and stationary lamps, and we now feel that we are comparatively safe.
>
> We shall not allow this opportunity to escape of testifying to the value and convenience of your process for lighting establishments of this kind. We are now using about seventy burners, and are satisfied that it is not only a very good but also a very cheap light.[13]

Sanborn Fire Insurance maps from the 1880s and 1890s indicate that the firm still used the Springfield system for lighting and metalworking some thirty years after it was installed.[14]

Similarly, the Dover Stamping Company used a Springfield system to both provide light and power production operations. The company, founded in Dover, New Hampshire, in 1833, moved to Cambridge, Massachusetts, in 1865. An 1896 history of Cambridge and its industries noted that the firm's principal "business was the stamping of tin plates into tinware of all kinds."[15] The company, which installed a three-hundred-light Springfield system, reported that the gas it produced provided excellent light and heat. In addition, the system needed little maintenance and cost considerably less than the charcoal and city gas used previously. "The light it furnishes is of superior quality," a member of the firm wrote to Gilbert and Barker,

> while our solder coppers are safely, neatly and rapidly heated. In first starting up we find an ordinary solder copper is heated in from five to eight minutes. . . . The care of it is merely nominal. We fill it say ten times a year [and] wind the pressure weight . . . ten minutes daily, and this covers all that has to be done.
>
> In 1874 our charcoal bill for soldering was $300. Our city gas bill for the same year for lighting was $527.82, making a total in that year for

> heating and lighting of $827.82. In 1875 we paid for gasoline to feed your generator, $396.08. We did all our work of lighting and heating that year with the products of the generator, the cost of the gasolene covering the whole expense. Our work was better done and our saving was $431.74.
>
> We cannot close this communication without recalling the peculiar warrant which we required of you, covering the ability of the machine to both heat and light "in a satisfactory manner." We allude to this that we may say . . . that very condition has been fulfilled absolutely and to the very letter.[16]

The Overman Wheel Company of Chicopee Falls, Massachusetts, also used the Springfield system for metalworking. The firm operated what it touted as the largest bicycle manufacturing facility in the world at the time, producing the famous "Victor" bicycle.[17] In *Picturesque Hampden*, Warner refers to the Springfield as "most extensively used at the Overman and other bicycle works through the country, for brazing purposes."[18] Overman almost certainly used the gas to fuel brazing apparatus to assemble frames for their "Victor" and "Victoria" models. An 1891 Sanborn map documents that the Chicopee Falls plant had electrical light, indicating that the firm used the Springfield system exclusively for production purposes.[19]

The "Electro-plate Ware" manufacturer Simpson, Hall, Miller and Company, which had a manufacturing plant in Wallingford, Connecticut, and principal sales office in New York City, employed the Springfield for both lighting and production. Founded in 1844 by Samuel Simpson, a "pioneer in the electro silver plate trade," the firm grew to become one of the three largest manufacturers of fine electroplated goods by the mid-1880s.[20] In addition to manufacturing a full line of William Rogers's cutlery, the company produced hollow wares such as tea sets, vases, cups, and goblets.

A company representative explained to Gilbert and Barker in 1873 that Simpson, Hall, Miller and Company used "some two hundred burners for illumination, and about thirty blow pipes for soldering purposes. It has been used by us about six years." This testimonial indicates that the firm purchased its Springfield system circa 1867 in the early days of Gilbert and Barker. By 1878, the firm wrote that it now used "double the burners than in 1873, with an enlarged Machine . . . and we are very much pleased with the working of it."[21] The company clearly continued to operate the system in 1885, as the Sanborn map from that year indicates that the plant derived its lighting from "gasoline gas." The Sanborn map illustrates a one-story gas house on the periphery of the company's industrial complex.[22]

The American Sterling Company of South Glastonbury, Connecticut, also used the Springfield system for both lighting and production. An 1878 testimonial letter reports that "we have had one of your (500-light) Gas Machines in use about seven years. . . . The light is strong, steady, and brilliant, and we have found it to work well, both for hard and soft metal soldering. The furnace for soldering and brazing is the best of the kind we have ever seen, and we cheerfully recommend its use to all those interested."[23] Interestingly, American Sterling makes particular note of the gas furnace sold by Gilbert and Barker, an appliance designed specifically for metalworking processes, such as soldering and brazing. This metalworking furnace is another clear example of Gilbert and Barker targeting a specific market for the Springfield system by creating specialized equipment that utilized air-gas.

Manufacturers of optical products, particularly eyeglasses, also began to use the Springfield system for soldering and brazing and heating precious metals in the late 1860s and 1870s. The largest and foremost optical factory in the United States, the American Optical Company of Southbridge, Massachusetts, purchased a system in 1869, the year it incorporated. American Optical resulted from the merger of several optical manufacturing firms, including Robert H. Cole and Company. Founded in the early nineteenth century, Cole and Company manufactured the first American steel eyeglasses; over the next thirty to forty years, Cole became the leader in producing inexpensive spectacles and, thanks to inventor and machinist George W. Wells, introduced a host of innovations in eyeglass manufacture. Wells purchased the largest textile mill in Southbridge, and along with Robert Cole and other investors, he converted it into the new American Optical Company's factory.[24] Not unlike Gilbert and Barker, American Optical obtained numerous patents; by 1909, Wells personally held twenty-six patents and the company another twenty-seven.[25]

American Optical's superintendent, H. C. Cady, wrote to Gilbert and Barker in 1893 that the firm had its "Springfield Gas Machine in successful use for the past twenty-four years, using the gas for illumination purposes, soldering, melting gold and silver, annealing, tempering, hardening, and for all kinds of heating purposes."[26] The 1888 Sanborn map, which illustrates the company's large factory, confirms the use of gas for lighting purposes; by 1894 the factory used electric light. The firm's new lens factory, completed in 1888, also used gas initially and then converted to electricity by 1894.[27] The application of the Springfield system for annealing, tempering, and other heating processes suggests that, like American Sterling, American Optical used gas furnaces fueled by the Springfield system. Air-gas, with its ability to provide a steady and easily controllable temperature without introducing impurities provided significant advantages for fine metalworking firms like American Optical; prior to the introduction of the Springfield system, such companies relied exclusively on dirty coal furnaces and fires.

Despite the Springfield system's suitability for metalworking, several firms used it exclusively for lighting. For example, Poole and Hunt, Founders and Machinists, used a Springfield Gas Machine to light its huge facility in Woodberry, Maryland. Founded in 1851 by Robert Poole and German Hunt, the company was one of the premiere foundries and mechanical firms in the country. Poole and Hunt produced a wide range of machinery, such as steam engines, boilers, pumps, water wheels and turbines, and gearing. It also made architectural castings, including the iron columns for the dome of the U.S. Capitol and cast-iron lighthouses for the U.S. government. The company's Woodberry facility featured a "foundry, pattern shop, machine shop, melting house, brass foundry, boiler yard sheds, tool house, and large and commodious offices and draughting rooms." Historian G. W. Howard noted in 1873 that the firm used a gas machine to light the entire complex: "The buildings are substantial edifices of brick and stone, heated by steam, and lighted by gas manufactured on the grounds."[28]

The same year, Poole and Hunt provided a testimonial letter to Gilbert and Barker that offers some additional details on the Springfield system: "We would state that the three hundred light Springfield Gas Machine we have recently put in to light our works, gives perfect satisfaction, and does everything you represented it to do." The company also specifically indicated that it had selected the Springfield system based on its satisfactory performance at Mr. Poole's Woodberry residence.[29]

The Springfield Gas Machine also functioned well in the canning industry, which required plants close to farms or fisheries to ensure freshness and economy. Because of their outlying locations canning and packing operations typically could not take advantage of the municipal gas works available in larger, more urban areas. An 1888 newspaper article, "New Fields for a Favorite Local Company," discussed Gilbert and Barker's expansion into this market: "The canning industry has made generous drafts on the Gilbert and Barker Company; factories are springing up all over the country in agricultural districts, where every surplus crop of vegetables and fruit is run into cans and saved where once it went to waste. Cheap heat for soldering is wanted and that is why the Springfield Gas Machine operates three plants which have a daily capacity of 250,000 cans each, and three sardine factories on the coast of Maine, besides many smaller canning factories east and west and north and south."

By 1906, Gilbert and Barker reported that "more than two hundred and fifty canners and can makers" used its system.[30] The last decades of the nineteenth

Figure 5.1. Poole and Hunt Foundry, Baltimore. Detail of advertisement from *The Monumental City, Its Past History, Etc.*, 207. Special Collections, University of Maryland Libraries.

century witnessed a huge growth in the production of canned foods in the United States. The production of canned vegetables grew from four million cases in 1870 to twenty million by 1889, while the number of canneries grew from about one hundred in 1870 to about eighteen hundred by the turn of the century. During the second half of the nineteenth century, the can-making industry went from an exclusively hand-crafted to a completely mechanized operation. In 1898 the Max Ams Company introduced a fully automated system for producing "sanitary" cans that machines made, filled, and sealed without the use of any lead solder. Prior to this innovation, workers hand constructed tin cans and sealed the seams with lead solder.[31] For example, in his study of the canning industry in Delaware, historian Dean Doerrfeld notes that "crews of tinsmiths working at the cannery during winter frequently manufactured the hole-and-cap cans prevalent in the nineteenth century."[32]

With mechanization, a typical cannery bought its supply of empty cans from a local or regional can-making firm. Workers then filled the cans and soldered on tin lids. A capping bench, usually located in the center of the cannery, would have several "fire pots" to heat the soldering "copper" or iron used to attach the lid. In many cases these fire pots were fueled by coal or kerosene and required a great deal of tending—thus slowing down the actual production rate. The adoption of portable gasoline gas machines like the Springfield system by the canning industry provided a dependable, low maintenance, and far safer heating source for soldering. Gilbert and Barker noted in its 1878 catalog:

> For heating soldering—coppers, it possesses these advantages over ordinary charcoal fire-pots: cleanliness, safety, convenience and cheapness; no time of men is lost in building or keeping up the fire; the flame is more fully under control; by turning on as much or little gas a very hot or low fire is obtained; when the workman has finished, the gas can be shut off at once, and the fire entirely stopped, thus economizing the fuel; again, the room is never filled with smoke and foul gas, as often happens when charcoal is used, caused by a bad draft of the chimney or contrary winds; nor are the floors and benches littered with coal, or dust and ashes.[33]

Similarly, can-making firms used the gas produced by the Springfield to solder can seams during construction. Although can makers could operate in urban areas served by municipal gas, they often elected to use a portable gas machine as a more flexible and cheaper alternative.

The E. F. Kirwan Manufacturing Company of Baltimore, a major wholesaler of canning supplies, illustrated an "idealized" canning plant in its circa 1890 catalog. Interestingly, the canning bench and fire pots in this model plant used gas produced by a gas machine located outside the building.[34] The presence of a "gasoline tank" and "air pump" suggests the plan called for a cold process gas machine system like the Springfield. The introduction of the self-contained gas machine into the cannery and can-making industries clearly increased the efficiency and rate of production and also greatly decreased the risk of fire and injuries.

Sanborn Fire Insurance maps of numerous cannery operations in Delaware and New Jersey confirm the widespread use of gas machines in the industry. For example, the 1885 Sanborn map shows that the D. Reise Company in Milford, Delaware, a fruit canner and evaporator, had lights and heating stoves powered by gas and that "gas [was] used in the tin shop," presumably to solder cans. This map also refers to "gasoline gas manufactured on premises" by a "novelty gas machine." The 1885 map of E. B. Seller's Evaporating Establishment in Middletown, New Castle County, Delaware, has a notation that the plant ran "night and day," although it operated "only through peach season."[35] The lights needed for working throughout the night were provided by "gasoline gas." Alternatively, the Stetson Elison Cannery in Camden, Delaware, used "lamps," likely kerosene, although, as the Sanborn map shows, a "gasoline oil tank underground" provided air-gas for the "can manufacturing" process. Finally, the Farmers' Preserving Company of Camden used gasoline gas to power the lights for its plant and for soldering in the "capping room." A separate gas house, shown on the map, stood just outside the "peeling" room, near the steam engine.[36]

The J. V. Sharp Canning Factory, a tomato processing firm located in Williamstown, New Jersey, that opened in 1883, utilized a Springfield Gas Machine. By the end of the nineteenth century, this company had the capacity to pack as many as twelve thousand cases of gallon cans of tomatoes per day.[37] The 1895 Sanborn map indicates that the plant used kerosene oil for lighting and gasoline for soldering. The map also illustrates a small, one-story building labeled "gasoline tank," located west of the main factory. By 1900, the Sanborn map shows a new "gas house" set farther west to accommodate the expansion of the main plant building. The map indicates that this one-story building had a "gasoline tank" and includes another building just to the south labeled "Springfield Gas Machine." Notes on the map specify that kerosene lamps still provided lighting, while gasoline powered soldering equipment. The 1905 Sanborn indicates the use of the gas machine for an increasing number of production operations. Detailed notes list "one gasoline gas soldering machine, 2 gasoline furnaces—1st, 3 gasoline furnaces—2nd" and explain that a "3/8 inch feed pipe under air pressure from one bbl. tank in yard" provided gasoline.[38] The furnaces, located on the first and second floors, are likely heating furnaces used for soldering and can production, possibly the specialized metalworking furnaces manufactured by Gilbert and Barker. Doerrfeld explains that "references in various insurance documents to 'gas generators' or 'can shop' on an upper floor . . . implies that the factory itself provided the preferred location for this activity [can making] rather than an ancillary building."[39]

The Springfield system made it possible for canning factories and can manufacturers to expand significantly prior to the introduction of automated can-making and canning machines. For these companies, most located in rural, agricultural regions of the country, coal stoves offered the only real alternative for soldering. Ultimately, gasoline gas provided more efficient and controllable heat with a much improved safety record and allowed for round-the-clock operation by offering good quality light for the plant during peak production seasons.

Mills

From the very beginning of its production, the Springfield system was used in a wide variety of mills. The machine typically provided lighting for mill complexes but was not used for production. In large mills especially, exterior windows alone could not provide enough good quality light for the entire interior production area. This was particularly true in the winter months due to low light levels and

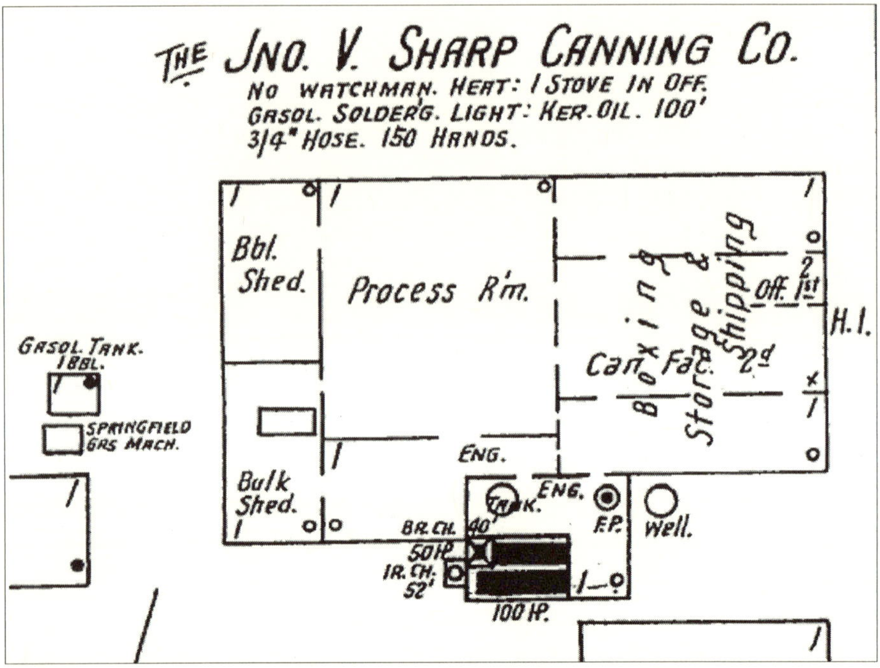

Figure 5.2. Sanborn Fire Insurance map of Jno. V. Sharp Canning Company, 1900. Sanborn Library, Sanborn Fire Insurance Company, Williamstown, New Jersey, 1900.

fewer daylight hours. Facilities that relied on kerosene lamps had to contend with attendant time-consuming maintenance and much higher fire risk. In addition, kerosene lamps gave off substandard light compared to gas light, and the introduction of the incandescent burner further exaggerated this difference. Many of these mills used extremely dangerous machinery and better lighting enhanced worker safety. Gas light also made it possible for mills to operate around the clock, which allowed owners to maximize the use of their facilities and machines and increase productivity and efficiency.

Gilbert and Barker catalogs document ninety-five mills that used the Springfield Gas Machine for lighting. Massachusetts had the largest number of these (n = 37, or 38.9 percent), followed by New York (n = 11, or 11.6 percent), Connecticut (n = 8, or 8.4 percent), and Pennsylvania (n = 6, or 6.3 percent). North Carolina and New Jersey each had five (5.3 percent), and the remaining states represented had fewer than 5. Not surprisingly, New England had the largest number of mills with Springfield Gas Machines (n = 48, or 50.5 percent), followed by the mid-Atlantic (n = 25, or 26.3 percent). Thus the Northeast had over 75 percent

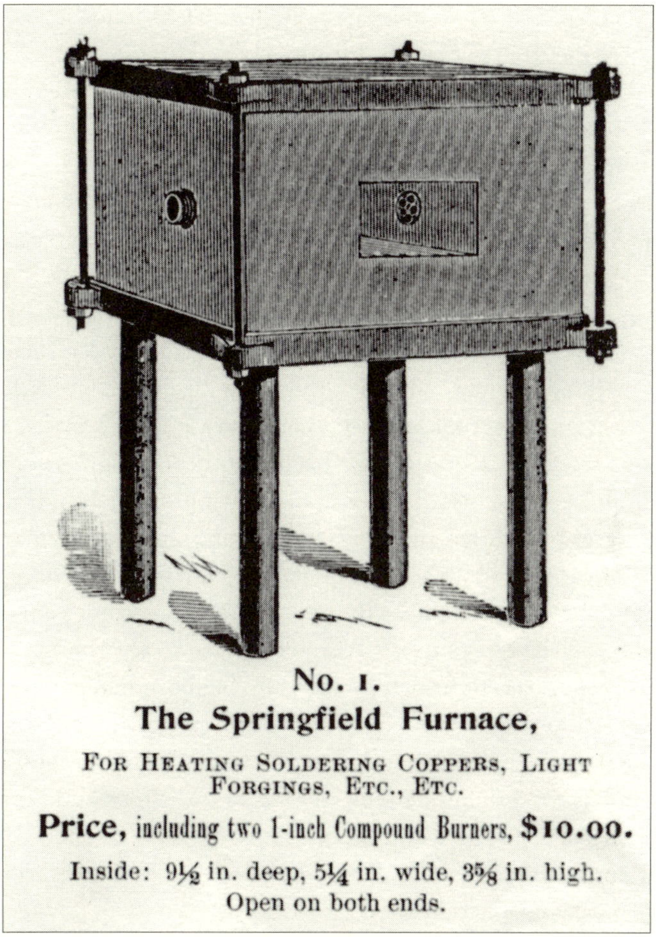

Figure 5.3. The Springfield Furnace, No. 1. From Gilbert and Barker Manufacturing, untitled advertising flyer, n.d., Shelburne Farms Collections, Shelburne, Vermont.

of the U.S. mills with Springfield systems, while the South had 12.6 percent (n = 12) and the Midwest had 10.5 percent (n = 10). Sixty-one of the ninety-five Springfield systems used to light mills could be readily categorized by type of mill: thirty-four were textile mills, twenty-two were paper mills, four were lumber mills, and one was a flour mill. Textile and paper mills were thus the principal users, perhaps not surprising given the complex machinery used by these industries.

The Philmont Hosiery Mills, located in Philmont, New York, used a large, five-hundred-light Springfield system to light its entire facility. Nelson P. Aken established the firm in 1861 and expanded it in 1865, building a new lower mill employing three hundred operatives. Historian Franklin Ellis noted in 1878 that the firm produced "four hundred dozen of underwear per day." Ellis also reported that "Mr. Aken is at present erecting a second mill also of brick, four stories high" that would employ another two hundred workers.[40]

Based on a testimonial letter to Gilbert and Barker, this second mill received a Springfield Gas Machine in 1878. Nelson Aken wrote, "I am pleased to write that the 500-light Machine you put in our new mill, with Hydraulic Blower, works well and gives us perfect satisfaction." Used with a hydraulic blower powered by the mill's water distribution system, the Springfield unit would have required virtually no maintenance beyond replenishing the gasoline as needed. Aken also reported that the machine "put in my house four years ago, pleases us very much, and gives us no trouble," indicating that his use of the machine at home probably influenced his decision to install a unit in his new mill. And he indicated that he had used other gas machines previously, suggesting that he may have installed other brands of gasoline gas machines in his 1865 lower mill: "I have seen many and used several gasoline Machines the past thirteen years, but yours is the only one that has stood the test of use."[41] Aken continued to use the Springfield system in the 1878 mill, as confirmed by the 1887 Sanborn map, which notes that the lights were powered by "gas manufactured on premises by Springfield Gas Machine." The map also illustrates a gas pipe running from the southeast corner of property into the main mill building; a note on the "Stock and Storage Bldg.," explains that it was forty-five feet from this building to the gas house.[42] Thus the gas machine and gasoline were located at the periphery of the mill complex, well away from principal production buildings.

Another major textile firm in Philmont, New York, a town known for its knitting industry, was the High Rock Knitting Mill, which reportedly manufactured "shirts and drawers."[43] Writing about his Springfield Gas Machine in 1905, owner B. F. Harder explained that "we have used your machine for lighting our mill since it was built in 1879, only renewing it once, in 1885, when we exchanged it for a larger size generator which has been in continuous use now over 20 years, and this 250-light machine has been carrying nearly 300 lights in our mill and about 100 lights in our residences. We have recently placed our order with you for a 500-light generator, which we hope soon to have installed, and feel confident that it will give us very satisfactory results."[44] This testimonial provides good evidence of extended use of the Springfield system for lighting. The 1887 Sanborn map confirms that the lights are powered by "gas, manufactured on premises by gas machine" and illustrates a one-story gas house with gas pipe running to main mill building; a notation below reads "Springfield Underground Gas Mach." Harder also indicates that the system lit the firm's "residences," either nearby workers' tenements or larger, single-family dwellings of overseers and the superintendent both shown on the 1887 Sanborn.[45] Given the larger single residence's distance from the generator, Harder may have referred to the adjacent tenements. If this is the case, it demonstrates an interesting use of the system for both the mill and the workers' housing, providing employees with high-quality lighting at work and home.

Like the knitting mills, silk thread factories constituted a large and innovative industry in mid-nineteenth-century New England. For example, the city of Florence, Massachusetts, became known, according to historian Elizabeth Sharpe, for "3 things: silk, abolitionism, and free thought."[46] The Nonotuck Silk Company, in nearby Northampton, Massachusetts, one of the region's largest silk firms, lit its mill with a Springfield system. Established in the early 1840s, the firm introduced its "machine twist" silk thread in 1852.[47] This thread, the first compatible with the new mechanical sewing machine, quickly attracted the attention of Issac Singer of the Singer Sewing Machine Company, who began ordering all that Nonotuck Silk could produce.[48] By 1881, Moses King reports in his Boston guidebook, the firm's mills had expanded to one hundred thousand square feet, employed seven hundred people, consumed 175,000 pounds of raw silk, and produced three thousand miles of finished silk each day.[49]

The company installed a Springfield Gas Machine to light the mill in 1872. Writing to Gilbert and Barker the next year, a company representative asserted that "the Springfield Gas Machine procured of you a year ago last October, has met fully our expectations. The 'Hydraulic Blower' works to a charm, and with little or no trouble in keeping it adjusted." Another testimonial in Gilbert and Barker's 1878 catalog indicates that Nonotuck Silk continued to use the Springfield machine for lighting: "Constant use of your Machine . . . enables us to say that the views expressed are not weakened by use and time."[50] The 1902 Sanborn map confirms that the mill still used its Springfield, at least as a backup system, as a notation reports that "electric and gasoline on premises" provided lighting.[51]

A number of large textile mills in Baltimore, Maryland, also used the Springfield Gas Machine for lighting. These companies represent the center of the manufacture of "cotton duck" fabric in the United States. Baltimore historian G. W. Howard wrote in 1873 that "under the enterprise of Mr. [Horatio] Gambrill, David Carroll, William E. Hooper & Sons, and William Kennedy, the great Cotton Duck interest of Baltimore at Woodberry, Mount Vernon, and adjoining points, has enormously increased, and become the principal manufacturing city of that article in the world." The firm of Gambrill Sons and Company, founded by

Horatio Gambrill, was a major player in the manufacture of cotton duck cloth in Baltimore. Gambrill purchased property in Woodbury in 1842 and had doubled the capacity of his mill by 1845.[52] In 1847, he bought Laurel Mills and subsequently built Clipper Mills, Park Mills, and Druid Mills.

Someone from the Gambrill firm wrote to Gilbert and Barker in 1870 about the use of the Springfield system in Druid Mills: "The Gas Machine purchased of you for lighting our factory, known as the Druid Mills at Woodbury, Baltimore Co., has to this time given us perfect satisfaction." The remainder of the letter makes clear that Gambrill and Sons considered insurance and safety issues paramount in selecting the system and indicates that the Springfield met the requirements of the company's insurers: "The machine was set in accordance with the regulations adopted by the National Board of Fire Underwriters to govern in such cases, and after a personal inspection by the agents of the various companies in which our Insurance is made, our arrangements were approved, and their consent to the use of the machine indorsed upon our policy."[53]

Textile mills posed a notorious fire hazard due to the accumulation of small fibers in the air and on the equipment, which could cause catastrophic explosions and fires. In addition, the vast quantities of both raw materials and finished products provided potential fuel for disastrous conflagrations. Thus textile mills required both fire protection systems and safe, reliable lighting. While mills used multiple windows for natural light, owners struggled to find ways to provide consistent light throughout the building in general and in the central sections in particular. The use of artificial lighting not only solved this problem but also provided the opportunity for running mills well into or even through the night. The owners of the Hampshire Manufacturing Company of Huntington, Massachusetts, wrote to Gilbert and Barker that their three-hundred-light Springfield system had "been in constant and successful operation in our mill for the past thirteen months, and it has never failed to furnish us with a brilliant and reliable light. We use about one hundred burners nightly."[54] Many other manufacturing firms using the Springfield Gas Machine reported that they lit and operated their facilities all night.

In 1873, Gambrill and Sons wrote another testimonial letter, which provides additional information about its Springfield system and lighting needs: "We have used the machine during the past three years to light our works, 'Druid Mills,' at Woodbury, Md, for which purpose over two hundred burners were required, with entire satisfaction when gasoline used was of the proper quality." Gilbert and Barker added a note in the 1874 catalog that "in the summer of 1873, Messrs. Gambrill having built a second mill, we exchanged the above machine for one of

500 burners' capacity, which now lights both mills." By 1878, Gambrill and Sons apparently used three separate machines to light its several mills. A letter from the company confirms that it viewed safety as a major concern: "We have been using three of your Springfield Gas Machines, about 600 burners, in our mills for the past eight winters. We consider them much safer than coal oil lamps. We would not be without them."[55] As the reference to using the Springfield system for "eight winters" confirms, winter months and their shortened daylight hours proved the most difficult period in which to ensure good lighting.

In the South, particularly in North and South Carolina, textile mills also employed the Springfield Gas Machine for lighting purposes. For instance, the Odell Manufacturing Company of Concord, North Carolina, used the system for its cotton mills. John Milton Odell and his brother James Alexander Odell founded the company in 1877.[56] Historian Wallace Jarrell reports that John Milton Odell "became one of the most successful textile industrialists in the state, being engaged in that business for more than 40 years."[57] While Gilbert and Barker did not list the firm in any of their catalogs, an 1885 Sanborn map confirms that a Springfield system lit Odell's Concord mill. The map shows a "covered gas machine" about fifty feet from the main mill building and notes "light: Springfield Gas Machines."[58]

Paper making, another industry that used the Springfield, received special marketing attention from Gilbert and Barker. In 1882, an article on the Springfield apparatus in the industry trade journal *Paper World* boasted that the system had "demonstrated [its] practicability in some of the largest paper mills in the country. They can be seen in operation at the mills of Chapin & Gould and Chester Paper Co., Huntington [Mass.]; Worthy Paper Co., Mittineague [Mass.]; Crane & Co., Dalton [Mass.]; and Smith Paper Co., of Lee, Mass.; W. C. Hodge, Rainbow, Ct.; Walloomsac Paper Co., North Hoosick [Mass.]; Hudson River Pulp and Paper Co., Jessup's Landing [N.Y.]; and the Remington Paper Co., of Watertown, N.Y. Many others have them and have testified to their efficiency." The same issue of the journal contains an advertisement from Gilbert and Barker that encourages readers to "light your paper mill with a Springfield Gas Machine."[59]

The Berkshire region of Massachusetts became a center of paper making during the first half of the nineteenth century. Over time, firms flocked to the area around Lee, Dalton, and Housatonic, for its plentiful water power.[60] The first paper mill in the area opened in Lee in 1806 and had at least twenty-four competitors by 1857.[61]

One of the major nineteenth-century paper-making firms of the region, the Smith Paper Company of Lee, Massachusetts, used the Springfield Gas Machine in several mills. Elizur Smith founded Plater and Smith in 1835 and subsequently

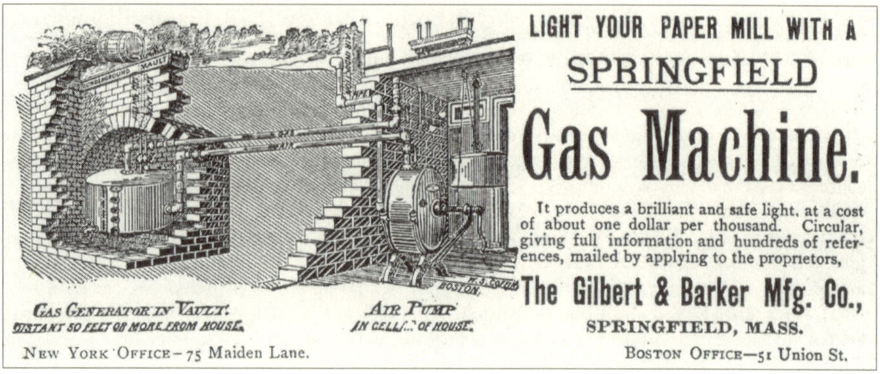

Figure 5.4. Advertisement for Springfield Gas Machine. From *Paper World*, Oct. 1882, 39. Author's collection.

owned and operated a number of paper mills around Lee. Organized in 1864, the Smith Paper Company became one of the largest paper-making firms in the world.[62] Smith Paper owed some of its success to its creation of the first paper manufactured solely from wood pulp; the firm maintained a monopoly on this innovation during the mid-1860s.[63]

Based on the company's testimonial letter to Gilbert and Barker, Smith Paper appears to have used the Springfield Gas Machine to light all of its principal mills in and around Lee. "It gives us pleasure," noted Treasurer Wellington Smith in 1878, "to write that all our five Gas Machines are working to our entire satisfaction in giving us good, bright light and causing us little trouble to look after them, and think your machine the safest and best in the market, and if we need any more would put in yours in preference to any that we are acquainted with."[64] The firm's use of Springfield systems in their Eagle and Columbia mills is documented by an 1884 Sanborn map. The map notes that Eagle Mill, located in Lee, used "gasoline gas" lighting and illustrates a gas machine close to the "Horse Shed" and "Fire Eng. House" and approximately seventy-five to one hundred feet from main mill buildings. Similarly, Columbia Mill, also located in Lee, had a gas machine next to a one-and-a-half-story "Storage" building and about seventy-five feet south of the main mill building.

The Springfield system also found some success with a host of industries, including paper making and the lumber trade, in the Midwest. For example, the lumber firm of Knapp Stout and Company of Menomonie, Wisconsin, used a Springfield system to light several of its saw mills. Established in 1846 by William Wilson, John H. Knapp, and Henry Stout, the company had become the city's principal industrial employer by 1870. The firm reportedly had multiple mills in

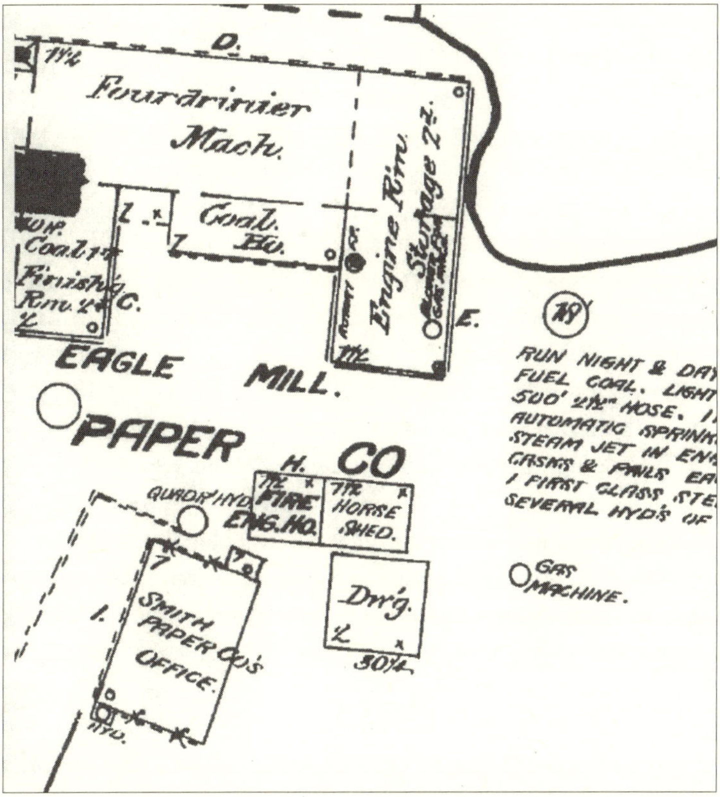

Figure 5.5. Sanborn Fire Insurance map of Smith Paper Company, Eagle Mill, 1884. Sanborn Library, Sanborn Fire Insurance Company, Lee, Massachusetts, 1884.

Menomonie and controlled all the timber rights along the Red Cedar River valley. During the height of operations in the last quarter of the nineteenth century, the firm employed over two thousand people and sent an average of eighty-five million board feet of lumber to St. Louis each year. Several of the firm's owners maintained elaborate homes in Menomonie, which also employed the Springfield for lighting. Their use of the Springfield system in their homes and mills would have been a novelty in rural Wisconsin during this period and no doubt further confirmed their "baronial" status as entrepreneurs and innovators.[65]

The company installed a Springfield system to light its mills in 1871, the beginning of its most productive period; thus the firm may have purchased the Springfield system as it expanded its facilities to accommodate increased production and perhaps introduced night shifts. Knapp Stout and Company wrote to Gilbert and Barker in 1872 that "the 150 light Gas Machine you put up in our saw and shingle mills last September, works well, gives a grand light, and is by far the best gas machine we have tried."[66] Like many other industrial users, Knapp Stout first used other types of gas machines to provide light but found the Springfield superior. It is unclear when Knapp Stout replaced its Springfield with electric light. According to an 1885 Sanborn map, the shingle mill used electric lights by that year.[67]

During the second half of the nineteenth century, flour mills also began to use the Springfield Gas Machine. After 1850, flour milling became regionalized and highly industrialized.[68] For example, the Patapsco Flouring Mills, owned by the C. A. Gambrill Manufacturing Company of Ellicott City, Maryland, had a production capacity of six hundred barrels per day in the late 1880s. An 1887 Sanborn map indicates that the mill used a Springfield Gas Machine with "all lights in mill enclosed"; the map places the gas machine west of the principal mill building with the notation "gasoline tank underground." A later Sanborn map indicates that the company used electric lights by 1894 but had "gasoline in reserve, Springfield Gas Machine." Patapsco made an early switch to electricity but kept its Springfield system operational as a backup to the relatively undependable new lighting source.[69]

Other Companies and Industrial Facilities

A host of other companies from tobacco firms in North Carolina to clock manufacturers in Connecticut employed the Springfield Gas Machine to light their factories and stores. For example, Blackwell's Durham Tobacco Company of Durham, North Carolina, used a Springfield system to light its extensive production facilities from the mid-1870s until the late 1880s. The firm, which began as J. R. Green and Company in the mid-1860s, held the trademark for "Genuine Durham Smoking Tobacco."[70] Renamed W. T. Blackwell Company in 1870, the company created the "first truly national tobacco brand," introducing "production, packaging, and marketing techniques that made Bull Durham a part of American industrial history and folklore."[71]

Blackwell constructed "a factory which would match their growth and ambition; this factory building (later known as "Old Bull") was completed in 1874."[72] At one time, this structure was "considered the world's largest tobacco factory."[73] The large Italianate-style masonry factory, a novelty in Durham at the time, used a Springfield Gas Machine to provide lighting. The 1875 Gilbert and Barker catalog lists Blackwell, suggesting that the system was installed sometime in 1874 or 1875 to coincide with the completion of the new structure.[74] The Springfield system still lit the building in 1884, when a Sanborn map indicates that "gasoline gas" provided its light. The map illustrates a one-story gas house in a courtyard behind the main factory building, surrounded by storage, cutting, stemming, and packing operations. A notation on the map indicating the presence of a "cut off valve at gas machine" was in keeping with insurance guidelines of the period.[75]

Another firm that only utilized the Springfield system for lighting was the hat manufacturer National Straw Works located in Westboro, Massachusetts. National Straw Works and its patent company, H. O. Bernard, were founded in 1868 and incorporated as H. O. Bernard Manufacturing Company in 1885.[76] By the turn of the century, the firm employed more than one thousand people and had a reputation as "one of the most extensive enterprises of its kind in New England."[77] The firm may have installed its Springfield system when it built its first factory in 1868; Gilbert and Barker catalogs begin listing it in 1870. The catalogs indicate that National Straw used a 150-light system, which suggests that it may have lit multiple buildings in its large factory complex. The firm continued to operate its Springfield system until at least 1888, when a Sanborn map indicates that lighting for the factory came from "gas made on premises."[78]

Several firms, including the New England Pin Company and William L. Gilbert Clock Company, in the small industrial town of Winsted, Connecticut, used Springfield Gas Machines to provide factory lighting. New England Pin produced an amazing fifteen million pins a day in its Winsted factory by the early twentieth century.[79] Based on catalog references, the company began using a Springfield system by 1870 and continued to use gas light in combination with electric lights into the early 1900s. An 1887 Sanborn map indicates that the firm used "gas made on premises" for its lighting, while the *American Electrical Directory of 1886* shows that the firm had installed an Edison electric generating system that powered four hundred lights.[80] The discrepancy between these sources may have resulted because of a lag in the publishing of the Sanborn map or because New England Pin had ordered but not yet installed the electric system in 1886. Alternatively, New England Pin may have used the gas for production purposes as well as lighting and thus retained the Springfield for metalworking functions even when lighting had switched to electricity. A 1904 Sanborn map indicates that the company then used both "Gas & Elec." to light its factory.[81]

Like New England Pin, the William L. Gilbert Clock Company used a Springfield system to provide lights for its large facility in Winsted, beginning circa 1873. Clockmaker William L. Gilbert established the company in 1871, and by 1873, the firm had completed construction of its factory and apparently installed a Springfield Gas Machine.[82] An 1887 Sanborn map indicates that the Gilbert Clock factory still used its Springfield machine for lighting. Like the New

England Pin factory, the Gilbert Clock Company added electric lighting by the time Sanborn published its 1904 map, which notes lights as "Gas & Electric." This suggests the presence of combination fixtures that could use gas when the electric lighting system failed.[83]

The Springfield system became essential equipment in a wide variety of industries from metalworking to textile production. Gilbert and Barker specifically marketed the machine for commercial and industrial applications from its introduction into the marketplace. The Springfield Gas Machine Company's 1867 catalog promotes its portable gas machines as "suitable for the lighting of mill, factories, [and] machine shops." In fact, the catalog explains that "manufacturers are particularly invited to examine our improved Apparatus. We make the lighting of manufacturing property a specialty."[84]

Gilbert and Barker produced machines suitable for lighting large structures and powering a variety of manufacturing activities. The company continually introduced new components designed especially for industry. These innovations not only increased the Springfield's suitability for factory use but also improved the quality and dependability of the light it produced and minimized maintenance and operating costs. Apparently referring to its "Air-Forcing Apparatus," the company notes in its 1867 catalog that "manufacturers whose mills are propelled by water should see us. We have just the thing for you."[85] The firm's 1870–71 catalog describes its Air-Forcing Apparatus as "designed especially for large mills, factories, and establishments where water power is obtainable." The apparatus, used in place of the weight-powered air pump, employed a small turbine water wheel to power a rotary fan blower and provide a steady supply of air pressure; Gilbert and Barker advised that a "2 or 3-inch pipe will carry water sufficient to operate this wheel and furnish gas for a thousand burners."[86] In its 1874 catalog, the firm notes that while many customers used the weight-driven air pump, "where water-power is attainable we have another instrument, which we often use in place of [pumps], especially in lighting mills and factories. Gilbert and Barker intended this new "hydraulic blower," patented in 1871, to improve upon the Air-Forcing Apparatus.[87]

Unlike the earlier Air-Forcing Apparatus, which employed a series of pulleys and belts to transfer power from the turbine to air pump, the hydraulic blower worked simply, so that "no mechanism intervenes between the power and the use made of it, no pullies, wheels, or other appliances." "The power of this instrument for producing a blast of air is wonderful. The largest mill of 500 or 1,000 burners," Gilbert and Barker enthused, "is as readily lighted as one of fifty."[88]

Gilbert and Barker also introduced an "Anti-Vibration" light fixture specifically designed for lighting factories and mills. As the company notes in its catalog for the Springfield Incandescent Gas Burner, this fixture made "the use of Incandescent Mantles in manufactories practicable and economical." The incandescent burner, a critical innovation that allowed gas lighting to successfully compete with electric lights into the early twentieth century, produced a much brighter and stronger light while using less gas than earlier burners. However, this burner used mantles that were very fragile and prone to break as a result of even the smallest bumps or vibrations. A fixture attached to the interior framing of a mill or factory building had to withstand high levels of vibration from machinery that could destroy the mantles quickly and repeatedly. Gilbert and Barker's anti-vibration fixtures were isolated from vibrations by enclosing them in wire frames that hung from springs, which provided a dampening action between the fixture and building. The fixtures received gas through flexible rubber tubes, which also helped dampen movement.[89]

Churches

Churches across the northeastern United States and as far afield as Wisconsin and Kentucky also utilized the Springfield system. At least fifty churches across the country used the Springfield system, with thirty-eight (76 percent) in the Northeast and twelve (24 percent) elsewhere. Thirty-two of these churches (64 percent) were in just three states: New York (ten), Connecticut (eleven), and Massachusetts (eleven). These thirty-two church buildings represent a range of denominations, including Baptist, Catholic, Episcopal, Methodist, Unitarian, Presbyterian, and Congregational. In several cases, multiple denominations in one town employed the Springfield system for lighting their churches. For example, in Rye, New York, both the First Presbyterian and Christ Episcopal churches used Springfield Gas Machines. An 1892 Sanborn map confirms that these buildings were lit with "gas, made on premises by the Springfield Gas Machine."[90] According to Gilbert and Barker's 1874 catalog, the First Presbyterian Church installed its two-hundred-light system by that year, and given that the congregation built the church in 1870, it may have installed the system during construction.[91] The 1892 Sanborn map locates the system approximately fifty feet north of the church; a note reads "underground gas machine." Interestingly, by 1892, Rye had municipal gas and electric service provided by Consolidated Gas and Electric Company. The continued use of gas machines, like the Springfield, speaks to their durability and functionality, as well as the costs related to both switching over to and operating electric lights. This may have held particularly true for churches, which utilized their buildings intermittently, and thus found the Springfield's ability to operate on demand and efficiently a bonus.[92]

Churches were the types of gathering places that brought the use of the Springfield Gas Machine system to the attention of community residents and businessmen; most churches lit with the Springfield were in towns without the availability of municipal gas or electricity. As will become clear, the installation of the Springfield system at Unity Church in North Easton, Massachusetts, in 1875, for example, led to its use elsewhere in town, notably the Ames family's country homes.

Unity Church

Unity Church is the earliest documented installation of a Springfield system (1875) in the town of North Easton, Massachusetts. This initial installation clearly caught the attention of the town's decision makers, as a number of other public buildings soon had their own Springfields. These buildings included Oakes Ames Memorial Hall (1879) and the Ames Free Library (1877), both designed by H. H. Richardson. The Ames family installed Springfield systems at a number of their homes in and around North Easton during the later nineteenth and early twentieth century (see chapter 4).

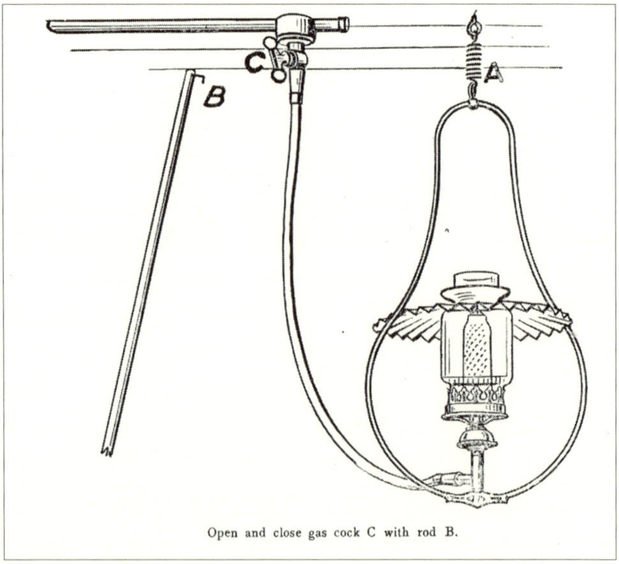

Figure 5.6. The Anti-Vibration Springfield Incandescent Gas Burner. From Gilbert and Barker Manufacturing, *Springfield Incandescent Burner*, ca. 1900. Shelburne Farms Collections, Shelburne, Vermont.

Built in 1875, Unity was the home church of the Ames family, one of North Easton's wealthiest and most powerful clans. Oliver Ames Jr. (1807–1877), the church's principal benefactor, was heavily involved in the construction process. John Ames Mitchell, nephew of the donor and founder of *Life Magazine*, designed the Gothic Revival–style structure.[93]

Documentary records and extant fixtures confirm the installation of a Springfield system during the church's construction. On June 23, 1875, with much of the building completed, Ames wrote in his diary that he "went to Boston and we selected Chandoliers [*sic*] for the Church and fixtures for the Basement. The Chandiliers [*sic*] for the Church I think will look very well." On June 29, with work on the church tower proceeding, Ames recorded that he "ordered gas Machine to be delivered early next week." He fretted on July 3, "Gas Pit not yet finished, Barney has made a poor job of it today." The gas machine arrived on July 8, along with a "man to put it in," with workers "getting in gas machine" on the following day. As the landscaping of the church grounds got underway in late July, Ames reported progress on "covering up the gas pit" and doing the "brick work for entrance to Gas Pit." As July ended, he noted that the work on the system was being finished and the "gas fixtures [were] up."[94]

Figure 5.7. Unity Church, built 1875, North Easton, Massachusetts. Photograph by Greg Galer.

Figure 5.8. Gas fixtures (now electrified), Unity Church sanctuary, North Easton, Massachusetts. Photograph by John Ventresco.

A brief inspection of the basement of Unity Church failed to turn up evidence for the actual Springfield system, but the original gas fixtures, now converted to electricity, remain throughout the building. The Ames diary entries confirm that the system had a generator, likely copper, installed in a belowground vault located some fifty to one hundred feet from the church building. A weight-driven air pump in the church basement would have initially powered the system, since it predates the introduction of the water-driven air pump in 1878. The system produced gas for six gasoliers and at least twelve wall fixtures and was probably a smaller unit, perhaps a No. 2 or No. 3, that could power between twenty-five and forty burners. A 1903 Sanborn Fire Insurance map indicates that the church used gas light, but it does not specify the location of the gas machine pit.[95] A letter from Gilbert and Barker Manufacturing to John S. Ames, dated May 14, 1914, indicates that the church still used its machine for lighting and that it had upgraded from a weight-driven to water-driven air pump at some point. The letter explains that a Gilbert and Barker repairman who performed maintenance had tightened the packing nut on the air pump shaft and "left everything in good order."[96] The company also suggested that the church replace the water in the "top holder of the pump," which sealed the expandable gas holder, with twenty gallons of oil to "do away with the rust and dirt which collect in it." This change would also eliminate any potential freezing problems and the need to regularly fill the water holder, making the system relatively automatic and reducing maintenance.

The endorsement of the Springfield Gas Machine system by one of the church's richest and most influential members no doubt impressed the congregation. This relatively early and very public installation seems to have encouraged the Ames family to install Springfield systems in subsequent public buildings in town, as well as in many of their private residences. Unity Church and these other public buildings attracted national notice—a leading architectural journal chose Ames Memorial Hall as the nation's tenth best building in 1885—adding to their cachet. Moreover, these buildings played a central role in the lives of North Easton's citizens. The marvelous interior spaces of these structures designed by Richardson and others would have appeared all the more spectacular and modern with lighting provided by the Springfield Gas Machine.

Educational and Governmental Institutions

A surprising number of college campuses and private and public schools also used Springfield Gas Machines. Approximately twenty-three colleges or universities, according to available records, used the system to provide lighting. For example, Swarthmore, founded in 1864 by the Religious Society of Friends in Philadelphia, Pennsylvania, reported to Gilbert and Barker that it presently used "over five hundred burners, and have no kind of difficulty in securing good light in every part of the college. The machine has never given us any trouble; neither have we had any accident of the most trifling character."[97] Some colleges, like Mount Holyoke and Swarthmore, also employed the system for heating, laundry, and kitchen purposes.

Others used it principally for laboratories, specifically to power Bunsen burners and furnaces like those used in metalworking firms. An early Springfield user, Amherst College, installed its system in 1866. W. S. Clark, the college's professor of chemistry, reported total satisfaction with the system installed in Williston Laboratory. He informed Gilbert and Barker that "during the past two months from ten to twenty burners have been in daily use, and the whole time consumed in supplying the Generator with the requisite fluid has not exceeded ten minutes." "In my opinion," Clark continued, "there is also much less danger in burning of Naphtha from gas pipes than in the employment of kerosene or fluid lamps. Not a day passes without serious accidents from the explosion of lamps or cans containing these fluids, and when we consider the dishonesty of manufacturers and dealers, and the carelessness and ignorance of consumers, we must regard them as unavoidable."[98] The University of Florida also utilized the Springfield system to power its laboratory burners. A testimonial letter to Gilbert and Barker in 1905 reports the "satisfaction that the Springfield Gas Machine has given in the chemical laboratory of the University of the State of Florida. . . . A year and a half ago we installed an Automatic Mixing Regulator which has performed its work perfectly and has become indispensable to us."[99]

Some colleges appear to have used the Springfield generator principally to enhance the coal gas produced by their own gasworks. For example, Dartmouth College, listed in Gilbert and Barker's 1906 catalog, had a gas works with a retort and large gasometer which clearly provided the gas for lighting.[100] However, an 1884 Sanborn map also shows, adjacent to the gasworks building, a small, one-story building marked petroleum tanks. This probably indicates gasoline storage and a generator. The company explains in its 1874 catalog that

> where buildings are lighted with coal gas, we can cheapen the cost of light very materially by the introduction of one of our gas generators. To do this we build a vault outside the building to be lighted; in this we place our gas generator and fill it with gasolene. The gas that supplies the building is conducted by pipes, after it has passed through the meter, to this generator. Here it becomes enriched with the carbonaceous vapors of the gasoline, and returned to the distributing pipe of the building. Coal gas is enriched by this process to that extent, that a two foot burner gives an equal light to five feet of ordinary coal gas.[101]

Gilbert and Barker had so much confidence that the Springfield system would lower the cost of coal gas that it guaranteed "a certain percentage of savings in the cost of lighting as a condition of payment."[102]

Mount Holyoke College

Mount Holyoke College in western Massachusetts had perhaps the largest and most impressive Springfield system used at an educational institution. One of the largest units ever in operation, the system could light thirteen separate buildings on campus. Use of the system by a prestigious women's college seems to have given Gilbert and Barker a valuable marketing opportunity by demonstrating its capacity, utility, and safety to well-heeled students and their parents. Documentary research led to the discovery of the intact generator unit, buried and forgotten for more than one hundred years, as well as the identification of several amazing fixtures long since converted to electricity.

Mount Holyoke College was founded as the Mount Holyoke Female Seminary in 1835. Founding principal Mary Lyon had a "vision of an education for women, comparable to that then available for men, [attracting] young women from the hill towns of the Connecticut Valley, from Vermont, Ohio, and from as far away as Hawaii."[103] The seminary grew to more than 250 students by 1849, graduating women who were much sought after as teachers and missionaries.

Mount Holyoke became a model as other women's educational institutions appeared. The original seminary building was continually expanded, ultimately forming a quadrangle, and improved by the addition of steam heat (1868), an artesian well (1879), and a twenty-passenger elevator (1880). The school constructed new buildings as well, including an observatory (1853), a library (1870), a natural science and arts building (Lyman Williston Hall, 1876), and a new observatory (1881). In 1881, the school received a collegiate charter and became known as Mount Holyoke Seminary and College. In 1890, Elizabeth Storrs Mead became the college's new president. In addition to academic and intellectual enhancements, Mead oversaw facilities improvements, and the construction of a new science hall (Shattuck Hall) in 1892. In 1896, a fire completely destroyed the Seminary Building. The trustees responded quickly with a plan to replace it with Mary Lyon Hall and over the next year also authorized the construction of four new dormitories.[104] Documentation indicates that the college installed a Springfield Gas Machine system in 1896, during its extensive rebuilding project. The system was designed to provide light to all existing campus buildings and eventually lit the college's new gymnasium (Blanchard Hall), dormitory (Mary Wilder Hall), and art building (Dwight Memorial Hall), all constructed in 1899.[105] By 1900, the college had completed the basic layout of the campus and boasted an enrollment of 550 students.

A significant amount of documentary evidence, including Sanborn Fire Insurance maps and Gilbert and Barker catalogs, provides a good sense of the size and scope of the Springfield system at Mount Holyoke.[106] Sanborn Fire Insurance maps of the campus, dating to 1905, 1910, 1918, and 1929, offer conclusive evidence of Mount Holyoke's use and eventual abandonment of gasoline gas lighting. The 1905 Sanborn map indicates that by this date, the college got its light from a combination of gasoline gas and electricity, the latter produced by a dynamo located in the Power House. The 1905 map also bears notations for "Gaso. Tank, Capy. 100 Gals" and a small one-story "gas house," just north of the Power House.

The 1910 Sanborn map notes that "Electric I.E.P. and Gasoline Gas" powered the lights on campus, indicating that the college had abandoned its dynamo in favor of power from the local power company.[107] By 1929, however, the Sanborn map notes "lights in all buildings: Elec. I.E.P.," clearly indicating that the college no longer used the gas machine system for lighting; the 1929 map doesn't feature the gas house, likely a sign that it had been removed.

Gilbert and Barker's 1906 catalog also provides good documentary evidence of the Springfield system at Mount Holyoke. The company chose to devote three pages of its catalog, *How Best to Light Our Country Homes and Resorts,* to the Mount Holyoke system as it "illustrates the size and scope of these machines." In addition to several paragraphs that explain the enormous capacity of the system, the catalog includes illustrations of eight of the principal campus buildings lit by the system. The machine, wrote Gilbert and Barker, had a "total capacity of eight thousand gallons of gasoline; the gas mains are eight inches in diameter; it supplies thirteen different buildings, seven of them being dormitories, an administration building, science hall, laboratory, gymnasium, art building and library." The system, the firm continues, supplied one thousand incandescent burners and one thousand flat-flame burners and all of the dormitories used "the gas for fuel

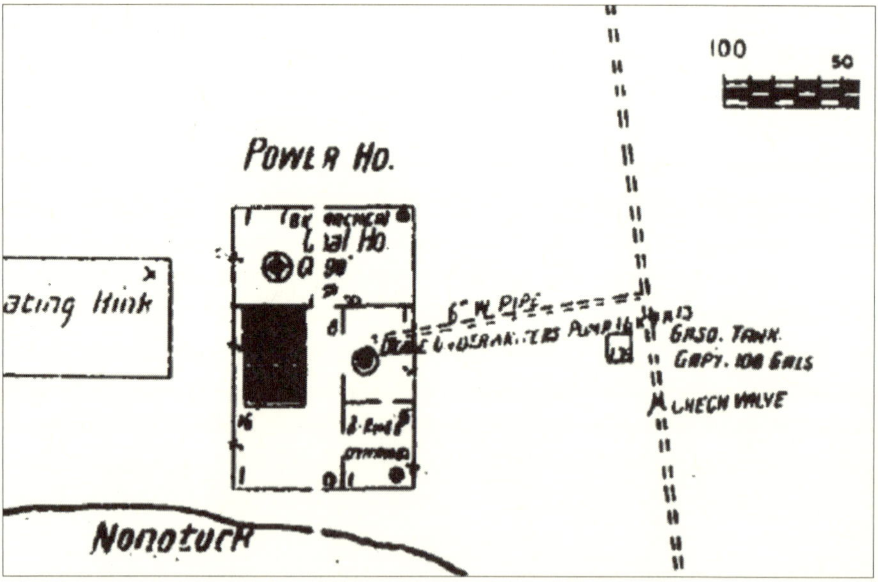

Figure 5.9. Sanborn Fire Insurance map of Power House, Mount Holyoke College, 1905. Sanborn Library, Sanborn Fire Insurance Company, South Hadley, Massachusetts, 1905.

purposes in their laundries." The system also supplied gas for Bunsen burners and other laboratory apparatus.[108]

A 1905 testimonial letter from college treasurer A. L. Williston validates many of the 1906 catalog's boasts and adds some information on the use of gas in the dining hall:

> The 2000-light Springfield Gas Machine which you furnished Mount Holyoke College in 1897, lighting seven dormitories, and administration building, science hall, laboratory, gymnasium, art building and library, has been in use for eight years, and is giving very good satisfaction.
>
> With your Automatic Mixing Regulator and incandescent burners, the quality of the light is good and well adapted to all various purposes of the institution and I am pleased to be able to so report to you.
>
> About a year ago you made us a special burner for use under chafing dishes, and we have installed several hundred of them to our satisfaction, and we find them much more economical and safer than lamps using alcohol.

Interestingly, by this date, the college was also using electric lighting in some buildings, produced by its own dynamo. For example, the "Library Note" in the *Mount Holyoke* published in 1904 indicates that the temporary library in Dwight Hall was lit with electric lights.[109]

While the college archives didn't have any information on the system, Russ Boudreau, a member of the Buildings and Grounds Department, recalled that a couple of trades supervisors had found several eight-inch pipes of unknown origin that he suspected might be from the Springfield system. In addition, Boudreau discovered a 1936 campus map that had references to an "old 8 inch Gas Main."

Using measurements from Sanborn Fire Insurance maps, Boudreau triangulated from the corners of the existing Power House to locate the "gas house" site. He identified a concrete slab in the vicinity of the area labeled on the maps as the "gas house." The exposed part of the slab measured approximately two feet square and capped the opening to the brick vault. Removal of the concrete lid revealed the presence of the Springfield machine's generator, installed in a manner similar to the Ashland unit. Gas pipes to the generator measured exactly eight inches, as reported in the 1906 catalog and by the trades supervisors.[110] The brick vault measures approximately eighty-four by forty-two inches, and the generator itself exceeded eighty-four inches in diameter.[111] A three-inch fill pipe runs from the surface into the top of the generator within the vault area.

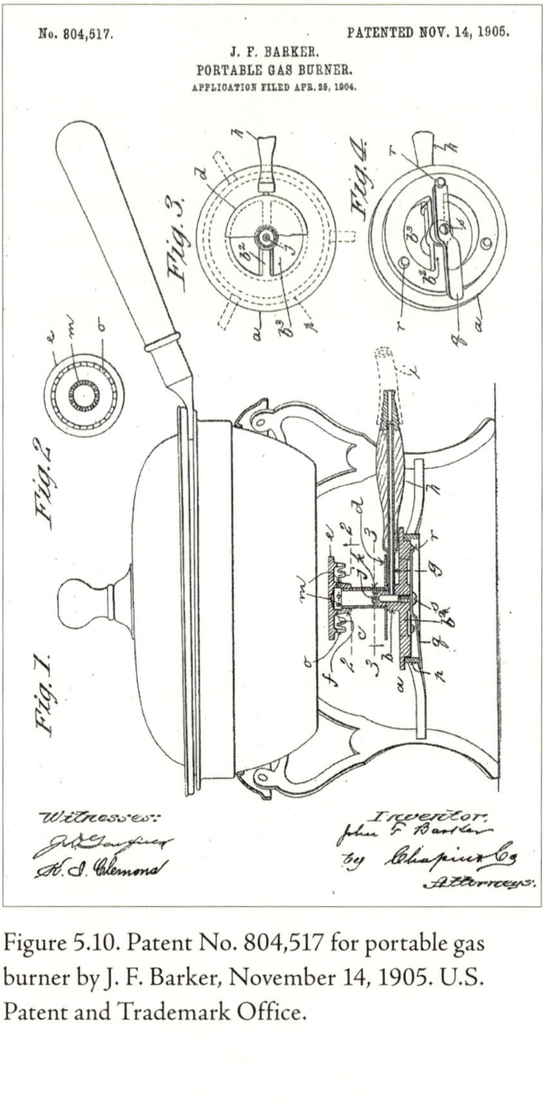

Figure 5.10. Patent No. 804,517 for portable gas burner by J. F. Barker, November 14, 1905. U.S. Patent and Trademark Office.

Remaining fixtures include two huge gasoliers located in the gymnasium (now Blanchard Hall). These circular gas fixtures, subsequently electrified, hang from the ceiling. Each has approximately eighty individual flat-flame burners, which adjust with a small screw and are probably of the burner design patented by John F. Barker in March 1881 (Patent No. 238,205). Barker noted that he intended to provide a "gas-burner through which the flow of gas to the tip may be readily controlled from the outside and while the jet is ignited."[112] The 160 burners in these two fixtures must have produced quite a spectacular light.

In July 2000, the college decided to remove the Springfield generator and vault to make way for a new underground oil tank that would supply the nearby Power House boilers.[113] Because of the potential hazards, the removal had to be performed by a certified environmental mitigation company and the machine couldn't be salvaged. Boudreau photographed and sketched the removal process to document the machine, which provided additional details about the system.

The gas generator measured approximately 108 inches in diameter and stood 84 inches tall; the unit contained nine evaporator pans that held approximately two thousand gallons of gasoline. Brass "pump pipes," which ran from each

Figure 5.11. Large gasolier, Blanchard Hall, Mount Holyoke College. Photograph by author.

evaporator pan to a control box at the surface, were used to pump gasoline out of the evaporator pans. The generator, constructed of riveted steel (likely Bessemer), was mostly buried, with only six inches of it protruding into the vault. A cylindrical, six-thousand-gallon storage tank, measuring twenty-nine feet, six inches long by six feet in diameter and also constructed of riveted steel, was buried just east of the generator vault. The generator and tank are almost identical in design to the New Feed Springfield Gas Machine pictured in the company's "Directions for the Care of the Springfield Gas Machine," dated May 15, 1897. Gilbert and Barker patented the New Feed process on December 6, 1892, and the New Feed machine on July 17, 1894; thus this system was their newest design when installed at Mount Holyoke in 1897.[114] The New Feed system allowed for the generator to be easily refilled or topped off as needed by drawing gasoline from the storage tank via a control valve.

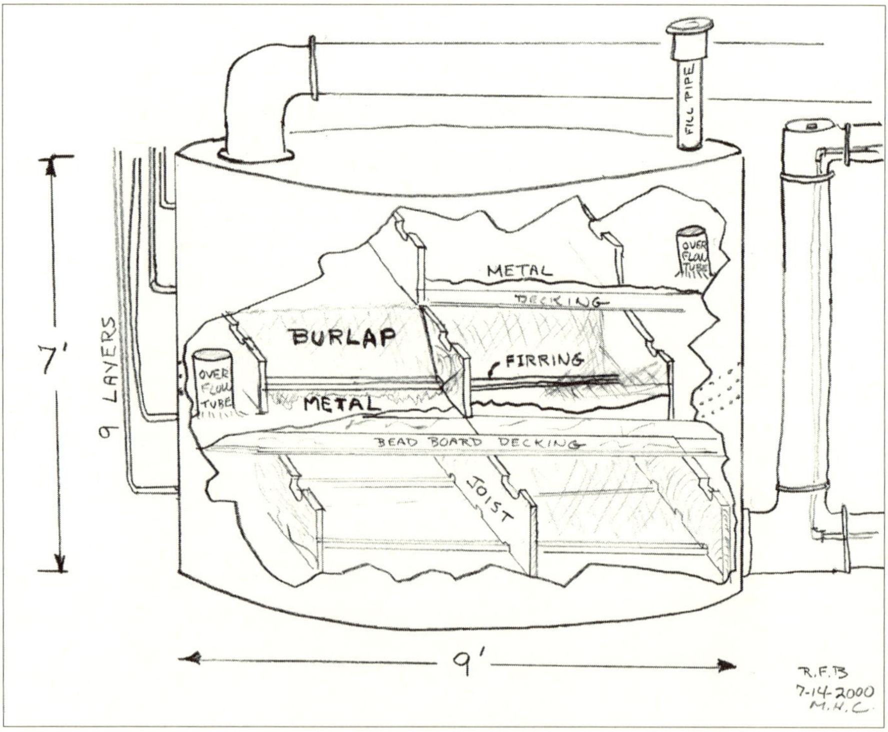

Figure 5.12. Sketch of Springfield Gas Machine generator, Mount Holyoke College. Delineated by Russell F. Boudreau. Courtesy of Facilities Management, Mount Holyoke College.

A 1902 "Map of Mt. Holyoke College Showing Electric Duct Line and Gas Pipes," discovered by Boudreau after the system's removal, illustrates the system in detail.[115] It identifies the vault, with generator and storage tank, and shows two eight-inch pipes running from the generator to the Power House; this piping configuration confirms that the air pump was located in the Power House. A third eight-inch pipe exits the Power House and turns west, running up the hill to Mary Lyon Hall, and smaller pipes branch off the eight-inch main to connect individual buildings to the system. A three-inch drain is shown running from the generator vault to Lake Nonotuck.[116]

Interestingly, this map also labels an "old gas machine" in Shattuck Hall and two "old tanks" located between Shattuck and Williston halls.[117] Shortly after the excavation of the Springfield system, another building project resulted in discovery of the "old gas machine." The second gas machine was located in the vicinity

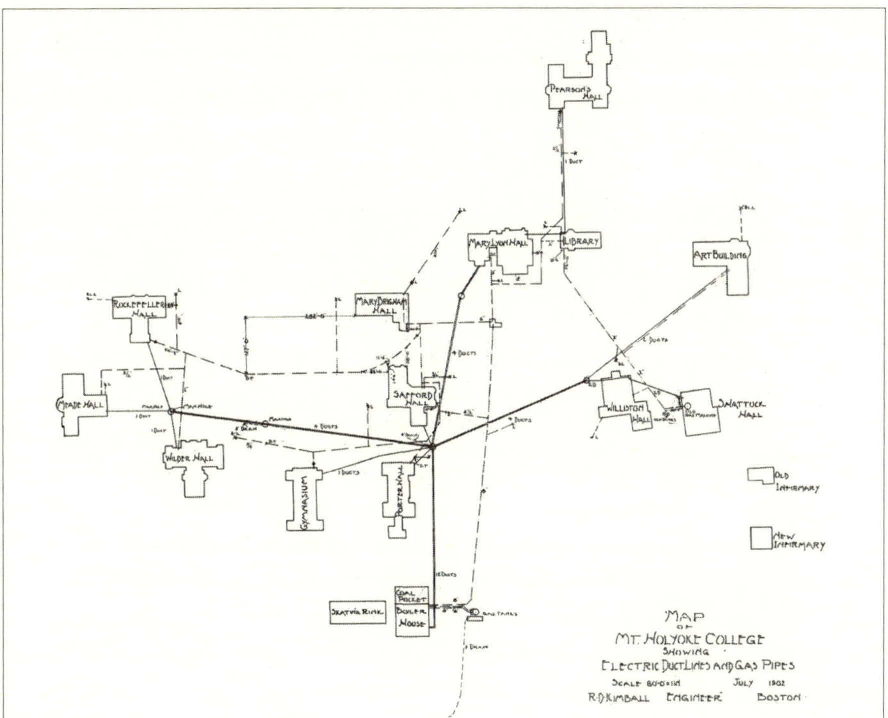

Figure 5.13. Map of Mount Holyoke College showing electric duct lines and gas pipes, R. D. Kimball, engineer, Boston, July 1902. Courtesy of Facilities Management, Mount Holyoke College.

of old Shattuck Hall, built as the physics and chemistry building in 1892 and torn down in 1954.[118] Based on photographs taken during its removal, this machine resembles a Springfield gas generator from the outside; it is round, approximately eighty-five inches in diameter, and stands about forty-four inches tall. However, the similarity ends there, as the interior of this machine has a carousel-like platen that revolves around a central pipe. The platen has small wings along its edge, as if it mixed or stirred as it rotated.

The "old gas machine" installed in Shattuck Hall obviously predates the installation of the Springfield system in 1897. The college may have installed it in 1892 when Shattuck Hall was built. Given its use in the physics and chemistry building, this machine may have exclusively provided gas for Bunsen burners and other scientific apparatus. It seems possible that the one-inch piping around the base of the excavated machine may have circulated hot water or some other liquid to gently heat the gasoline in the generator. The carousel-like mixing apparatus would have agitated the liquid gasoline in order to further speed the evaporation process. The college appears to have abandoned this machine with the purchase and installation of the Springfield system in 1897.

The Springfield Gas Machine system at Mount Holyoke introduced the modern convenience of gas lighting to the campus at a time of expansion and improvement. One of the larger units ever in operation, the system reportedly produced the astonishing amount of three and a half million cubic feet of gas during the year 1905.[119] In addition to providing gas to some two thousand burners in thirteen buildings across the campus, the college used the gas to power an intriguing array of appliances from the laundry to dining hall.

In addition to colleges and universities, a large number of private and public schools in New England, the mid-Atlantic, and the Midwest utilized the Springfield Gas Machine for lighting. The Northfield Seminary, founded in 1879 by A. G. Moody to provide "better Biblical education" and to "meet the demand for trained women who would devote themselves to missionary work," used a Springfield for lighting and to power appliances on its large campus in Northfield, Massachusetts.[120] Sometime later, Moody opened Mount Hermon, a school for training boys, which, like Northfield, apparently used a Springfield system early in its history. By 1906, Moody's "Northfield Schools" used seven machines to supply gas for lights and appliances. Moody's testimonial clearly shows that he found the introduction of the mixing regulator a major improvement: "We wish to give you our testimony for the efficiency of the Mixing Regulator which goes with your Springfield Gas Machine. The first machine which you put in for any of the Northfield schools was without that adjunct, but since then you have put in for us seven of your Gas Machines, all with the Mixing Regulator attached. These seven machines have not given us so much trouble as the old one without the Regulator."[121] While not a ringing endorsement of the early Springfield Gas Machine, this testimonial highlights the importance of the mixing regulator to the company in terms of improving the overall operation of its system.

The Rye Female Seminary, founded in 1869 in Rye, New York, also used a Springfield system to provide its lighting. A 1892 Sanborn map indicates that the school had lights powered by "gas, made on premises by the Springfield Gas Machine." The Hotchkiss School, a small private school in Lakeville, Connecticut, used a Springfield to provide fuel for lighting, cooking, and laboratories. Marie Hotchkiss founded the school in 1891 to prepare young men for entrance into Yale.[122] The school's headmaster reported to Gilbert and Barker that Hotchkiss had used a Springfield system "in combination with the Automatic Mixing Regulator for several years with perfect satisfaction. The gas is used for lighting purposes continuously, for cooking, for heating laundry machines, and for fuel in our Physical Laboratory."[123]

Public schools also installed Springfield systems. The Murdock High School in Winchendon, Massachusetts, begun in 1884 and completed and dedicated in 1887, employed one for light and laboratory purposes. The school, constructed in the Richardsonian Romanesque style, was a modern facility that contained "four recitation rooms, a library, art studio, physical and chemical laboratories, cloak rooms, basement, assembly hall, trustee's and committee's room and a principal's office." Considered "the finest and best equipped school building in the state" at the time of its dedication, Murdock High included the most up-to-date features then available, such as steam heat, gas lights, and a supply of running water.[124]

Ephraim Murdock Jr., a local entrepreneur who built a large wooden-ware business, bequeathed the money to construct the new school. Murdock utilized a Springfield Gas Machine in his personal residence in Winchendon from 1869 until his death in 1882.[125] He also utilized a Springfield system to light a group of commercial buildings in Winchendon known as the Murdock Block. Given Murdock's familiarity with the Springfield system, he may have specified its use in his bequest for the school.

Although no longer used, an examination of the building revealed several intact parts of the Springfield Gas Machine, including the system's air pump and mixing regulator in the basement. The standard water-powered air pump is identical to those found at Sagamore.[126] Interestingly, the serial number on the water-powered air pump at Murdock School, 6325, is slightly more than half of that on the Sagamore air pumps (10287 and 11037), suggesting that the company sold

Figure 5.14. Murdock High School, built 1884–87, Winchendon, Massachusetts. Photograph by author.

roughly four thousand units during the decade between the installation of the system at Murdock High School and the installation at Sagamore circa 1897–1902. The school added a "Combination Mixing Regulator," patented February 23, 1897, sometime in the 1890s.[127]

In addition to the main components of the Springfield system, the building contains several fixtures and burner parts with patent dates. A flat-flame burner marked "Gilbert and Barker Mfg., Springfield, MA, Pat. Mch—1881" was discovered on the first floor, as was a hanging fixture with an incandescent burner marked "Gilbert and Barker Mfg. Co., Pat. Jan 19, 1897, Nov. 21, 1899." While the flat-flame burner likely represents the type of burner installed with the original system, the later element confirms that the school made upgrades, including more efficient incandescent burner technology that could compete against increasingly popular electric lighting systems. This upgrade may have coincided with the addition of the Combination Mixing Regulator, first available in the early 1890s. Historic photographs from about 1910 confirm the use of four-burner gasolier fixtures in the school's science and physics lab, along with Bunsen burners mounted on the laboratory tables. These photographs also document the school's use of the Springfield well into the twentieth century.[128]

Resort Hotels and Tourist Sites

Gilbert and Barker catalogs document the use of the Springfield Gas Machine system at a long list of hotels and tourist destinations across the United States in the late nineteenth and early twentieth centuries. The list included railroad hotels in Maryland, the massive summer resorts of the White Mountains, small mineral springs hotels in the Midwest and South, and hotels along the Florida coast, the "Winter Newport." Many of these hotels served a burgeoning tourist trade in the second half of the nineteenth century. Victorians traveled to visit the country's historical sites and natural beauties, to seek rest and relaxation, to enjoy luxury and opulence, and, often, to overcome the many debilitating diseases supposedly helped by the fresh, clean air of the mountains or the humid and salty air of the seacoast.

The second half of the nineteenth century also witnessed a democratization of travel and tourism, brought about by the ease and economy of rail travel and the growing economic power and aspirations of the middle classes. As the number of tourists grew, the number of hotels to accommodate them also increased. These hotels often served as destinations in themselves, offering a high level of service,

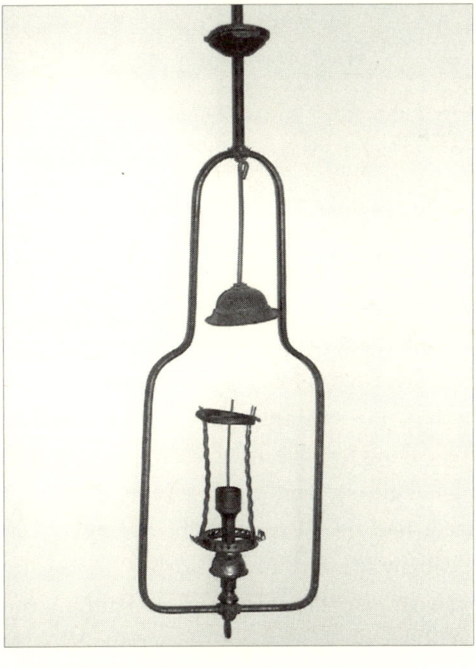

Figure 5.15. No. 95 Springfield Incandescent Burner in hanging fixture, Murdock High School, Winchendon, Massachusetts. Photograph by author.

Figure 5.16. Laboratory Bunsen burner, Murdock High School, Winchendon, Massachusetts. Photograph by author.

comfort, and convenience, particularly by utilizing the latest technological innovations, such as the Springfield Gas Machine.[129]

As train travel expanded across the country, railroads began to build hotels and resorts along their lines that themselves became tourist destinations and way stops for long-distance travelers. For example, the Baltimore and Ohio Railroad built a series of grand hotels along its line in Maryland. These hotels offered various amenities, including gas lighting from Springfield Gas Machines. The company reported in 1873 that it had five-hundred-light Springfield systems at the Deer Park Hotel in the Allegheny Mountains and at the Queen City Hotel in Cumberland, Maryland, along with a smaller two-hundred-light machine at its new "Relay Station and Hotel" at Relay or Washington Junction, Maryland.[130] The railroad also used a 650-light Springfield unit to light its wharves, piers, shops and offices at Locust Point.

The White Mountains of New Hampshire established a popular tourist industry as early as the 1840s. The expansion of rail and steamship travel to remote regions of the country in the 1840s and 1850s created a national interest in resorts. Industrialization and urbanization rapidly became dominant social forces, prompting Americans to look to "the idyllic American past for cultural inspiration, ideas and values." Those individuals who had the wealth to do so sought to escape the fast pace and competitiveness of the urban industrial lifestyle by traveling to places of historic significance and natural beauty. As train access to unspoiled places like the White Mountains attracted travelers who wished to escape the din of the city and factory, resort hotels began to spring up to accommodate them. The resort hotel culture of New England appealed to both old-money families and the new entrepreneurs of the industrial age.[131] The truly grand hotels, such as the Mount Washington Hotel, drew the country's rich and famous whose share of wealth and privilege increased exponentially during this period of unbridled commerce. Many smaller and less ornate resorts in the White Mountains catered to middle- and upper-middle-class industrialists and businessmen who aspired to amass greater wealth and power. All of these hotels and resorts, regardless of their size and degree of opulence, sought to employ the most up-to-date technologies to attract and retain customers.

Architectural historian Bryant Tolles Jr. notes that White Mountain hotels continually undertook renovations, not just architectural, that "involved interior decoration or modernizing such as the updating of heating, lighting and mechanical systems." Tolles argues that to "stay competitive with their rivals and continue to attract a large enough clientele, all the hotels in the region regularly devoted portions of their profits to these improvements and aggressively promoted them in guidebook and serial advertising." The 1874 edition of Bachelder's *Popular Resorts* notes of the renovated Fabyan House, for example, that "there are two hundred and fifty sleeping-rooms, high, airy, light, and newly and handsomely furnished. The entire house is lighted by gas, and warmed by steam."[132] Gas lighting with the Springfield Gas Machine was one of these modern conveniences and luxuries that helped to establish an atmosphere of glitter and fantasy that so many tourists of the period sought to consume.

In their 1888 catalog, *How Best to Light Country Homes and Resorts,* Gilbert and Barker boasted that "all hotels in the White Mountains are lighted by Springfield Gas Machines."[133] While it is difficult to confirm this claim (the firm omitted the word "all" in its 1906 catalog), Gilbert and Barker did supply gas lighting systems to a large number of resorts. These included most of the best known and most highly prized facilities, such as the Pleasant House, Profile House, the

Arlington Hotel, Flume House, Forest Hills House, Sinclair House, Crawford House, Maplewood Hotel, and the internationally famous Mount Washington Hotel. Nestled among picturesque mountain ranges in the northern part of New Hampshire, most of these resorts could be reached by rail lines connected to the major northeastern population centers.

Hotels also appeared in the vicinity of popular tourist sites, such as Howe's Cave, discovered in the 1850s in Schoharie County about forty miles from Albany, New York. Visitors could reach the cave, billed as second only to Mammoth Cave of Kentucky, easily by train from both New York City and Boston. The site quickly became a popular destination, complete with accommodations for overnight guests. The Howe's Cave Association, which managed the cave and the Cave House hotel, explained in an 1878 testimonial letter to Gilbert and Barker that it had used "for about five years one of your large Gas Machines with several hundred burners, for lighting a portion of the cave and the Cave House at the entrance of the cave. The machine has answered every purpose which it was intended and beyond my own anticipations. I cheerfully commend it to the public use."[134] This letter indicates that the association had the Springfield system installed in the "third" Cave House, built in 1873 to replace the hotel that had burned in 1872. This system lit not only the hotel but also the cave itself. In fact, the association touted Howe's Cave in advertisements as "the only cave in the world lighted with gas!"[135] Clearly, the association deliberately selected gas lighting to enhance the tourist experience and to emphasize the unique quality of the site. A *New York Times* article published in 1874 noted that "to increase the novelty means have recently been taken to have [the cave] lighted with gas as far as the lake, this being the only cave in the world that is lighted with gas."[136] The Gilbert and Barker staff added a note to the association's testimonial letter indicating that the "gas is carried into this cave from the machine a distance of over half a mile." In its 1874 catalog, the company reports that "this cavern is always cold and damp, the circumstances are most favorable to condensation, yet no inconvenience from this cause is experienced; a good light is always obtained."[137] Gas continued to be utilized at the site into the 1890s, as referenced in the Reverend Horace Hovey's 1896 publication *Celebrated American Caves*. Based on a visit to the cave in 1880, Hovey reported to his readers that the cave featured "modern improvements" like gas. He did not consider gas lighting an unqualified success, however, and recommended its replacement with electric lighting:

> [It provides] a pleasing effect ordinarily, though far less picturesque than torches and not free from danger. This appeared on the occasion of my first visit, which was in a company of a party of 400 excursionists, many

> of whom caught hold of the pipes overhead to steady themselves along difficult paths. This procedure disturbed the flow of gas. A number of jets were extinguished; and although frequently relighted, they could not be kept burning.
>
> The next day we examined critically the whole system of light in the cave in the company with Dr. Lewis, the chemist of the Boston Gas Works, our conclusion being that it is safe enough, if the pipes and jets are not tampered with nor allowed to be eaten through by rust. We recommended the substitution of electric lights, which are now used.[138]

While many nineteenth-century middle-class travelers chose active and improving visits to sites like Howe's Cave, wealthier tourists preferred extended sojourns to luxury resorts during the winter "season." Florida's elaborate hotels expressly catered to members of the upper classes who expected luxury and opulence and to members of the middle classes who sought to better themselves and experience a taste of exclusive refinements. This socially important form of travel provided news and hearsay for the hungry gossip sheets of the mid-nineteenth century. As newspapers and magazines vied to deliver the latest scoop, Americans of all classes found themselves fascinated by the lifestyles of the rich and famous; if some people couldn't experience the winter season first hand, they could at least read about it in their favorite newspaper or magazine.

Gilbert and Barker catalogs list fifteen Florida hotels that used the Springfield Gas Machine: the Brock House, Fort George Hotel, Waltham, Magnolia Hotel, Ocala House, Hotel Ormond, Royal Poinciana Hotel, Indian River Hotel, Sanford House, Silver Springs Hotel, Hotel San Marco, Florida House, Leon Hotel, Hotel Punta Gorda, and Seminole.[139] The guest registers of these hotels read like a who's who of American business, ranging from titans of industry such as Rockefeller, Vanderbilt, and Morgan, to their associates and management employees, to local and regional entrepreneurs.

Tourist Hotels of the White Mountains

In 1874, travel writer Henry Burt described the arrival of a traveler at the Profile House after dark: "Fatigued by an all-day's ride, he leaves the coach and enters this magnificent hotel, surprised at its extent and the strange wildness of the scene around it. . . . Retracing his steps, he enters the hotel and joins, perhaps, in the merry-making in the parlor, where a gay and bewitching scene presents itself."[140] The Profile House, built about 1852–53, underwent a major expansion beginning in 1866 that almost doubled its size.[141] It seems that the renovation included the installation of a Springfield Gas Machine. In a testimonial letter to Gilbert and Barker, owners Taft and Greenleaf report that "the gas-works put up for us nine years ago have given us perfect satisfaction. We have about 200 burners in our halls, parlors, etc. The pressure is sufficient to give us a very steady, and at the same time a strong and most beautiful light. We were in doubt about trying the machine, but can truly say now that we would rather part with almost anything in connection with our hotel than with your splendid Machine."[142] Several interior photographs from the 1870s and 1880s confirm the use of gas lighting in the public rooms at Profile House. An image of the drawing room from about 1875 illustrates a fashionable Victorian parlor lit with gasoliers outfitted with relatively plain eight-burner fixtures.[143] In 1872, the resort added a five-hundred-seat dining room with twenty-three-foot ceilings, which was lit with gas. Another photograph, circa 1885, shows the huge dining room illuminated by two rows of gasoliers. These fixtures, more ornate than the gasolier pictured in the 1875 image, had six burners each and provided the elegant space with good, bright light.

The Arlington, located in Bethlehem, New Hampshire, the major hotel center of the White Mountains, utilized a Springfield Gas Machine not only for

Figure 5.17. Dining room, Profile House, Franconia Notch, New Hampshire, ca. 1885. Courtesy of Richard Hamilton, Littleton, New Hampshire.

lighting but also in their dining room operation. In a 1905 testimonial letter to Gilbert and Barker, owner F. C. Abbe notes that "the Springfield Gas Machine which you put in for me in 1895 did its work to perfection, not only lighting my hotel with satisfaction, but *furnishing fuel for heating my carving table and griddle cake baker.* The Automatic Mixing Regulator, which you installed with the plant, kept the lights always brilliant and steady with not the least blowing, so common with some gas machines. It will give me pleasure to recommend your machine to anyone desiring a good and economical lighting plant."[144]

The Maplewood Hotel, another major Bethlehem resort hotel, also used a Springfield system. Isaac S. Cruft, a transplanted Boston merchant, purchased the Maplewood as a modest inn in 1871 and within eight years transformed it into a massive three-story hotel crowned with a Mansard roof. An ell added in 1877–78 had a "capacity of over 350 people" and "contained several new dining rooms and parlors . . . and over 130 bedrooms, each lit with gas and connected to the office by ingenious speaking tubes." The 1888 and 1906 Gilbert and Barker catalogs confirm the use of a Springfield Gas Machine at Maplewood, as does a circa 1910 photograph that shows a gas fixture on the newel post of the main staircase in the building's lobby.[145]

The Crawford House, established on the west side of Mount Washington in the early 1850s, was one of the earliest Mount Washington hotels. The original hotel burned in 1859, but a new building was constructed and opened within a year. This building underwent renovations several times in the 1870s and 1880s and probably acquired its Springfield Gas Machine system during one of these expansions. While the hotel clearly used gas light in its public rooms, it is not known if individual rooms also had gas lighting. A photograph dated about 1865 shows an ornately decorated Victorian parlor with several large gasoliers with multiple burners. Tolles notes that the opening of the Portland and Ogdensburg Railroad in 1875 "brought a new, greater, influx of tourists to the area," which may have induced the resort to switch to a Springfield by this time to stay abreast of the competition and meet guests' expectations.[146]

The Forest Hills House did not open until 1883, a relative latecomer to the resort hotel industry of New Hampshire. Overlooking the picturesque Gale River Valley, the beautifully sited hotel "immediately caught on with the summer resort clientele." With a capacity of 150 guests, the resort "boasted all the latest technological advances—gas, steam heat, hot and cold running water, spring beds, electric bells, a telegraph and 'model plumbing.'"[147]

The best known New Hampshire resort lit with a Springfield Gas Machine was the Mount Washington Hotel. Gilbert and Barker included a photograph of

Figure 5.18. The Crawford House, White Mountains. Advertisement engraving from Thursty McQuill, *From New York to the Summer resorts of New England,* 1880. Courtesy of Bryant F. Tolles Jr., Concord, New Hampshire.

the hotel in its 1906 catalog and noted after a testimonial letter from the managers of the Hotel Ormond in Florida that "Messrs. Anderson and Price are also using one of our largest gas machines during the summer months in the Mount Washington Hotel."[148] Of the many resort hotels built in the White Mountains, the Mount Washington Hotel stands out "in the eyes of most critics [as], the most ambitiously conceived, elaborately appointed and conspicuously palatial" of the group. Unlike many of its predecessors, it did not expand over time but was originally designed and built as a massive, ultra-luxurious resort. With its completion in 1902, the hotel quickly "became the centerpiece of the entire White Mountains hotel industry."[149]

The hotel, built in the Spanish Renaissance Revival style, could accommodate five hundred guests on opening day. While little is known of the Springfield system at the Mount Washington Hotel, it seems likely that it was installed during the original construction. Gilbert and Barker's 1906 catalog entry highlights it as one of the firm's largest systems capable of lighting thousands of burners. It is unclear exactly when the Springfield system gave way to electric light, but the remoteness of the facility would have required an onsite dynamo until transmission lines could reach the complex. As often seen, the two systems probably ran concurrently for a time, with the gas machine providing backup power during electric power outages.

Figure 5.19. Crawford House parlor, Crawford Notch, New Hampshire, stereo views, ca. 1865. Courtesy of Bryant F. Tolles Jr., Concord, New Hampshire.

The Red Lion Inn

The fabulous resorts and hotels of the White Mountains and Florida were not the only tourist establishments to use the Springfield Gas Machine for lighting, cooking, and other purposes. Several hundred hotels, inns, and resorts across the country, especially in remote or rural locations without municipal gas, employed the Springfield system. One particularly noteworthy New England hotel was the Red Lion Inn, located in Stockbridge, a longtime fixture in Massachusetts' Berkshire Valley. The Berkshires, another important tourist destination beginning in the mid-nineteenth and continuing into the early twentieth century, attracted travelers primarily from Boston and New York. The area also became an important locale for country houses and estates in the later part of the nineteenth century.

The 150-light Springfield system at the Red Lion Inn was installed as part of the rebuilding of the inn following a devastating fire in 1896. Previously, the inn used the kerosene lamps typical for the period. In a large, wood-framed public building, like an inn, the use of portable oil lamps posed a great danger and may have caused the Red Lion's fire.

Established in Stockbridge, Massachusetts, in 1773, the Red Lion had several owners by the mid-nineteenth century. As seen with the White Mountains region, the introduction of rail lines in the area guaranteed the inn's success. The Housatonic Railroad, completed in 1842, made Stockbridge a destination for wealthy Boston and New York families. Simultaneously, a literary colony began in the Lenox and Stockbridge areas. These developments brought increasing numbers of visitors, leading the inn, then called the Stockbridge House, to expand. In 1873, Charles H. Plumb acquired the inn and renamed it Plumb's Hotel.[150] In 1886, author Clark W. Bryan wrote that "public accommodations for those who had no private entertainment by friends were given by the Stockbridge House, which stands, today, under the proprietorship and management of C. H. Plumb, one of the choice hotels of the world."[151] In the early 1890s, Plumb turned over the management of the Stockbridge House to his nephew Allen T. Treadway, who managed and owned the inn until the 1940s. Treadway had worked as clerk at the inn after his graduation from Amherst College in 1886, which at the time was itself using a Springfield Gas Machine to fuel its chemistry lab.[152]

In 1893, with business expanding, Treadway added a new wing and dining room to the inn. On August 31, 1896, however, a fire caused heavy damage to the inn.[153] Treadway quickly reopened a new and enhanced hotel, renamed the Red Lion Inn, on April 30, 1897.[154] Treadway furnished the newly rebuilt facility with "modern" and elegant features like gas lighting. The Red Lion Inn, the *Valley Gleaner* effused, "never looked so handsome as it did after the entire building has been lighted up and many people were out in the streets to see the pretty sight. Just eight months from the time the old inn went up in smoke and ashes the elegant new one was opening." Another local paper, the *Courier,* noted that "the new hotel . . . is a graceful structure in the colonial style, simple yet elegant and appointed after the luxurious fashion of a gentleman's country house." The *Valley Gleaner* illustrated the new building and described its many features, noting that "public rooms are heated by hot water and lit by gas."[155] Several photographs from early brochures depict gas light fixtures in the inn's public rooms and kerosene lamps in the private rooms.[156] The 1898 Sanborn map notations indicate that the lights in the structure were powered by "gas & K[erosene] Oil," confirming that the public rooms were lit with gas and the private rooms with kerosene lamps.

A 1905 letter from Treadway to Gilbert and Barker unequivocally corroborates that he had a Springfield system installed in the rebuilding following the fire. Treadway writes on November 13, 1905, that "it is now nearly nine years since my experience began with the Springfield Gas Machine, and I am very glad to be able to say a word not only in praise of the machine and its satisfactory workmanship, but also as to the marked courtesy invariably shown in all business dealings with your company." Treadway also notes that he had two Springfield machines, one at the Red Lion Inn and one at his other Stockbridge inn, Heaton Hall. "My outfit," he writes, "consists of two machines with Waterwheels and Automatic Mixers." Treadway records that the light "is used as an auxiliary, and fuel is also supplied for certain uses in the kitchen." These comments further support newspaper reports that gas lighting was used in the public rooms only or as "auxiliary light" but also point out that the inn used its system to run equipment in the kitchen. Both machines, Treadway reports, "have always given entire satisfaction, are very cheap to maintain and perfectly simple in construction and care."[157]

The new hotel, clearly as popular as the old, became a hub of social and cultural activity in the area. Joseph Hodges Choate, ambassador to the Court of St. James, reportedly said that "if anyone sat long enough on the Red Lion veranda he would see pass everybody worth knowing."[158] Open only during the social "season," the hotel became the favorite of New York's elite and hosted the visits of five U.S. presidents from Grover Cleveland to Franklin Roosevelt.[159]

Figure 5.20. Springfield Gas Machine, water-powered air pump (left), and Combination Mixing Regulator (right), basement of main dining room addition, Red Lion Inn. Photograph by author.

Treadway became an important local figure as both the proprietor of the Red Lion Inn and as a prominent politician representing the region. He served as a member of the Massachusetts House of Representatives, the state senate, and the U.S. House of Representatives.[160] It seems clear that Gilbert and Barker traded on the name recognition of both Treadway and his inn, providing yet another opportunity for them to point to the use of the Springfield by rich and powerful travelers and tourists.

A Sanborn map of 1911 provides evidence of the abandonment of the Springfield lighting system at the Red Lion in the form of notations reading "Lights: Elec. I.E.P."[161] Period photographs confirm this evidence.[162] During the initial inspection of the basement of the Red Lion Inn, one room gave way to another without even a hint of the Springfield system. Surprisingly, the "newer" cellar under the dining room addition, presumed to postdate the installation of the Springfield system, contains the Springfield air pump and mixing regulator.

Based on the location of two 2 1/2-inch pipes exiting the cellar's east wall, it seems likely that the generator is buried in the courtyard between the dining room addition and a small, one-story carriage house south of the kitchen addition. The size of the piping suggests that the inn had a No. 7 Springfield system, which could have powered approximately 140 to 175 lights or burners. Given the date of installation, it is possible that the machine was a fully or partially buried unit.[163]

As noted by Treadway in his 1905 testimonial, the Springfield system at the Red Lion used a water-driven air pump and an automatic mixing regulator.[164] A one-inch pipe provided water to turn the air pump's small water wheel. The entire pump unit sits on a stone and brick base.[165] The pump, except for its much larger size and capacity, is identical to Gilbert and Barker's water-driven air pumps at Sagamore and several other sites. The mixing regulator at the Red Lion is also identical to the mixer in the Play House at Sagamore.[166] Like the air pump, the mixing regulator is raised on a brick platform. The presence of the mixing regulator indicates that Treadway either purchased the system with the "latest technology" in 1896–97 or very quickly upgraded it by adding a mixing regulator.

The Springfield Gas Machine system at the Red Lion Inn enhanced an already prominent structure and business. The device played an important part in the rebuilding and reopening of the inn in 1896–97, as the owners sought to improve and modernize it. By this date, tourists enjoyed many more options with the development of resort areas like Florida. In addition, many well-to-do city residents began eschewing travel to resorts in favor of their country homes and estates.

Documentary records make it clear that the inn exclusively used kerosene lamps for lighting prior to the fire of 1896. It seems likely that Treadway decided against the installation of gas lighting in the private rooms because of the cost and/or because the hotel then operated only from April 16 to December 1, when the demand for in-room lighting would not have been as great. Guests would most likely have retired to their rooms primarily for sleeping, while using the parlor, library, and sitting rooms, all lit by gas as of 1897, for reading and socializing.

The change from murky kerosene to bright gas light clearly warranted mention, such as the *Valley Gleaner*'s reports that the inn "never looked so handsome as it did after the entire building has been lighted up." The *Courier* noted that "the new hotel [was] simple yet elegant and appointed after the luxurious fashion of a gentleman's country house."[167] During the hotel's heyday in the late nineteenth and early twentieth century, the Red Lion Inn became one of the Berkshire's most desirable destinations. Gas light had become a sine qua non for country houses by 1896, and travelers began to expect gas light in all the fashionable establishments they patronized. Both trends increased demand for portable gas lighting systems around the turn of the century. Gilbert and Barker, as innovative and as attuned to emerging and growing markets as ever, met this demand with exceptional success.

The Flagler Hotel Empire

Some of the finest of the country's resort hotels utilizing the Springfield system resulted from the energy, business interests, and fierce competition of two men: Henry M. Flagler and Henry Plant. Flagler, a New Englander by birth, found himself in partnership with John D. Rockefeller after several failed business ventures. Established in Cleveland, Ohio, this partnership resulted in the formation of Standard Oil in 1870. By 1872, Standard Oil had become the leading oil refinery in the United States.[168] With the formation of the Standard Oil Trust in 1882, Flagler withdrew from any direct management responsibility at Standard and began to look for other investments.[169] Following the death of his first wife, Flagler remarried in 1883 and, with his new wife Ida Alice Shourds, traveled to St. Augustine, Florida, in 1885. The couple stayed at the Hotel San Marcos, owned by New England businessman and resort tycoon Isaac Cruft and designed by the architectural firm of McGuire and McDonald.[170] Flagler "was so impressed with St. Augustine and with the work of [hotel manager] Seavey, [and architects] McGuire and McDonald that he hired all three men to build an even grander hotel," the Hotel Ponce de Leon, in the same city.[171] Realizing the need for good transportation to his envisioned hotel empire in Florida, Flagler purchased the Jacksonville, St. Augustine and Halifax Railroad, the first company in what would become the Florida East Coast Railway.

Not long after the Ponce de Leon opened in 1888, Flagler began to extend his hotel and railroad empire south. He built a rail bridge that provided access to the southern part of the state and purchased the Hotel Ormond north of Daytona.[172]

Owners John Anderson of Maine and Joseph Price of Kentucky sold the hotel to Flagler in 1891 but continued to manage it into the twentieth century. A 1905 testimonial letter from Anderson and Price, "Managers," indicates that the Hotel Ormond used a Springfield system with automatic mixer with "eminent satisfaction to ourselves and guests." Gilbert and Barker proudly added a note after the testimonial that "Messrs. Anderson and Price are also using one of our largest gas machines during the summer months in the Mount Washington [New Hampshire] Hotel."[173]

The Hotel Royal Poinciana

Just two years after purchasing the Hotel Ormond, Flagler began constructing the Royal Poinciana in Palm Beach, Florida. The hotel would become known as the "largest wooden structure in the world" and the center of Winter Newport, temporary home of the country's wealthiest tourists. The Royal Poinciana also featured a Springfield Gas Machine to provide the dependable high-quality lighting expected by the hotel's well-heeled guests.

According to Flagler, the three-thousand-burner Springfield Gas Machine in the Royal Poinciana may have been the largest Springfield system in the world at the time. When constructed in 1893–94, the Royal Poinciana epitomized a true luxury resort in every sense. Flagler no doubt chose a Springfield Gas Machine for the Royal Poinciana because Gilbert and Barker had become a subsidiary of Standard Oil of New Jersey and one of the largest retailers of gasoline in the nation. Flagler clearly had personal experience with the Springfield system, noting in 1905 that he had used one in his Mamaroneck, New York, home for twenty years.[174] Again, it is no coincidence that Flagler, a director of the Standard Oil Trust, would have installed a Springfield system in 1885, as Gilbert and Barker had been bought just two years earlier by the Pratt Manufacturing Company, itself owned by Standard Oil.

In the early 1890s, Flagler crafted a plan to extend his Florida East Coast Railway system and build a luxury hotel on the barrier island that was to become West Palm Beach. The architectural firm of McGuire and McDonald designed and built the Royal Poinciana, a Colonial Revival structure with six floors and five hundred rooms. The massive, wooden hotel opened in February 1894, and in March of that year, the Florida East Coast Railway began regular train service to West Palm Beach. The main section of the five-hundred-room hotel consisted of a long, double-loaded block with shorter, unequal wings on each end. A two-story rotunda or lobby rose from the center of the main block. Topped with a skylight and sporting a second-story gallery for the ladies, this space, replete with gas light fixtures, served as the building's "most pivotal public space for physical and social circulation."[175]

Guests at the hotel found it equipped with the latest technology, including gas light from the Springfield Gas Machine. Architectural historian Susan R. Braden notes that while guests enjoyed the products of this up-to-date technology, "the machines and mechanical equipment remained hidden from view," typically in separate buildings behind the hotel. French visitor Paul Bourget "marveled at the very American exaggerated contrasts that he observed . . . where electricity and expensive ball gowns could be found in the wilderness of semitropical Florida." The hotel's 1894 brochure boasted of its "romantic 'Arabian Nights' quality" and referred to the complex as something out of the science fiction novels of Jules Verne. "Again and again," writes Braden, "the hotel's publicity underscored the luxury, technology, and sheer scale of the Royal Poinciana." Guests at the hotel, notes Braden, included "New York society figures and some of the wealthiest families in the country." Many of these families arrived in their own Pullman cars, which could be parked at the hotel.[176]

Flagler made major additions to the hotel in 1899, 1900, 1901, and 1929, the first of these including a new dining room and an additional 150 rooms with private baths. The Royal Poinciana actually suffered from the opening of Flagler's

Figure 5.21. Postcard of the Hotel Royal Poinciana, Palm Beach, Florida, ca. 1890s. Author's collection.

nearby Breakers Hotel in 1896, which by the early twentieth century had become the "epitome of modern luxury." By 1929, the hotel was suffering from the financial collapse that ended the 1920s tourist boom. The deteriorating hotel was finally torn down in 1935.[177]

Despite the lack of physical evidence, several sources provide a clear picture of the Springfield system at the Royal Poinciana. The most revealing document is a testimonial written by Henry Flagler to Gilbert and Barker on October 4, 1905. Flagler explains that "when we built the Royal Poinciana Hotel at Palm Beach, we lighted with gas, installing the Springfield Gas Machine, which, if I remember correctly, had nearly 3000 burners. At that time I think it was the largest private gas machine in the country. We are still using the machine to its utmost capacity for laundry and other heating purposes."[178] Given the size of the machine, the system probably provided light for public areas, such as the rotunda and hallways, as well as for individual guest rooms. The facility's long double-loaded corridors needed good artificial lighting. An 1894 photo of the rotunda from a souvenir brochure provides solid evidence of gas lighting fixtures in this important public space; these fixtures included two- and four-burner gasoliers and two-burner wall sconces. It seems curious that Flagler installed gas lighting in his new hotel in 1893, given that he had installed electric lights in the Ponce de Leon in 1888. The Royal Poinciana had at least some electric service, perhaps provided by battery, as the 1894 brochure touts the hotel's electric annunciator system for contacting the front desk.[179]

While the Springfield system originally provided lighting for the facility, Flagler reports that by 1905 it only powered ancillary activities like the laundry. The abandonment of gas lighting in favor of electricity isn't surprising since electricity represented the height of modern technology at the time. The conversion to electric lighting may have occurred during one of Flagler's expansion projects in 1899, 1900, or 1901. Flagler may have chosen gas light at first because the hotel's initial power plant could not produce enough power to light the massive complex and municipal power did not yet exist on Florida's frontier. Subsequent improvements to the generating system no doubt allowed for the conversion to electric light. Of course, Flagler had access to gasoline and controlled the railway that could deliver it to the hotel, an important factor in running a three-thousand-light system would have required approximately six thousand gallons of gasoline or almost 150 barrels at each filling.

In addition to providing accommodations to the very rich, Braden writes, the "romantically evocative appearance of Flagler's hotels . . . charmed and delighted visitors." The hotels "decorated their cities and functioned as landmarks and points of interest even to those not staying at the hotels." These magical attractions provided entertainment and wonder to a whole range of visitors, years before the Disney brand would arrive. Even those who could not afford to stay at these hotels made the pilgrimage to follow the rich and famous a trend. With their ornately decorated interiors illuminated by gas light, these luxury palaces drew in thousands of visitors who for a few minutes could consume the splendor of life at the Royal Poinciana.

The Plant Hotel Empire

New Englander Henry Plant had a long career in transportation before he began his railroad and hotel empire in Florida. An employee of the Adams Express Company, an express mail and cargo business, Plant bought the firm's southern holdings at the beginning of the Civil War and reorganized them into the Southern Express Company.[180] With the acquisition of the Texas Express Company after the war, the firm grew and prospered throughout the South. In the late 1870s and early 1880s, Plant and his partners acquired several railroads that eventually became the South Florida Railroad, which stretched from Georgia to Jacksonville and ultimately to Tampa along the state's western coast. In 1888, Plant began construction of the Tampa Bay Hotel, a huge "Islamic Revival structure that became the flagship for the Plant System hotels."

Over the next five years Plant bought or built a series of impressive hotels on Florida's western coast that rivaled Flagler's holdings on the state's eastern coast. In 1891, Plant acquired the Seminole in Winter Park, Florida. Designed and built by the McGuire and McDonald firm, the five-story wood-frame structure opened in 1886. By 1888, the hotel had a Springfield system in operation, according to the Gilbert and Barker catalog. Another major resort, the Hotel Punta Gorda in Trabue, Florida (later renamed Punta Gorda), also employed a Springfield system. Built in 1887 by the Florida Commercial Company as the southern terminus of the Florida Southern Railroad, the hotel probably used the Springfield system for lighting from its opening, as it too is listed in the 1888 catalog. The hotel's guest list for its first season included W. K. Vanderbilt, Andrew Mellon, and John Wannamaker. Henry Plant acquired Florida Southern Railroad and the already popular hotel in 1894. Another resort built in 1884 and acquired by Plant in 1895 was the Ocala House. This three-story hotel in the small commercial town of Ocala in central Florida had two hundred rooms. Plant advertised it "as if it were a health spa. Publicity for the hotel advocated a stay in Ocala as being good for sufferers from gout, chronic rheumatism, or feeble nerves."[181]

While Plant's Florida resort hotels never had the cachet of Henry Flagler's, Plant found success in his operation. In general, the Plant hotels were more modest than the Flagler hotels; they offered the same standard amenities but lacked the luxury and opulence of the Ponce de Leon, Royal Poinciana, or Breakers. As a result, the Plant hotels did not attract guests from the highest ranks of leisure-class families. Nevertheless, Plant's clientele of prosperous local and regional entrepreneurs comprised a very important market segment for Gilbert and Barker. Guests enjoyed the luxury of good light, and Gilbert and Barker made sure they knew it emanated from the Springfield Gas Machine.

Mineral Springs Hotels

Along with places of natural and historical significance, mineral springs drew thousands of visitors in the second half of the nineteenth century. Hotels and inns near these springs also installed Springfield Gas Machines. Mineral springs hotels grew in popularity during the Victorian era, as Americans increasingly sought relief from the illnesses and fatigue associated with urban and industrial life. The small town of Grand Haven, on Michigan's west shore, for example, became a destination for travelers from as far away as Detroit and Chicago. The Cutler House reigned as the town's premier mineral springs hotel, while the Pomona House in nearby Fruitport was another popular place to "take the waters."

Dwight Cutler built the Cutler House in downtown Grand Haven, directly across the street from the newly established W. C. Sheldon's Magnetic Mineral Spring. Opened in 1871, Sheldon's mineral springs soon attracted thousands of visitors to "take the cure" by drinking the sparkling water reportedly "highly charged with mineral substances and magnetism."[182] The ornamental sanitarium buildings and tastefully landscaped grounds of Sheldon's establishment provided the ideal resort setting for travelers, with the Cutler House offering the finest accommodations in the area. When it opened on July 4, 1872, the Cutler House immediately attracted visitors to Sheldon's mineral springs and quickly became responsible for establishing the town's reputation as a popular summer retreat for Detroit and Chicago residents. The five-story Second Empire–style hotel was "the most conspicuous object confronting any person entering the city."[183] Built at a cost of two hundred thousand dollars, the three-hundred-room hotel offered all the "modern improvements," including a steam passenger elevator, hot and cold running water, steam heat, and gas light provided by a Springfield Gas Machine.[184]

Gilbert and Barker's 1874 catalog confirms that the hotel had a five-hundred-burner Springfield system, probably installed during the hotel's construction.[185] An 1883 Sanborn map notes that almost ten years later the Cutler House still used "gasoline gas." The map shows an underground pipe running from the back of the hotel down an alley to a one-story building marked "gasoline tank" that stood across the alley from the hotel. This structure clearly housed the Springfield Gas Machine's gas generator. By the publication of the 1906 Sanborn map, the original hotel had burned and been replaced by the "New" Cutler House, which had "electric and gas" lighting. The gas house down the alley no longer exists.[186] The Grand Haven Gas Light Company, which began service in 1885, probably provided the gas lighting at this point.[187]

In addition to the Cutler House, the Grand Haven area had several large mineral springs hotels, earning the area the sobriquet "Saratoga of the West."[188] Nearby Spring Lake had the well-known Spring Lake House, while the town of Fruitport had a large and popular hotel known as the Pomona House. Owned by the Fruitport Magnetic and Sulphur Springs Company, the elegant, four-story Pomona House opened in 1871, soon after the discovery of mineral water in the

Figure 5.22. Woodcut of Grand Haven, Michigan, 1876, showing Cutler House and Mineral Springs. Photo courtesy of the Loutit District Library, Grand Haven, Michigan.

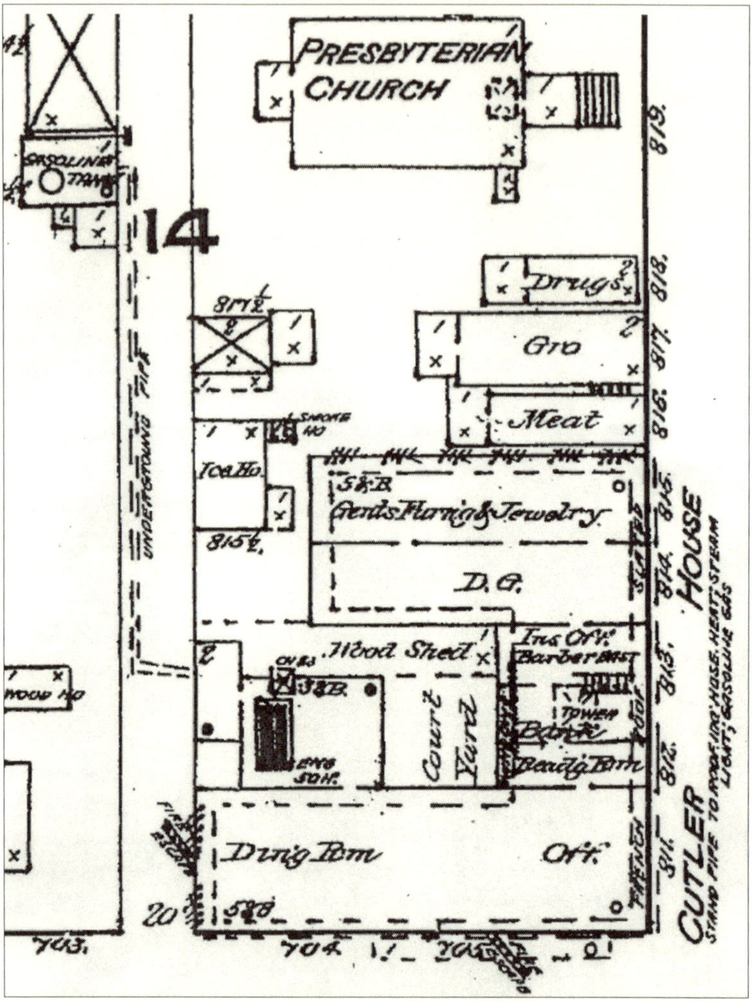

Figure 5.23. Sanborn Fire Insurance map of Cutler House, 1883. Sanborn Library, Sanborn Fire Insurance Company, Grand Haven, Michigan, 1883.

vicinity.[189] An 1872 advertisement in the *Grand Haven Herald* characterized the accommodations at the Second Empire–style brick hotel as "superior to those of the Saratoga hotels, where the charges are much higher." The same ad made the claim that "the water had proved efficacious in numerous cases of the following diseases: paralysis, incipient consumption, rheumatism in all its forms, deafness, anemia, dyspepsia, neuralgia, female irregularities . . . dropsy, hemorrhoids, piles, general debility, or chronic diarrhea . . . and all diseases of the skin."[190] The hotel's manager, B. F. Starr, wrote to Gilbert and Barker in 1872 about the Pomona's five-hundred-light gas machine, reporting that "it gives me much pleasure to say that it fully meets my expectations in all respects. It gives a brilliant, strong light, superior to the best coal gas, and at less than half the price of city gas."[191] As with the Cutler House, having modern conveniences like the Springfield Gas Machine allowed the Pomona House to advertise its interior comfort, in addition to "furniture and upholstering . . . of the most elegant and appropriate design, comfort and good taste," to its middle- and upper-class guests.

The Jekyll Island Club

While these mineral springs hotels and the grand resort hotels of the White Mountains, the Berkshire Valley, and Florida catered to middle- to upper-class businessmen and industrialists, one resort location served the most exclusive clientele of all. The Jekyll Island Club in coastal Georgia, a private resort designed for the most powerful and well-to-do individuals of the country, became the private winter retreat for the Rockefellers, Morgans, Astors, Vanderbilts, and other fabulously wealthy American families.[192]

The Jekyll Island Club's use of the Springfield Gas Machine system is important for several reasons, not the least of which was that the club represented the crème de la crème of leisure retreats of the late nineteenth and early twentieth century. Another significant factor is that the Jekyll Island Club provides one of the few negative assessments of the Springfield Gas Machine system that has yet come to light.

John Eugene du Bignon, a Brunswick, Georgia, entrepreneur, and his brother-in-law Newton S. Finney of New York City conceived of the idea of selling Jekyll Island as a hunting club for wealthy northerners.[193] In 1885, du Bignon and Finney sold the island for $125,000 to the principals of the soon-to-be-formed Jekyll Island Club. The club, incorporated in 1886 with a limit of one hundred shares of stock, then sought fifty individuals to purchase two shares each.[194] Members included the likes of Marshall Field, William Rockefeller, J. P. Morgan, Joseph Pulitzer, Vincent Astor, and William K. Vanderbilt, all men at the heart of "the technological and economic revolution that was transforming America from an agrarian society to an industrial world power."[195] Significantly, these men controlled major sectors of the oil and steel industries that ultimately supplied Gilbert and Barker. In fact, by 1886, Gilbert and Barker Manufacturing had become part of Pratt Manufacturing Company, a subsidiary of Standard Oil of New York and thus a part of the Standard Oil Trust. Jekyll Island Club member William Rockefeller served as the Trust's senior executive and New York representative.

Figure 5.24. Jekyll Island Clubhouse, Brunswick, Georgia. From *American Architect and Building News*, January 1887. Special Collections, University of Maryland Libraries.

Reporting on the founding of this club and its well-heeled members, a *New York Times* reporter wrote that the "fifty gentlemen composing the club include some of the highest social and wealthiest citizens, and many members of the Union, Yacht, and other clubs. It is predicted that the Jekyll Island Club is going to be the 'swell' club, the *crème de la crème* of all, inasmuch as many of the members are intending to erect cottages and make it their Winter Newport."[196] Most of the club's initial members came from New York and almost half of them already belonged to the tony Union Club of New York. A small but influential group, including Marshall Field, hailed from the Chicago area.

The club hired the well-known landscape architect Horace William Shaler Cleveland to design the island development. Cleveland's plan sought "'a natural paradise' without the 'artificial decoration . . . in common use by landscape gardeners.'"[197] He would, Cleveland wrote, strive for a "style of severe simplicity" and avoid anything that "even in appearance involves the idea of continuous care and labor." The club's building committee engaged Chicago architect Charles A. Alexander to design the club house. In May 1887, work began on the large Queen Anne–style building. When completed in November 1887, the clubhouse stood four stories tall with a corner tower and included "dining facilities; sixty guest rooms; rooms for reading, card playing, and billiards; a barber shop; and a pavilion."[198]

In December 1887, a journalist with the *Advertiser and Appeal* described the clubhouse as an "elegant structure . . . unlike anything else in this section. . . . From a distance it looks like some English castle with its square shaped windows and its lofty tower. The building is constructed for comfort—the ceilings low, but each room is well ventilated, and the entire building can be heated from cellar to garret by means of a heater in the basement. . . . In fact there is no attempt at display, and there is nothing gaudy about the building—everything is substantial." The club house opened for regular use in January 1888. Interior photographs taken during one of the first seasons prominently feature an ornate gasolier fixture in the entrance hall, an element that is quite restrained for the period and in keeping with the simplicity of the club's overall design.[199]

As the first season's guests arrived, Superintendent Richard Ogden reported that "hot water was insufficient for kitchen, laundry, and baths; gas fixtures provided inadequate lighting; and the stables were too small for all the members' horses." An overwrought and defensive Ogden wrote that "all these annoyances that could not be foreseen were charged to the incompetency of your superintendent." In the end, however, the first season proved a success and Ogden ultimately found everything "almost unanimously satisfactory."[200]

Sanborn Fire Insurance Maps confirm the installation of the Springfield system at the clubhouse. The 1893 Sanborn map notes that "gas and Ker[osene] Oil" provided light and that a "furnace and open grates" heated the facility. The map shows a small, approximately ten-by-twenty-foot, one-story structure bearing the notation "gasoline tank underground" approximately fifty to seventy-five feet east of the clubhouse building. The 1898 Sanborn map of the island complex indicates that the gas house had been moved from its location in back of the clubhouse. This change accommodated the expansion of the clubhouse, which added a billiard room and covered veranda. The 1908 Sanborn shows the installation of the dynamo/power plant, powered by coal, and lists the lights as electric.[201]

The Springfield Gas Machine at the Jekyll Island Club represents the system's use at arguably the most exclusive resort, with the most demanding clientele, in the United States at the time. It stands out as one of the few installations known to have received any negative comments, namely, Henry B. Hyde's criticism of the gas light provided by the Springfield as unhealthy, smelly, and poor in quality. While Hyde, known as the "Czar of Jekyll," clearly disliked the system, it is unclear whether members and guests found it as problematic, as it took almost ten years for the club to switch from gas light to electricity.

Figure 5.25. Entryway with gas light fixture, Jekyll Island Clubhouse, Brunswick, Georgia. Courtesy of the Jekyll Island Museum Archives.

The club acknowledged periodic requests to install electric lighting on the island in an 1892 stockholder's meeting, at which they also approved expansion of the dining room and new roads to the beach. Hyde wrote to an associate that the installation of electric lights and other improvements "will make it a very charming place and I hope I shall have the pleasure of enjoying your society there next winter." Despite this optimism, a yellow fever epidemic and financial panic in 1893 caused considerable delay. In 1895, Hyde wrote the New York office of the club to report that the "gasoline lamps and fixtures in use throughout the club house [were] dangerous to health as well as disagreeable" and provided poor light. Hyde's wife apparently complained continually about the smell and "he agreed that it was 'positively sickening.'"[202] Hyde urged the board to act on the plan to install an electric plant on the island. His admonition still did not get immediate attention, no doubt to the satisfaction of Standard Oil representatives on the club's board.

By 1900, the plan to build an electric plant for the island again came under consideration. Ironically, the gasoline lighting system disappeared as the first "gasolene automobile" arrived on the island. In 1902, the club awarded George A. Williams a contact to build an electric power plant. Completed in December 1902, the plant transformed the "1903 season into a brightly lit wonderland, its splendid new incandescence flickering in the island darkness." Superintendent Ernest Grob wrote Joseph Pulitzer that his recently electrified cottage and grounds "were all lighted up, and it certainly makes a fine appearance. Your house and grounds are the best lighted of any on the island."[203]

Commercial and Institutional Lighting with the Springfield Gas Machine

As with residential lighting, commercial and institutional lighting improved markedly with the introduction of gas light in the early nineteenth century. Gas lighting remained an urban phenomenon until mid-century, when firms like Gilbert and Barker began to develop and market portable systems for lighting buildings beyond the reach of city gas works. Throughout its long history, Gilbert and Barker specifically pitched its systems as "suitable for the lighting of mills, factories, machine shops, hotels, public halls, churches, blocks of stores."[204] Ever responsive to the demands of the market, the company manufactured systems capable of lighting huge structures and complexes, such as the Royal Poinciana Hotel in Florida and Mount Holyoke College in Massachusetts. Gilbert and Barker also developed and sold components, such as the hydraulic blower, anti-vibration fixtures, and heating furnace, specifically designed for industrial applications. This segment of the market became particularly important to the company's overall sales strategy following Gilbert and Barker's purchase by Standard Oil affiliate Pratt Manufacturing in 1882.

The introduction of the Springfield self-contained lighting system clearly had an impact on the world of work and labor. High-quality gas light was an important innovation for several reasons related to both owners and workers. On one hand, gas lighting presented the possibilities for an extended work day and multiple shifts for workers, and on the other, the improved lighting conditions provided a generally safer work environment, particularly in manufacturing settings like mills. In these industrial applications, the gas light of the Springfield had a significant impact on workers' environments. While it is difficult to fully appreciate and assess the addition of gas lighting for workers, who didn't as a rule leave significant written accounts, they certainly would have perceived the stark contrast between the workplace illuminated with reasonably bright, plentiful, and safe gas light and their homes lit with smoky, smelly, and dangerous kerosene lamps. This striking difference served to further signal the power of management and owners and their ability to control technology to literally change night to day, and to control the very ways in which workers and their families and communities structured their individual and communal schedules. As Cowan reminds us, the "daily schedule was set by the factory manager, not by the worker—and for the manager the best use of the machinery was the overriding concern, not the best use of the worker."[205]

Like domestic Springfield systems, commercial and institutional installations represented concentrations of extraordinary economic and social capital and power. Workers and visitors were witness to the literal and figurative power of good lighting provided by the Springfield system, whether used in a factory, mill, or magnificent hotel. For instance, luxury hotels that catered to people with money and power had to provide the best and latest in accommodations and amenities. Improved lighting also supported design changes that allowed hotel structures to be expanded so that central halls accessed rooms on both sides of the building; without good lighting these central halls would have been dark and dangerous for guests. These facility expansions and improvements in services like lighting and heating created a sense of opulence and luxury as many hotels and resorts grew to become virtually self-contained pleasure palaces. Gilbert and Barker also introduced new appliances and heating devices that enabled factories to power machinery more easily, consistently, and economically. Similarly, colleges, schools, and especially hotels could provide more extensive laundry and kitchen services.

The various uses of the Springfield allowed Gilbert and Barker to vigorously cross-sell its products. Satisfied residential customers often installed Springfields in their places of business. Selling systems to hotels and sites in important travel and luxury centers around the country also offered wonderful marketing opportunities, allowing Gilbert and Barker to target upper-middle- and upper-class business customers. By publicizing sales to wealthy homeowners and successful businesses, the firm established its Springfield brand as the "Cadillac" of gas machine systems. In a period when Americans both revered and rejected technology and engaged in indulgent consumerism, the purchase of an innovative system like the Springfield Gas Machine unequivocally demonstrated a company's mastery of technological advances and purchasing power.

Numerous manufacturing firms utilized Springfield Gas Machines for both lighting and production of various kinds. While most used the Springfield system principally for lighting, some Springfield customers used the system for both lighting and the industrial production operations. The gas produced by the

Springfield system, Gilbert and Barker noted, was employed "in the Mechanical Arts in the various ways in which coal gas is applied: by manufacturers of Britannia and Silver-plated ware, and by those who make tin and japanned goods, and by workers in metals generally."[206] The self-contained and "portable" nature of the system made it an ideal choice for industries, like food processing and canning, that required fuel for soldering processes but also needed to be located in rural areas in order to quickly preserve perishable agricultural produce.

Like industrial users, churches, schools and colleges, public halls, hotels, and resorts all found advantages in using the Springfield system for lighting. The huge system at Mount Holyoke lit the campus in ways and at levels impossible with kerosene lamps. The Springfield provided good quality lighting in the college's thirteen buildings and the grounds and fueled ancillary uses in laboratories, laundries, and kitchens. In its 1899–1900 catalog, the college touted its new gymnasium as "a fine building furnished with all the modern improvements for light, heat, and ventilation." These "modern" conveniences clearly made a difference in the daily activities of students, faculty, and staff. For the parents of prospective students, the college's use of such an innovative system demonstrated the school's commitment to progress and cutting-edge technology, as well as safety. Mount Holyoke regularly advertised its use of modern technology in its catalog, boasting to potential students and their parents that "all the [dormitory] halls are heated by steam and lighted by gas, and the students' rooms are appropriately furnished."[207]

For most industrial and commercial users, the Springfield provided the best possibility for improved lighting, lighting that rivaled or exceeded that offered by municipal gas works or fledgling electric plants in urban areas. In this sense, the Springfield provided a transitional technology between the ubiquitous kerosene lamp with all of its attendance problems, including light quality, safety, and maintenance, and the arrival of dependable and affordable electrical lighting. As a transitional system, Springfield gas lighting technology was restricted to businesses and public customers who could afford the substantial initial investment. Beyond the initial investment, however, business users found the cost of operating the system comparable to or even cheaper than city gas and kerosene. Businesses also saved because of the reduction in accidents and machinery malfunctions due to poor lighting; the ability to run multiple and longer shifts, particularly in winter; and the reduction of time spent in tending coal fires for industrial heating applications and kerosene lamps for lighting. In a period when efficiency was the mantra of management and Taylor's time study experiments were all the rage, the Springfield system allowed manufacturers and businessmen to get the most out of their facilities and workers.[208]

While the Springfield Gas Machine system successfully competed with early forms of electric lighting for several decades, it eventually lost out as electricity, always more flexible, became more reliable and spread out from urban areas into the countryside. Gilbert and Barker and its Standard Oil investors clearly realized the potential impact of electric lighting on the lighting business and by the late 1880s introduced other products that utilized oil and gasoline. This transition in the late nineteenth and early twentieth century allowed the firm to survive at a time when most gas machine companies went out of business. Chapter 6 examines the slow demise of gas lighting, as the industry attempted to compete against the spread of electric lighting, and explores Gilbert and Barker's response to the changing marketplace.

CHAPTER 6

Gas Lighting Gives Way to Electricity

The turn of the twentieth century saw the beginning of the end of "the age of the gaslight" as the gas lighting industry received its first real competition from electric light.[1] As one electric company advertisement in a local newspaper declared, "A lamp that requires no oil to burn, no matches to light it, and does not smoke, is the kind to have."[2] Even with these advantages, however, electric lighting progressed slowly at first because it took time and money to build the infrastructure necessary to generate and distribute electric power.[3] Furthermore, the earliest electric lamps burned far too brightly for domestic use. In addition, early electric systems often proved undependable, prompting both gas and electric system manufacturers to develop combined gas and electric fixtures that let customers resort to gas light when the electricity went out. The critical problems of electrical infrastructure and dependability allowed gas lighting systems, like the Springfield Gas Machine, to endure into the early years of the twentieth century.

Like early municipal gas lighting, electric lighting was initially an urban phenomenon. It remained one well into the 1930s due to high infrastructure costs and low density of potential customers outside of cities and suburbs. Strasser notes that "while almost half of the nonfarm dwelling units in America had electric service" by 1920, "farm homes lagged far behind." Even by 1930, electricity had reached "only about 10 percent" of farm homes. American farms and rural communities did not start to feel the full impact of electricity until 1935, when the newly created Rural Electrification Administration began subsidizing the expansion of the electrical grid to the country's rural counties.[4] As electric lighting historian David Nye has written of Muncie, Indiana, in 1931, "this bright urban center contrasted sharply with the dark countryside, where long and costly electric lines were needed to reach but a few potential customers." Nye goes on to explain that it wasn't until after World War II that rural electric service matched urban service.[5]

Early problems with lamps powered by electricity also slowed the technology's progress. While inventors struggled to develop a practical electric lamp in the 1840s and 1850s, a workable version of what became known as the arc lamp did not appear until the 1870s.[6] Inventor Charles Brush patented the first commercially successful arc lamp in 1879.[7] Early arc lamps never gained widespread use, however, because they actually burned too brightly to use in the home, although they did serve well as street lamps and factory lighting and in lighthouses.[8] As with later portable gas machine lighting, the potential of arc lighting encouraged start-up firms who found that the promise of the technology often greatly exceeded its actual commercial success.[9]

The electric filament, or incandescent, lamp soon replaced the arc lamp, and inventors quickly touted it as the superior technology. Hilaire Belloc described the contrast in his "Newdigate Poem" about the "the benefits which the Electric Light Confers on us, especially at night":

> As for the light, they hang about the town,
> Some praise them highly, others run them down.
> This system (technically called the Arc),
> Makes some pages too light, others too dark.
>
> But in the house the soft and constant rays
> Have always met with universal praise.
> For instance: if you want to read in bed
> No candle burns beside your curtain's head,
> For from some distant corner of the room
> The incandescent lamp dispels the gloom,
> And with the largest print need hardly try
> The powers of any young and vigorous eye.[10]

Consisting of a carbon filament enclosed in a gas vacuum chamber, or lightbulb, the filament lamp became possible only after William Crookes's improvements to the vacuum pump in the early 1870s.[11] The first practical filament lamp, introduced simultaneously by Sir Joseph Wilson Swan and Thomas Edison in the late 1870s, "divided the light" in order to cut down the brightness and to enable the lamp to be used in homes. According to lighting historian William O'Dea,

> There is little doubt that the credit for the practical introduction of electric lighting in the home must be shared between Swan in England and Edison in America, the Crookes-Sprengel pump being the touchstone. Swan was actually the first, in 1878, but for thirty years he had been experimenting until the vacuum pump of 1875 provided the factor that made success possible. Edison followed in 1879 after only eighteen months or

> so devoted to experiment, and both his lamps and his ideas of electricity distribution to feed them were more advanced than those of Swan. A quarrel over the dry dust of precedence developed, but happily these two great men composed their differences and, in the Edison-Swan Electric Company in Great Britain, established a monument to their co-operation that is still in existence today.[12]

The first three decades of the twentieth century would see continuous research on and improvement of essential electric lighting elements like bulbs. While Edison and Swan's research provided a viable new lighting system, early lamp filaments were problematic because evaporation from the filament surface blackened the bulb. In 1913, chemist Irving Langmuir solved this problem by creating a coiled filament that lessened heat loss from convection and made the lamp more efficient. In 1934, a doubly coiled filament, the "coiled coil," appeared, increasing the efficiency of lightbulbs by up to 20 percent.[13]

Interestingly, when Edison first envisioned suitable, reliable, and widely available electric lighting for homes, he took domestic gas lighting design as his model. Historians Cross and Szostak note that "Edison and his laboratory staff began to focus their energy on the problem of electric lighting in 1878 [as] the popularity of urban gas lighting convinced him of the commercial potential."[14] Gas industry historian Schivelbusch points out that for all its extraordinary qualities, "Edison's incandescent electric light is, in fact, nothing but a methodical imitation of gaslight in a new medium."[15] Edison recorded in his laboratory notebook that the success of his electric light would lie in its ability to mimic gas lighting, writing that his "great effort" was "not to make a large light or blinding light, but a small light having the mildness of gas." His object was to "effect exact imitation of all done by gas, so as to replace lighting by gas by light by electricity."[16]

Soon after his work on incandescent electric lighting had begun in 1878, Edison incorporated the Electric Light Company. With physicist Francis Upton, he developed a commercial dynamo for the generation of electricity in 1879 and later the same year perfected the carbon incandescent lamp. In 1880, he incorporated the Edison Lamp Works.[17] Ironically, Gilbert and Barker, well aware of the growing popularity of electric light, would install Springfield systems in the Edison Machine Works, the Edison Lamp Works, and even Thomas Edison's personal "experimenting laboratory" in Llewellyn Park, New Jersey.[18]

The gas lighting industry closely followed the rise of electric lighting but initially downplayed the competition's potential. At the 1878 meeting of the American Gas Light Association, Professor Charles Morton assured the gathered gas industry executives that numerous technical problems still existed with electric lighting and that until they were solved, "electricity cannot enter into competition with gas."[19] The association's president, in his closing remarks, calmed the audience by prematurely reporting that the industry faced "no real danger from Edison's electric light bulb."[20]

While the gas industry continued to minimize the importance of Edison's innovations, electric lighting continued to build its market share and its reputation at the expense of gas lighting. An 1881 program from the Paris Electricity Exposition reported, "Every individual light shines like gaslight, but this is a type of gas that has not yet been invented—a gas that gives a completely steady light but nevertheless shines vividly and brightly. . . . But then—how different from gas! Electric light leaves no combustion residues in the house—no carbon dioxide and carbon monoxide to pollute the air, no sulphuric acid and ammonia to damage paintings and fabrics. Electricity does not raise the air temperature, and does not give off the uncomfortable and fatiguing warmth associated with gas lighting." Schivelbusch notes that "by the late 19th century, gaslight seemed dirty and unhygienic—at the beginning of the century it was thought clean and pure." He further explains that "the deterioration in the quality of the air not only caused headaches and sweating, it also had an unpleasant effect on the interior decoration of rooms. As it burned, gas gave off small quantities of ammonia and sulphur, as well as carbon dioxide and water." Unlike gas, "electric light did not use up oxygen, and left the chemical composition of the air unchanged."[21]

With the successful establishment of the Edison Illuminating Company's Pearl Street generating station in New York City in 1882, electrical-generating companies began springing up across the United States to provide electric lighting to homes and businesses.[22] While it quickly found advocates, Nye notes that in the 1880s electric light "bordered on the supernatural, and the Edison light, the Brush arc light, any form of artificial light, astounded people because it violated the natural order."[23] In 1899, the *New York Times* quoted businessman J. J. Moore of Springfield, Massachusetts, as saying, "I may state it as my opinion that the great improvement in electrical machinery is fast bringing the generation and distribution of electricity to as simple and as easy a basis as the manufacture and distribution of gas."[24] Writing in 1894 in support of electric light, engineer and entrepreneur Robert Hammond detailed the six conditions necessary to produce "the perfect light for the home. . . . It should not: 1) rob the air of our rooms of oxygen; 2) add noxious fumes to the air; 3) be a source of danger in the house; 4) be an unpleasant light; 5) be difficult to control; 6) be costly." Hammond then proceeded to demonstrate how the "electric filament lamp met all these conditions admirably."[25]

Although gas industry leaders initially failed to see the potential in Edison's invention, they eventually realized that they could no longer take a "nonchalant attitude toward the electric light." As stock in Edison's companies rose "to all-time speculative high[s]," gas stocks declined precipitously.[26] In response, the gas industry began to argue that gas provided superior lighting and was safer, healthier, and cheaper compared with its competition. Industry agents reported that "electric light wiring was one of the fire hazards encountered in modern building construction because of the danger from defective insulation. This is a danger wholly absent from gas lighting."[27] Strasser explains that "ignorance [of electricity], combined with the genuine dangers of fire and electrocution, engendered fear" of the new technology. For example, "Mrs. Cornelius Vanderbilt, who in 1881 went to a costume ball dressed as an electric light, demanded that her Fifth Avenue house's entire new electrical system be removed a short time later because it caused a small fire."[28]

Stressing the "hygienic" value of gas, engineer Vivian Lewis wrote in 1907 that the "gas lighted room will nearly always be more pleasant and healthy to live in than one lighted by the newer form of illuminant."[29] The use of gas lighting, its proponents avowed, increased the ventilation in a room, "sterilizing the air and keeping the atmosphere at the breathing level pure and wholesome."[30] The gas industry also promoted gas as more reliable than electricity. Because gas was stored in bulk, it provided a ready supply to the customer even when production at the gas plant shut down, whereas the failure of electric generators led to an immediate loss of service.[31] The gas industry also developed "newer and more beautiful gas light fixtures" to appeal to consumers' aesthetics.

During this period of transition, as Edison sought to "mimic" the qualities of gas light, the gas lighting industry tried to imbue its product with the advantages of electricity. In many ways, the development of the Welsbach incandescent gas burner in the late 1880s allowed gas light to successfully compete with electric lighting for as long as it did. The incandescent burner, one of the greatest advances in gas lighting technology, used the gas flame to heat a refractory substance, or mantle, which then glowed with light. In addition to vastly improved light quality, these burners used gas in a very efficient manner to provide more candlepower.[32] The *Financial Times* hailed incandescent gas light as "electric light without electricity."[33] This innovation, according to Schivelbusch, was "the result of an attempt by the older gas technology to survive by mimicking the more recent, promising technology—and it was successful for some time." Lighting historian Roger Moss suggests that without the Welsbach burner and many later versions of incandescent burners, electric light would have spread much more rapidly than it did.[34]

An advertisement for the Welsbach incandescent gas light in the *Journal of Gas Lighting* in 1890 boasted, "Electric Light Surpassed. All the advantages of Electric Light and none of its drawbacks," a clear attempt to discredit electric light before it really got established.[35] The ad touted the many improvements that would be realized by using a Welsbach burner, which would

> consume the ordinary illuminating gas in such a way as to make the electric light look positively yellow.
>
> produce a given amount of light with less than one-half the gas used in the ordinary burner.
>
> [offer] extremely steady light; no flickering; delightful to read or work by.
>
> [require] no change in brackets or chandeliers—simply unscrew the old burner and screw on the Welsbach.

The ad ends by noting that the Welsbach burner "can be used with the Springfield or any other gas machine—in fact it can be used with any kind of vapor or gas that will produce heat in combustion."[36] Gilbert and Barker, always poised to enhance its market position, patented its own version of the burner, known as the Springfield Incandescent Burner, in 1897.

Gilbert and Barker's introduction of a mixing regulator in the 1880s also increased the company's ability to compete with electricity for as long as it did. By providing a consistent ratio of gasoline vapor to air, the regulator offered increased light quality and decreased fuel consumption. In combination, the mixing regulator and incandescent burner significantly lowered the cost to operate the unit, while producing light that was, the company argued, both brighter and cheaper than electricity.

In its 1906 catalog, Gilbert and Barker declared that the Springfield system equipped with a Springfield Incandescent Burner produced the "cheapest artificial light in the world." The system, according to the catalog, cost less than municipal coal gas, acetylene light, or even electricity. As the company explained, "It costs from one-half to one cent per hour to maintain an electric lamp giving the light of sixteen candles, while with the Springfield Incandescent Burner gasoline gas produces the light of sixteen candles, while consuming one cubic foot of gas per hour, costing one-tenth of one cent. Hence, it will be seen that electric light costs from five to ten times as much as light obtained by [the Springfield system]." Gilbert and Barker must have appreciated a testimonial by customer John Sheldon of Wakefield, Rhode Island, who wrote that in terms of quality and cost, he considered the Springfield "gas plant" in his store "far superior to the electric light service previously used."[37]

Echoing the earlier claims of gas system manufacturers, Boston Consolidated Gas Company trumpeted the superiority of "modern gas lighting" over "other recent types of lighting" as late as 1916:

> It is the comparatively large light-giving surface of the gas mantle which produces the soft and pleasing effect so noticeable with gas lighting, especially when compared with the harshness and glare of other recent types of lighting.
>
> Gas fixtures have recently been greatly improved in appearance and convenience. Artistic treatment of the modern gas fixture permits a selection conforming to individual requirements, and the chain pull pilot lighting attachment makes it very easy to find in the dark.
>
> Besides the superior quality of the modern gas light for reading and general purposes, the cost is about one third the cost of "Mazda" electric lighting. This is an important consideration in planning the home because of the advance in the cost of living.
>
> That these advantages of modern gas lighting have appealed to the home-builders is shown by the fact that 87.7 per cent of all the new homes last year used gas for lighting either wholly or almost exclusively.[38]

In a letter to customer J. A. Ames of Boston, Massachusetts, a man who ironically served on the board of directors of several new electric companies, Boston Gas boasted that "the superiority of gas light in preserving the eyes and in the soft restful illumination is generally conceded."[39]

Gas companies also sought to maintain their market share by expanding the range of domestic applications for gas. "A national advertising campaign, jointly sponsored by gas companies and appliance manufacturers in 1912–1913," writes Strasser, "promoted many uses for gas; by 1918, its future was assured, despite the serious competition [from electricity]."[40] Boston Consolidated Gas underscored the fact that "it is unusual now to speak of a kitchen without meaning gas. Prospective tenants will pay more for a modern gas kitchen than for almost any other improvement in the house." In addition to kitchen ranges, Boston Consolidated Gas also heavily marketed the use of gas water heaters, incinerators, irons, mangles, clothes dryers, gas logs, and auxiliary house heating.[41] The use of these new gas appliances, along with gas furnaces, quickly became the focus of the gas industry as it lost its market for lighting to electricity.

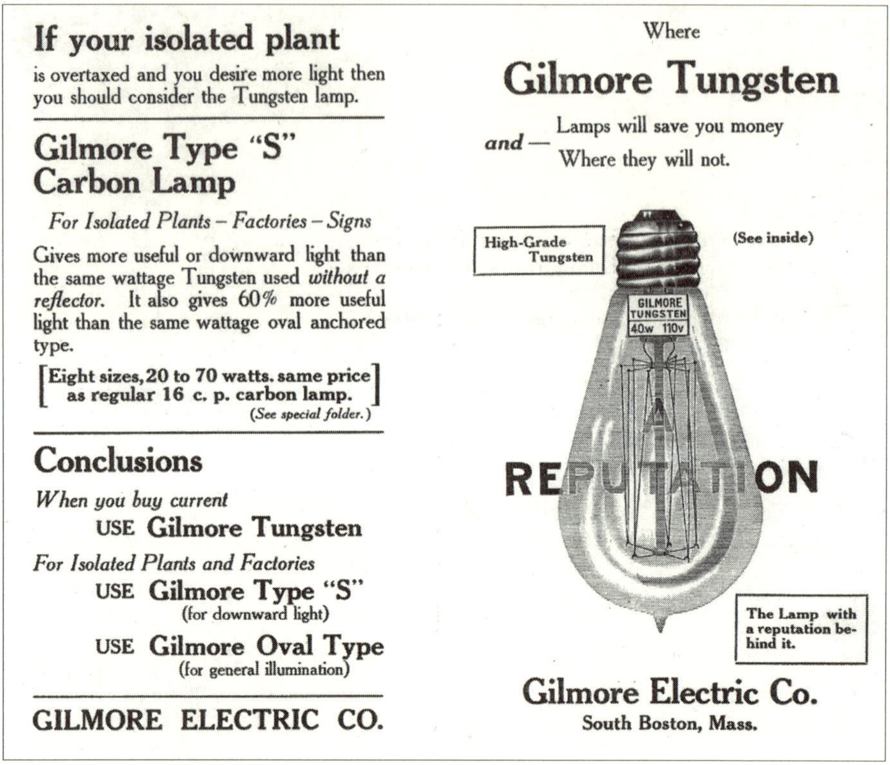

Figure 6.1. Advertisement for the Gilmore tungsten lightbulb, Gilmore Electric Company, Boston. Stonehill Industrial History Center, Stonehill College, North Easton, Massachusetts.

Gilbert and Barker, ever responsive to the consumer, actively promoted the broader uses of the Springfield Gas Machine. For example, a 1911 letter sent to "owners of gas machines," noted that "your gas machine is worth more to you today than ever before because you can do so much more with it. Besides furnishing you with an exceedingly economical and brilliant light the gas may also be used as a fuel in the kitchen and laundry, where you would find it a great convenience and comfort in the summer time." "Why not," the letter continued, "entirely do away with the use of coal and the dust and dirt that accompany it in these two important departments of the house and substitute in place of it a fuel that may be delivered directly to the cooking appliances through a pipe?"[42] The company developed a series of marketing pamphlets titled *Are You Aware?* that explained the various uses of gas and the expanding catalog of appliances and products designed to work with the Springfield system.[43]

As the use of incandescent electric lighting grew, the gas industry and manufacturers of gas machines kept trying new approaches to reverse the erosion of their ever shrinking market. Playing to concerns about monopolistic utility companies, manufacturers advertised the value and freedom that accrued from making one's own gas, noting that users of these systems did not have to depend on the whims of large gas or electric utility companies. An anti-monopolistic sentiment quickly developed within the public, spurring a variety of responses; for example, "antimodernists focused on electrification's capacity to increase the powers of subjection and destruction."[44] Embracing these concerns, advertisements for portable systems such as Tirrill's Equalizing Gas Machine exhorted readers to "make your own gas—be independent of any gas or electric company." Electric corporations, Nye explains, tried to respond to these fears "through the styling of products, so that the earliest appliances often appeared to be not radically new but rather familiar, even visually identical with the nonelectric competition. Many early electric light fixtures were designed to look like crystal chandeliers, candles, or gas jets."[45] Gilbert and Barker also emphasized the independence and freedom that accrued to owners of a Springfield Gas Machine, even as it was itself a small cog in the larger corporate machine known as Standard Oil.

Many Springfield users continued to maintain their gas systems in order to provide backup light in the rather regular event of electric failure. For example, the Patapsco Flouring Mills of C. A. Gambrill Manufacturing Company, near Baltimore, reported in 1887 that its lights were provided by the Springfield Gas Machine. By 1894, the company had switched to electric lights but had "gasoline [lighting]" generated by its Springfield Gas Machine "in reserve."[46]

Despite the gas industry's intensive marketing effort, by the 1920s electricity replaced most gas lighting systems in urban domestic and commercial buildings. Electricity finally won the battle to become the prevailing type of lighting for many reasons, including aggressive marketing by electric companies, installation of private generators, the increase in the number of local utility companies, and, finally, the intervention of the federal government in the 1930s. For most Americans, it epitomized the modern marvel. The nighttime lighting system at the 1893 Chicago Exposition, which used two dynamos to power ten thousand incandescent lights, according to author Erik Larson, gave many visitors "their first encounter with electricity. Hilda Satt, a girl newly arrived from Poland, went to the fair with her father. 'As the light was fading in the sky, millions of lights were suddenly flashed on, all at one time,' she recalled, years later. 'Having seen nothing but kerosene lamps for illumination, this was like getting a sudden vision of Heaven.' Her father told her the lights were activated by electric switches. 'Without matches?'

she asked."[47] In addition to the heavenly vision offered by this new technology, electricity provided a good quality, bright, and generally economical light. Like gas companies, electricity providers could easily regulate and measure individual household consumption. Wiring could be run to various outlets throughout the house, allowing people to move electric lights and appliances easily, unlike permanently fixed gas lights.

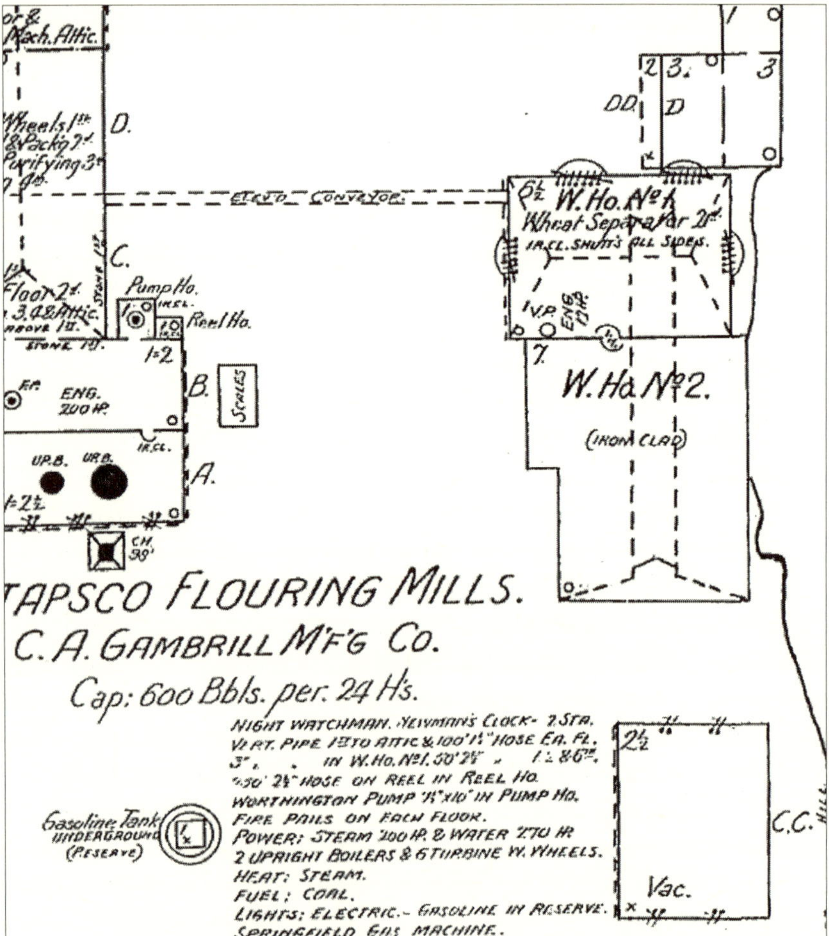

Figure 6.2. Sanborn Fire Insurance map of Patapsco Flouring Mills, 1894. Note that lights are electric with gasoline in reserve. Sanborn Library, Sanborn Fire Insurance Company, Ellicott City, Maryland, 1894.

In the early twentieth century, many of Gilbert and Barker's residential and commercial customers began to make the transition from gas to electric light. By 1905 Henry Flagler reported that the Royal Poinciana Hotel had begun using electricity to light the resort, although "we are still using the [Springfield] machine to its utmost capacity for laundry and other heating purposes."[48] Mount Holyoke College installed a dynamo for generating its own electricity by 1905, but for at least thirteen years it used both its Springfield and electricity for lighting; by 1929, however, "all b[ui]ld[in]gs" had electric light.[49] The McDowell family of Ashland switched from their Springfield Gas Machine to electric light provided by a local utility in about 1907.[50] Alfred Vanderbilt installed a hydroelectric plant at his Sagamore great camp in about 1914, but he selected combination gas/electric fixtures so that the family could use the gas machine as a backup lighting system, which it did for several years.

In terms of domestic applications, electric lighting ultimately offered consumers advantages of safety, cleanliness, and convenience over gas light. While electric lighting was initially sought for its novelty value, the public slowly came to appreciate its broad advantages, particularly as other uses and electrical appliances became available. A range of appliances from fans to carpet sweepers to refrigerators and even toys increasingly found their way into middle-class homes. In fact, electrical utilities, taking a page out of their gas company competitor's marketing plans, conducted door-to-door campaigns to sell appliances to "increase their use of electricity."[51] The gas industry fought back with its own appliances, but these didn't provide the same level of convenience as their electric competitors.

Like domestic applications, the use of electric lighting and other products in factories and commercial establishments provided clear advantages over the use of gas. Nye explains that "electricity has certain advantages in industrial production that gas cannot match. Both are sources of light and heat, but unlike gas, electricity can drive a multiplicity of small motors; it can produce high temperatures without consuming oxygen; it can link a series of machines through automatic feeding devices, scanners, and moving belts; and it can regulate the system of production with temperature gauges, meters, warning bells, automatic shut-off devices, heat sensors, and other electrical control devices." Nye notes that "although electric lighting greatly improved the luminous environment, reduced the danger of fires, and did not pollute the air, it also brought less desirable changes in working conditions." In addition to expanding the type of round-the-clock shift work made possible by improved gas lighting, Nye explains that in Muncie, Indiana's factories, "electricity eliminated both costly skilled labor, like glass blowing, and brute work, like carrying heavy loads. Increasingly factories required semiskilled workers, whose tasks could be learned in less than a month."[52]

Competition grew harsh as electric firms aggressively marketed their product to gas lighting customers. As early as 1905, Springfield customer John S. Ames regularly received solicitations from the Gilmore Electric Company of Boston that read, "As the short days are now coming on I thought possibly you might need some incandescent lamps to prepare you for the dark days. We should like very much to supply you."[53] For gas lighting manufacturers like Gilbert and Barker, dark days came on quickly in terms of gas lighting sales. Nevertheless, the Springfield managed to compete with electric lighting into the 1920s and 1930s thanks to the automatic and self-contained nature of the system and the improvements realized with the addition of the mixing regulator and incandescent burner. The slow encroachment of electrical distribution systems in most rural areas of the country extended the viability of self-contained systems. In fact, the Sears and Roebuck's catalog, a consumer lifeline for rural households in the early twentieth century, continued to offer kerosene and gas lamps until the 1930s, although their catalogs slowly introduced electric lighting and appliances as well. While the company's 1908 catalog contained mostly kerosene and gas lamps, their 1923 publication advertised a separate catalog for electric lighting fixtures and supplies.[54]

Gilbert and Barker reacted to the increasing competition from electric lighting by diversifying its product line in the late 1880s.[55] A history of the firm published in the employee newsletter conceded that "the introduction of electricity with low rates to the consumer seriously affected the gas machine business and Gilbert and Barker began to manufacture equipment for dispensing oil and gasoline."[56] The company began to produce fuel-oil-burning equipment, primarily burners for furnaces, in 1889 in an effort to expand the market for Standard Oil's fuel oil. It introduced the "first practical gasoline pump" for the growing automobile industry in 1910.[57] A letter from Gilbert and Barker to attorney E. F. Gebhardt at Shelburne Farms indicates that the company continued to produce gas machine systems until about 1916, when it "retired from the gas machine and gas appliance business, disposing of our interests to the Detroit Heating and Lighting Company of Detroit, Michigan."[58]

Gilbert and Barker entered a new era when it began producing its fuel-oil system, developing one of the earliest industrial fuel-oil burners in 1889.[59] The company later reported that it installed its first such system, a high-pressure type burner, at the Boston and Albany Railroad Shops and claimed that its introduction "marked the first successful use of oil as a fuel in industrial furnaces." Building on the success of its initial oil burning system, the company pioneered the development of the low-pressure industrial oil burner in the late 1890s. This technology allowed "thousands of the country's largest organizations . . . to convert coal and coke furnaces to oil firing."[60] The use of fuel oil for heating in the metal trades marked a major improvement because, unlike coal and coke furnaces, an oil burner provided both precise temperature control and quick startup and shutdown cycles.[61] The firm's low-pressure oil burner system was adopted by automobile manufacturers such as Ford, Stevens-Duryea, Locomobile, and Pierce-Arrow and a host of other companies, including Pratt & Whitney, General Electric, Allis Chalmers, Bethlehem Steel, Smith and Wesson, and Corning Glass.[62]

In 1916, the company developed a combination oil and natural gas firing system that further solidified its position as a leader in fuel-oil burners; this new system allowed manufacturers to switch easily from one fuel type to another to take advantage of the lowest fuel costs and avoid shortages in supply.[63] The company also modified its earlier gasoline gas-powered furnaces and burners to burn municipal and natural gas. The new products proved useful in the metal heat-treating process and became an important part of the firm's business. The increase in demand for heat-treating furnaces necessitated the construction of a new building for manufacturing furnaces in 1918; these furnaces played a critical role in the manufacture of munitions and ordnance during World War I.[64]

Based on its experience with oil burning systems and continued innovation during the 1920s, Gilbert and Barker introduced a household oil burner in 1930. Standard Oil's president, Walter Teagle, had initially resisted moving into the domestic fuel-oil market due to projected oil shortages, but supplies remained steady enough to allay these fears, and eventually he and other top executives became convinced of the potential of the home heating-oil market. Standard's move into this market increased demand for Gilbert and Barker's domestic oil burners and storage tanks.[65] A 1932 advertisement in the *Washington Post* announced Standard Oil's "5 year triple service heating plan." As the quarter-page advertisement explained, the company would install the Gilbert and Baker Flexible Flame Oil Burner, supply "carefully refined and carefully tested" Standard Oil fuel, and provide five years' service—"cleaning, oiling, adjusting, and replacing any parts."[66] With marketing packages like this, the Flexible Flame Oil Burner system caught on quickly; by the early 1940s over 100,000 customers used Gilbert and Barker's domestic oil burning systems.[67]

The other new direction for Gilbert and Barker, one that has carried it into the twenty-first century, was the manufacture of gasoline pumps for automobile service stations. The firm characteristically expanded into a new and rapidly growing market by taking advantage of its experience with petroleum-based equipment. In so doing, Gilbert and Barker once again helped to expand its parent company's market for gasoline. The firm introduced its first gas pump, the T-1, in 1910.[68] As the company's newsletter, the *Vent*, noted, "The volatile character of gasoline, the danger involved in its storage in open vessels and the carelessness of its handling in garages, service stations, etc., led to the demand for some other means of storing and handling." As the automobile market created demand for both gasoline and oil storage and delivery apparatus, Gilbert and Barker positioned itself at the forefront by developing related products such as "underground storage tanks and pumps, to cover all requirements for the storage and handling of gasoline and lubricating oils in the automotive field."[69] While the T-1 pump used a very simple design to transfer gasoline from storage tanks to an automobile, the company's later T-6 pump added the capacity to measure gas as it was pumped.[70] In 1911, the extent of Gilbert and Barker's gasoline pump and storage tank business had become evident. Their current factory could not handle the demand, so the company purchased eight acres of land across the Connecticut River in West Springfield. There it built a new factory, several times larger than the old, into which the company moved in August 1912.[71]

In May 1911, with Gilbert and Barker contemplating a major expansion in its facilities, the U.S. Supreme Court upheld a federal circuit court (Eastern District of Missouri) decision of 1909 that found Standard Oil of New Jersey, or Jersey Standard, guilty of violating the Sherman Antitrust Act.[72] While the court ordered the split-up and disaffiliation of many of the Jersey Standard companies, Gilbert and Barker Manufacturing Company remained as a subsidiary of Jersey Standard.[73] Oil Industry historians Gibb and Knowlton note that Jersey Standard's "manufacturing and marketing divisions were vitally dependent on Gilbert and Barker equipment and profited from the appreciable improvement in that equipment in these years and later."[74]

Automobile service stations were few and far between in 1912. Recognizing a remarkable opportunity, the newly reorganized Jersey Standard and other oil companies began to open experimental company-owned service stations over the next few years. As road systems expanded and roadside filling stations began to spring up, Gilbert and Barker worked to improve and refine its gasoline storage and handling systems.[75] Between 1912 and 1917, sales of Gilbert and Barker's gasoline equipment rose nearly 200 percent, while the firm's net earnings from manufacturing rose from $25,000 in 1912 to $1,113,000 in 1917. By 1918 "the use of barrels for storage and dispensing purposes had virtually been eliminated, and the trade was being educated to the advantages of Gilbert and Barker tanks and pumps."[76]

During World War I, the company supplied petroleum handling systems for military operations in France, introducing the firm's products to a new, large market in postwar Europe and England.[77] By 1925, the West Springfield factory had grown to eighteen acres, with a redoubled plant capacity.[78] This expanded plant principally manufactured five- and ten-gallon visible-measuring gasoline pumps. These pumps displayed the fluid in a glass cylinder because "purchasers expressed a desire to see the gasoline they were purchasing."[79] By 1926, the company's net earnings had reached almost $1,750,000 and its payroll had grown to over sixteen hundred employees.[80]

In 1929, Gilbert and Barker introduced the first electric gasoline meter pump. In 1935, it created its famous Model 70 Calco-Meter pump, which revolutionized the gas station pump by registering "on a dial not only the quantity of gasoline delivered but its cost in dollars and cents." The firm continually improved and introduced new gasoline pumps, becoming a world leader in their manufacture. It also offered a wide range of other products for automotive service stations, including a line of lubricating equipment, car lifts, and air compressors. By 1930, Gilbert and Barker Manufacturing Company had become a wholly owned subsidiary of Gilbert and Barker Company of America, which itself was controlled by Jersey Standard. The *Wall Street Journal* reported in December 1930 that the company had "taken a plant [in Toronto, Canada,] with 20,000 square feet of space and is now installing new machinery."[81]

During World War II, fuel shortages caused sales of Gilbert and Barker's fuel-oil and gasoline equipment to decline rapidly. The company instead focused on supplying military equipment to the army and navy. Prescient as always, Gilbert and Barker had begun converting to wartime production as early as 1939; its first products were fuse adapters for fourteen- and sixteen-inch projectiles for the U.S. Navy. The following year the firm began to manufacture fire-control equipment for antiaircraft guns. Like many manufacturing companies during the war, Gilbert and Barker discontinued the "manufacture of its regular products and converted entirely to war work," operating its factories around the clock by 1942. The firm's facilities became a center for the production of fuse setters and a range of other products, including "cooling equipment and oil gears for antiaircraft guns, steering engines for torpedoes, sights for 6-inch guns on light cruisers, and control parts for Navy Hell Cat and Avenger planes."[82] The company reportedly supplied more than $69 million in equipment to the armed forces during the war years.[83] In 1940, Gilbert and Barker operated factories in West Springfield, Toronto, and

Sydney, maintained sales offices in London, Paris, Buenos Aires, and Capetown, and marketed its products to 104 countries around the world.[84] The company's reputation for quality products earned it continued military contracts for new sighting equipment in the decades following the war. Gilbert and Barker resumed and expanded its petroleum handling products as well in the postwar years.

In 1966, the firm moved from its West Springfield plant to a new half-million-square-foot facility in Greensboro, North Carolina. The move reportedly took seven hundred tractor trailers to transport over ten thousand tons of materials to the new location.[85] In 1967, with operations in North and South America, Europe, and Australia, the company shortened its name to Gilbarco, Inc. General Electric Company of Britain purchased Gilbarco, Inc., then the "biggest U.S. maker of gasoline pumps," for $250 million in 1987.[86] The company continues to market gasoline pumps and petroleum handling products around the world, including its popular multi-dispenser, credit-card-reading gas pump and control console.[87] Millions of Americans encounter this product each time they pull up to a fuel pump, giving little if any thought to the company that made it or to its fascinating beginnings just after the Civil War.

While sales of the company's Springfield Gas Machine had faded in the first quarter of the twentieth century, a surprising number of systems continued to operate decades later. A newspaper article in the *Springfield Sunday Union and Republican* reported in 1940 that "many Gilbert and Barker gas machines are still in use in remote sections of the country. Many of these machines are 30, 40 and even 50 years old."[88] For example, the company celebrated the "old indestructible," a gas machine in the Monson Academy building that operated from 1889 until 1953, a span of time that witnessed extraordinary changes in American technology and society. In 1889, the machine and two Gilbert and Barker installers had arrived in Monson, a small Massachusetts village, via train and horse-drawn stage. The machine originally lit the entire building and provided fuel for laboratory burners. Although the building had used electric light since the early twentieth century, students continued to use the lab burners supplied by the gas machine until a fire, not related to the Springfield system, destroyed the building in 1953.[89]

As the last remaining Springfield systems were abandoned and the company reoriented its business to oil burners and gasoline pumps, the Springfield Gas Machine was quickly forgotten. The remnants of these systems continue to intrigue the owners and curators of historic houses and sites across the country. Gilbert and Barker arguably stand as the most reliable and innovative purveyors of gas machine systems. Gas lighting remained a technology that was accessible largely to the upper middle and upper classes. While the light of the Springfield certainly had an impact on the lives of lower- and middle-class workers, the advent of electricity really democratized lighting, bringing cheap, reliable, and good quality illumination into the homes and workplaces of all Americans. Nevertheless, the Springfield Gas Machine served as an important transitional technology that expanded gas lighting into rural and suburban homes and a multitude of industries across the country.

NOTES

Introduction

1. Daniel Yergin, *The Prize: The Epic Quest for Oil, Money, and Power* (New York: Simon & Schuster, 1990).

2. Daniel Mouer, "RE: Residential Gas Lighting Systems," Sept. 1, 1998, Historical archaeology LISTSERV mailing list, HISTARCH@asuvm.inre.asu.edu.

3. Barbara Carson, *Ambitious Appetites: Dining, Behavior, and Patterns of Consumption in Federal Washington* (Washington, DC: American Institute of Architects Press, 1990), 59.

4. Springfield Gas Machine Company, *Circular of the Springfield Gas Machine Company of Springfield, Mass.: Manufacturers of Portable Gas Machines and Contractors for the Erection of Gas Works . . .*, trade catalog (Springfield, MA: Springfield Gas Machine, 1867), Trade Catalog Collection, Baker Library, Harvard Business School, Cambridge, MA; Gilbert and Barker Manufacturing Company, *The Springfield Gas Machine, Gilbert and Barker Manufacturing Co.*, catalog (Springfield, MA: Gilbert and Barker Manufacturing, 1870–71); Gilbert and Barker Manufacturing Company, *The Springfield Gas Machine*, catalog (New York: Excelsior Printing, 1874); Gilbert and Barker Manufacturing Company, *The Springfield Gas Machine*, catalog (New York: Wheat & Cornett, 1878); Gilbert and Barker Manufacturing Company, *How Best to Light Our Country Homes and Resorts*, catalog (New York: Press of Fleming, Brewster & Alley, 1888), Box 6, Folder 648a, Catalogue Collection, Farm Managers' Papers Collection, Shelburne Farms Archives, Shelburne, VT; Gilbert and Barker Manufacturing Company, *How Best to Light Our Country Homes and Resorts*, catalog (Springfield, MA: Gilbert and Barker Manufacturing, 1906).

5. Leo Marx, *The Machine in the Garden: Technology and the Pastoral Ideal in America* (New York: Oxford Univ. Press, 1964).

6. Clifford E. Clark Jr., "Domestic Architecture as an Index to Social History: The Romantic Revival and the Cult of Domesticity in America, 1840–1870," in *Material Life in America, 1600–1860*, ed. Robert Blair St. George, (Boston: Northeastern Univ. Press, 1986), 535–49; Ruth Schwartz Cowan, "The 'Industrial Revolution' in the Home: Household Technology and Social Change in the Twentieth Century," in *Material Culture Studies in America*, ed. Thomas J. Schlereth (Nashville, TN: American Association for State and Local History, 1982), 222–36; Marilyn Ferris Motz and Pat Browne, eds., *Making the American Home: Middle-Class Women and Domestic Material Culture, 1840–1940* (Madison: Univ. of Wisconsin Press, 2002).

7. Gervase Wheeler, *Homes for the People in Suburb and Country: The Villa, the Mansion, and the Cottage* (1855; rpt., New York: Arno Press, 1972); John R. Stilgoe, *Borderland: Origins of the American Suburb, 1820–1939* (New Haven: Yale Univ. Press, 1988); Alan Gowans, *The Comfortable House: North American Suburban Architecture, 1890–1930* (Cambridge, MA: MIT Press, 1986).

8. Dona Brown, *Inventing New England: Regional Tourism in the Nineteenth Century* (Washington, DC: Smithsonian Institution Press, 1995).

9. Springfield Gas Machine, *Circular of the Springfield Gas Machine Company;* Gilbert and Barker Manufacturing, *Springfield Gas Machine,* 1870–71; Gilbert and Barker Manufacturing, *Springfield Gas Machine,* 1874; Gilbert and Barker Manufacturing Company, *The Springfield Gas Machine,* catalog (New York: Excelsior Printing, 1878); Gilbert and Barker Manufacturing, *How Best to Light Our Country Homes and Resorts,* 1888; Gilbert and Barker Manufacturing, *How Best to Light Our Country Homes and Resorts,* 1906.

Chapter 1

1. Richard Rhodes (2002) quoted in Mark P. Grisi, "The Automatic Gas Machine Era, 1853–1923: Analysis of Improvements in Patented Illuminating Devices and Processes," master's thesis, Goucher College, Baltimore, 2003, 6.

2. Ibid., 6–7.

3. Ruth Schwartz Cowan, *A Social History of American Technology* (New York: Oxford Univ. Press, 1997), 165.

4. Margaret Hindle Hazen and Robert M. Hazen, *Keepers of the Flame: The Role of Fire in American Culture, 1775–1925* (Princeton: Princeton Univ. Press, 1992), 168.

5. Ibid., 176, 175, 168, 176.

6. Arthur H. Hayward, *Colonial Lighting* (New York: Dover, 1962), 3; Helen Brigham Hebard, *Early Lighting in New England* (Rutland, VT: Charles E. Tuttle, 1964), 15.

7. Hayward, *Colonial Lighting,* 14.

8. Hebard, *Early Lighting in New England,* 15.

9. Hayward, *Colonial Lighting,* 11.

10. Hebard, *Early Lighting in New England,* 27.

11. In 1634, candles cost four pence apiece, "a sum which was considered the height of extravagance in those days." Hayward, *Colonial Lighting,* 75–76. Their expense stemmed from the initial scarcity of tallow due to a lack of domesticated animals and from the relatively high level of effort and thus cost required to make them.

12. Ibid., 76.

13. Hebard, *Early Lighting in New England,* 41.

14. William T. O'Dea, *A Short History of Lighting* (London: Her Majesty's Stationary Office, 1958), 7.

15. Ibid.

16. Hayward, *Colonial Lighting,* 133. Whaling began in the United States in 1671 and lasted for over two hundred years, providing a host of important products, including better fuels for lighting.

17. Loris S. Russell, *A Heritage of Light: Lamps and Lighting in the Early Canadian Home* (Toronto: Univ. of Toronto Press, 1968), 68–69.

18. Hebard, *Early Lighting in New England,* 62; Russell, *Heritage of Light,* 70.

19. O'Dea, *Short History of Lighting,* 7; Hayward, *Colonial Lighting,* 123; Russell, *Heritage of Light,* 75–78.

20. Russell, *Heritage of Light,* 77–78.

21. Hayward, *Colonial Lighting,* 123.

22. Brian Bowers, *Lengthening the Day: A History of Lighting Technology* (Oxford: Oxford Univ. Press, 1998), 28.

23. C. Malcolm Watkins, "Artificial Lighting in America: 1830–1860," in *Annual Report of the Board of Regents of the Smithsonian Institution, 1951*, by Smithsonian Regents (Washington, DC: Government Printing Office, 1952), 397; Leroy L. Thwing, *Flickering Flames: A History of Domestic Lighting Through the Ages* (Rutland, VT: Charles E. Tuttle, 1958), 78. For example, the Carcel oil or "mechanical lamp," invented in France in 1800, used an "elaborate clockwork" to pump fuel into its Argand burner, which made it one of the "most efficient lamps for burning viscous fuels."

24. Russell, *Heritage of Light*, 93–94; O'Dea, *Short History of Lighting*, 7.

25. Harold F. Williamson and Arnold R. Daum, *The American Petroleum Industry: The Age of Illumination, 1959–1899* (Evanston, IL: Northwestern Univ. Press, 1959), 33.

26. Nadja Maril, *American Lighting, 1840–1940* (Atgien, PA: Schiffer, 1995), 17.

27. Thwing, *Flickering Flames*, 61.

28. *Franklin Institute Journal* (1831) quoted in Watkins, "Artificial Lighting in America," 399.

29. Russell, *Heritage of Light*, 94.

30. Ibid., 93.

31. Watkins, "Artificial Lighting in America," 399.

32. Russell, *Heritage of Light*, 103–4.

33. Ibid., 111.

34. Hebard, *Early Lighting in New England*, 62; Russell, *Heritage of Light*, 111.

35. Russell, *Heritage of Light*, 113–14.

36. Watkins, "Artificial Lighting in America," 401.

37. Russell, *Heritage of Light*, 123.

38. Watkins, "Artificial Lighting in America," 403. The Ufford lamp also gained popularity during the 1850s and 1860s; made of iron and tin, it had a broad, flat wick and a matching smoke shade (Hebard, *Early Lighting in New England*, 62).

39. Watkins, "Artificial Lighting in America," 405.

40. These deposits were initially accessed via surface seeps that flowed like natural springs.

41. Hebard, *Early Lighting in New England*, 68; Russell, *Heritage of Light*, 136–39.

42. Russell, *Heritage of Light*, 139.

43. Ibid., 134, 187, 190–91, 188.

44. Watkins, "Artificial Lighting in America," 405.

45. Christopher J. Castaneda, *Invisible Fuel: Manufactured and Natural Gas in America, 1800–2000* (New York: Twayne, 1999), 42.

46. Samuel Hughes and Henry O'Connor, *Gas Works: Their Construction and Arrangement and the Manufacture and Distribution of Coal Gas* (London: Crosby Lockwood and Son, 1904), 6–7. In the 1660s, Dr. John Clayton identified "inflammable air" coming from a well shaft and began experiments into the distillation of coal; he noted that "a spirit [gas] arose" and caught fire at the flame of a candle.

47. Ibid., 10.

48. Ibid., 13.

49. D. Hale, "Diary of an Industry," *American Gas Journal* 186, no. 11 (1959): 22; Hughes and O'Connor, *Gas Works*, 13–14.

50. Hughes and O'Connor, *Gas Works*, 14, 15.

51. Ibid., 15.

52. Castaneda, *Invisible Fuel*, 6.

53. Hughes and O'Connor, *Gas Works*, 17. The Royal Society, the national academy of science of the United Kingdom and the Commonwealth, established the Rumford Medal in 1800. It is awarded in "recognition of an outstandingly important recent discovery in the field of thermal or optical properties of matter, or their applications, made by a scientist working in Europe" (http://www.royalsociety.org/).

54. Castaneda, *Invisible Fuel*, 7. Winzer anglicized his name to Winsor upon arriving in England.

55. Hughes and O'Connor, *Gas Works*, 16, 20, 22, 24.

56. Ibid., 94–97, 101, 155, 166, 216, 235.

57. Ibid., 318-19.

58. William E. Worthington Jr., *Beyond the City Lights: American Domestic Gas Lighting Systems*, exhibit at the National Museum of American History, Oct. 17, 1985–Apr. 20, 1986 (Washington, DC: Smithsonian Institution Press, 1985), 5.

59. Castaneda, *Invisible Fuel*, 56.

60. Hughes and O'Connor, *Gas Works*, 24, 26–27.

61. O'Dea, *Short History of Lighting*, 8.

62. Hughes and O'Connor, *Gas Works*, 337.

63. O'Dea, *Short History of Lighting*, 8; Hughes and O'Connor, *Gas Works*, 338.

64. Russell, *Heritage of Light*, 287.

65. Castaneda, *Invisible Fuel*, 9, 36.

66. William Richards, *The Gas-Consumer's Guide: A Hand-book of Instruction* (Boston: Alexander Moore, 1871), 20, 21.

67. Wolfgang Schivelbusch, *Disenchanted Night: The Industrialization of Light in the Nineteenth Century* (Berkeley and Los Angeles: Univ. of California Press, 1988), 42.

68. Ibid., 42, 43–44.

69. Frederick Christian Accum, *A Practical Treatise on Gas-Light* (London: R. Ackerman, 1815), 104.

70. Richards, *Gas-Consumer's Guide*, 21, 22, 23.

71. Oscar E. Norman, *The Romance of the Gas Industry* (Chicago: A. C. McClurg, 1922), 86–87.

72. Castaneda, *Invisible Fuel*, 35–36.

73. Watkins, "Artificial Lighting in America," 393.

74. Russell, *Heritage of Light*, 290.

75. Watkins, "Artificial Lighting in America," 393.

76. Melville (1813) quoted in Worthington, *Beyond the City Lights*, 3.

77. Watkins, "Artificial Lighting in America," 393; Castaneda, *Invisible Fuel*, 15–21.

78. Norman, *Romance of the Gas Industry*, 44.

79. Hale, "Diary of an Industry," 2; Watkins, "Artificial Lighting in America," 393.

80. Norman, *Romance of the Gas Industry*, 45–47, 116.

81. Hale, "Diary of an Industry," 22.

82. Norman, *Romance of the Gas Industry*, 46.

83. Schivelbusch, *Disenchanted Night*, 40; Hale, "Diary of an Industry," 26, 22.

84. Watkins, "Artificial Lighting in America," 393.

85. Hale, "Diary of an Industry," 22–26; John D. Wright, *Lexington: Heart of the Bluegrass* (Lexington, KY: Lexington–Fayette County Historic Commission, 1982), 81; Worthington, *Beyond the City Lights*, 8; Norman, *Romance of the Gas Industry*, 47.

86. Worthington, *Beyond the City Lights*, 7.

87. Reynolds (1854) quoted in Norman, *Romance of the Gas Industry*, 50.

88. Schivelbusch, *Disenchanted Night*, 29.

89. Alan Trachtenberg, *The Incorporation of America: Culture and Society in the Gilded Age* (New York: Hill and Wang, 1982), 4; Castaneda, *Invisible Fuel*, 52; Susan Strasser, *Never Done: A History of American Housework* (New York: Henry Holt, 1982), 69; Schivelbusch, *Disenchanted Night*, 74.

90. Trachtenberg, *Incorporation of America*, 130.

91. Cowan, *Social History of American Technology*, 150.

92. Schivelbusch, *Disenchanted Night*, 28.

93. Ibid., 29, 161–62.

94. Worthington, *Beyond the City Lights*, 1.

95. Ibid., 4.

96. Ibid., 7.

97. Maryland Portable Gas Company, *Gas Apparatus of the Maryland Portable Gas Company*, catalog (Baltimore: Murphy, 1855), 4, Baker Library, Harvard Business School, Cambridge, MA.

98. Donald W. Linebaugh, "Gas Lighting at the President Lincoln and Soldiers' Home National Monument," University of Maryland, School of Architecture, Planning and Preservation, College Park, technical report submitted to the National Trust for Historic Preservation, Washington, DC, 2006. The campus of the Military Asylum was renamed the Soldiers' Home in 1858 and is now known as the Armed Forces Retirement Home. The Soldiers' Home campus contains the "Lincoln Cottage," as it is now colloquially known. President Abraham Lincoln used the Soldiers' Home as a seasonal retreat from June to November in 1862–64.

99. Worthington, *Beyond the City Lights*, 12; Grisi, "Automatic Gas Machine Era," 81–114.

100. Worthington, *Beyond the City Lights*, 12.

101. James W. Bryant, ed., *Mining and Petroleum Standard and American Gas-Light Journal*, Nov. 2, 1864, 136.

102. Worthington, *Beyond the City Lights*, 12.

103. Grisi, "Automatic Gas Machine Era," 81, 82.

104. "Mace's Benzole Gas Apparatus," *Scientific American* 10, no. 1 (Jan. 27, 1855).

105. Charles Cunningham, "Benzole Vapor Apparatus," Patent No. 12, 535, Mar. 13, 1855 (Washington, DC: U.S. Patent Office, 1855).

106. Hiram Maxim, "Improved Gas-Machine," Patent No. 81,922, Sept. 8, 1868 (Washington, DC: U.S. Patent Office, 1868).

107. Dr. William H. Wahl, reprinted in "Gas Machines," *American Gas Light Journal and Chemical Repertory* 17 (1872): 130.

108. Gilbert and Barker Manufacturing, *Springfield Gas Machine*, 1874, 51.

109. M. C. Stebbins, "Improved Portable Gas-Apparatus," Patent No. 83,419, Oct. 27, 1868 (Washington, DC: U.S. Patent Office, 1868); George H. Kitchen and Scotto C. Nash, "Improved Portable Gas Apparatus," Patent No. 84,283, Nov. 24, 1868 (Washington, DC: U.S.

Patent Office, 1868); Iva Prichard, "Improvement in Gas-Machines," Patent No. 74,132, Feb. 4, 1868 (Washington, DC: U.S. Patent Office, 1868).

110. Prichard, "Improvement in Gas-Machines," 1868.

111. William Foster Jr. and George P. Ganster, "Improved Portable Gas Apparatus," Patent No. 84,941, Dec. 15, 1868 (Washington, DC: U.S. Patent Office, 1868).

112. Bryant, *Mining and Petroleum Standard and American Gas-Light Journal,* 136. An intact Drake system, still in its underground vault, was documented in suburban Philadelphia; it was installed at William Lockwood's Loch Aerie Estate in Chester County in 1865 or 1866 (Richard Veit, *Historical Documentation of the Loch Aerie Gasworks, Home Depot—East Whiteland and East Whiteland Township, Chester County Pennsylvania* [Highland Park, NJ: Cultural Resource Consulting Group, 1997]).

113. *New York Times,* Sept. 22, 1869, 1; *Washington Post,* Mar. 11, 1878, 1.

114. Excelsior Gas Machine Company, *The Excelsior Gas Machine, Described and Compared,* catalog (New York: Less & Skeen, 1872), 12–13.

115. Maxim Gas Machine Company, *The Maxim Steam Gas Machine for Lighting Buildings, Towns and Cities,* catalog (New York: Maxim Gas Machine, 1873), 10–11, fig. 4, 14–15.

116. Frank W. Ofeldt, "Improvement in Gas Apparatus for Carbureting Air," Patent No. 162,848, May 4, 1875 (Washington, DC: U.S. Patent Office, 1875).

117. Thomas B. Fogarty, "Improvement in Gas Machines," Patent No. 179,652, July 11, 1876 (Washington, DC: U.S. Patent Office, 1876).

118. Worthington, *Beyond the City Lights,* 16, 17.

119. William Paul Gerhard, *The American Practice of Gas Piping and Gas Lighting in Buildings* (New York: McGraw, 1908), 123–24.

120. Maxim Gas Machine, *Maxim Steam Gas Machine,* 18.

121. John A. Bassett, "Improved Apparatus for Carbureting Gas," Patent No. 39,541, Mar. 18, 1863 (Washington, DC: U.S. Patent Office, 1863).

122. Gerhard, *American Practice,* 185.

123. Worthington, *Beyond the City Lights,* 18.

124. Miscellaneous patents for acetylene gas machines, 1895–1903 (Washington, DC: U.S. Patent Office, 1895–1903).

125. Sunlight Gas Machine Company, *Acetylene: The Catalogue of the Sunlight Gas Machine Company* (New York: Sunlight Gas Machine, 1906), 5, Baker Library, Harvard Business School, Cambridge, MA.

126. Russell, *Heritage of Light,* 300.

127. Gerhard, *American Practice,* 186; Worthington, *Beyond the City Lights,* 19, 186.

128. General Acetylene Company, *Daylight by Night: The General Acetylene Generator* (New York: Press of Robert L. Stillson Co., ca. 1905), 5; Sunlight Gas Machine, *Acetylene,* 15, 19–21.

129. Sunlight Gas Machine, *Acetylene,* 20; Gilbert and Barker Manufacturing Company, *Gilbert & Barker M'f'g Co's Price List, July 1, 1882,* pamphlet, July 1, 1882 (New York: Gilbert and Barker Manufacturing, 1882).

130. Gilbert and Barker Manufacturing, *How Best to Light Our Country Homes and Resorts,* 1906.

131. Sunlight Gas Machine, *Acetylene,* 20.

132. Gerhard, *American Practice,* 186.

133. Worthington, *Beyond the City Lights,* 20.

134. General Acetylene, "Daylight by Night," 8.

135. Gerhard, *American Practice,* 142.

136. Denys Peter Myers, *Gaslighting in America: A Guide for Historic Preservation* (Washington, DC: Denys Peter Myers, 1978), 25–30.

137. Gilbert and Barker Manufacturing, *Springfield Gas Machine,* 1874, 70.

138. Russell, *Heritage of Light,* 294.

139. Ibid.

140. Richards, *Gas-Consumer's Guide,* 48.

141. Ibid., 57–59.

142. Gerhard, *American Practice,* 113.

143. Myers, *Gaslighting in America,* 14–15.

144. Gerhard, *American Practice,* 113; Myers, *Gaslighting in America,* 15.

145. Richards, *Gas-Consumer's Guide,* 65, 67.

146. Gerhard, *American Practice,* 120.

147. Ibid., 118, 119.

148. Richards, *Gas-Consumer's Guide,* 61.

149. Gerhard, *American Practice,* 120, 122.

150. Ibid., 123.

151. Ibid.

152. Roger W. Moss, *Lighting for Historic Buildings* (New York: Preservation Press, John Wiley & Sons, 1988), 100.

153. Myers, *Gaslighting in America,* 206.

154. Schivelbusch, *Disenchanted Night,* 49; Moss, *Lighting for Historic Buildings,* 102–3.

155. Richards, *Gas-Consumer's Guide,* 70–72. The reflectors typically consisted of "bright metal, porcelain, silvered glass, enameled metal, etc."

Chapter 2

1. Cowan, *Social History of American Technology,* 119.

2. Ibid., 130; Trachtenberg, *Incorporation of America,* 4–7, 137.

3. Ruth Schwartz Cowan, *More Work for Mother: The Ironies of Household Technology from the Open Hearth to the Microwave* (New York: Basic Books, 1983), 14.

4. Merritt Ierley, *The Comforts of Home: The American Home and the Evolution of Modern Convenience* (New York: Three Rivers Press, 1999), 139.

5. Gilbert and Barker Manufacturing, *Springfield Gas Machine,* 1874, 33.

6. Biographical Review, *The Leading Citizens of Hampden County, Massachusetts* (Boston: Biographical Review, 1895), 402–5, 406–7.

7. Cowan, *Social History of American Technology,* 125.

8. Biographical Review, *Leading Citizens of Hampden County,* 406–7.

9. Cowan, *Social History of American Technology,* 140.

10. Trachtenberg, *Incorporation of America,* 53–54.

11. Biographical Review, *Leading Citizens of Hampden County,* 407.

12. Dick Garvey, "Big Name in Business Born in Local Rooming House," *Springfield Union,* Nov. 3, 1994, 19.

13. "Charles Gilbert Dies at Home at Great Barrington," *Berkshire Courier,* Jan. 18, 1934.

14. Charles Gilbert, Alumni Records, Beloit College, Oct. 19, 1928, Beloit College Archives, Beloit, WI.

15. Gilbarco, "These Are the Men Who Started It All," *Gilbarco News* 14, no. 1 (1955): 8.

16. Ibid.

17. R. G. Dun and Company, Massachusetts, vol. 41, p. 233, Oct. 9, 1866, R. G. Dun and Company Collection, Baker Library, Harvard Business School, Cambridge, MA.

18. Biographical Review, *Leading Citizens of Hampden County,* 407; Gilbarco, "These Are the Men," 8.

19. J. N. Cummings, "From Appomattox to Versailles," *Vent* 3, no. 11 (Nov. 1919): 2–3.

20. Gilbert and Barker Manufacturing, *The Springfield Gas Machine,* 1878, 2. This catalog notes that in December 1878 it had been "some 12 years since we began to manufacture the Springfield Gas Machine."

21. R. G. Dun, Massachusetts, vol. 41, p. 233, Oct. 9, 1866.

22. J. Holton Messick, "Gas Machines Started Local Concern on 75 Years of Expanding Production," *Springfield Sunday Union and Republican,* Jan. 28, 1940, 3E.

23. Ibid. It is unclear if Maxim originally came to work for the New England Portable Gas Works Company or for the subsequent Springfield Gas Machine Company.

24. See chapter 1.

25. Messick, "Gas Machines Started Local Concern," 3E; Gilbarco, "These Are the Men," 8. Maxim later became famous for his invention of the Maxim Machine Gun and other military equipment; he was eventually knighted by the British government for his invention of the machine gun.

26. R. G. Dun, Massachusetts, vol. 41, p. 233, Oct. 9, 1866.

27. Biographical Review, *Leading Citizens of Hampden County,* 407.

28. Gilbert and Barker Manufacturing, *Springfield Gas Machine,* 1874, 46.

29. Ibid., 44–46, 58.

30. R. G. Dun and Company, Massachusetts, vol. 41, p. 253, Sept. 10, 1867, R. G. Dun and Company Collection, Baker Library, Harvard Business School, Cambridge, MA; R. G. Dun and Company, Massachusetts, vol. 41, p. 233, Sept. 24, 1867, R. G. Dun and Company Collection, Baker Library, Harvard Business School, Cambridge, MA.

31. Biographical Review, *Leading Citizens of Hampden County,* 407; Gilbert and Barker Manufacturing, *Springfield Gas Machine,* 1870–71. On September 10, 1867, R. G. Dun was reporting on the newly formed Springfield Gas Machine Company (R. G. Dun, Massachusetts, vol. 41, p. 253, Sept. 10, 1867).

32. Springfield Gas Machine, *Circular of the Springfield Gas Machine Company,* title page.

33. Gilbert and Barker Manufacturing, *Springfield Gas Machine,* 1874, 46.

34. R. G. Dun, Massachusetts, vol. 41, p. 253, Sept. 10, 1867.

35. Charles N. Gilbert, John F. Barker, and E. N. Ives, Assignors to the New England Portable Gas Works Company, "Improvement in Apparatus for Carbureting Air," Patent No. 61,004, Jan. 8, 1867 (Washington, DC: U.S. Patent Office, 1867).

36. Biographical Review, *Leading Citizens of Hampden County,* 407.

37. Springfield Gas Machine, *Circular of the Springfield Gas Machine Company,* 4; Gilbert and Barker Manufacturing, *How Best to Light Our Country Homes and Resorts,* 1906, 26. The 1867 catalog reports that the company had "recently lighted the Water Shops," while the 1906 catalog gives the date as 1868.

38. R. G. Dun and Company, Massachusetts, vol. 41, p. 253, Sept. 14, 1868, R. G. Dun and Company Collection, Baker Library, Harvard Business School, Cambridge, MA.

39. Williamson and Daum, *American Petroleum Industry*, xx.

40. Springfield Gas Machine, *Circular of the Springfield Gas Machine Company;* Messick, "Gas Machines Started Local Concern," 3E. The timing of the endorsement and Barker's trip is a bit unclear. Messick actually reports the trip as occurring in 1869, but the catalog and other references suggest that the firm clearly had an endorsement by 1867. It may be that the trip did occur in 1869 in response to a later motion by the underwriters to deny coverage on any property using a gas machine due to the high incidence of fires and injuries associated with other gas machine systems.

41. Springfield Gas Machine, *Circular of the Springfield Gas Machine Company*, 3; Gilbert and Barker Manufacturing, *Springfield Gas Machine*, 1870–71, 5.

42. Gilbarco, "These Are the Men," 8. The structure measured just eighteen by forty-two feet.

43. Springfield Gas Machine, *Circular of the Springfield Gas Machine Company*, 8. The machine sizes and prices are handwritten in the catalog by George Sedgwick.

44. Ibid., 3, 4.

45. Ibid., 5. The company even offered this carburetor free of charge with the provision that the customer "return the value of the amount of Coal Gas saved in a certain length of time."

46. Ibid.

47. Biographical Review, *Leading Citizens of Hampden County*, 407.

48. Gilbarco, "These Are the Men," 8.

49. "The Springfield Gas Machine," display ad, *Appletons' Journal of Literature, Science and Art*, May 29, 1869, 286; "The Springfield Gas Machine," display ad, *Scientific American*, May 29, 1869, 352.

50. R. G. Dun, Massachusetts, vol. 41, p. 253, Sept. 14, 1868.

51. Gilbert and Barker Manufacturing, *Springfield Gas Machine*, 1870–71; R. G. Dun and Company, Massachusetts, vol. 41, p. 253, Sept. 27, 1869, R. G. Dun and Company Collection, Baker Library, Harvard Business School, Cambridge, MA.

52. Gilbert and Barker Manufacturing, *Springfield Gas Machine*, 1870–71. It seems likely that W. S. Gilbert was a relative of Charles Gilbert, thus the new firm was entirely controlled within the Gilbert and Barker families. It is possible the W. S. Gilbert was William Stanley Gilbert, a cousin of Charles, who attended Beloit College in the late 1850s and early 1860s (Beloit College, 1859–60 catalog, Beloit College Archives, Beloit, WI).

53. R. G. Dun, Massachusetts, vol. 41, p. 253, Sept. 27, 1869.

54. Gilbert and Barker Manufacturing, *Springfield Gas Machine*, 1870–71; R. G. Dun and Company, Massachusetts, vol. 41, p. 253, Mar. 27, 1869, R. G. Dun and Company Collection, Baker Library, Harvard Business School, Cambridge, MA.

55. Trachtenberg, *Incorporation of America*, 82.

56. Gilbert and Barker Manufacturing, *Springfield Gas Machine*, 1870-71; R. G. Dun, Massachusetts, vol. 41, p. 253, Mar. 27, 1869.

57. Gilbert and Barker Manufacturing, *Springfield Gas Machine*, 1870-71.

58. Ibid.; Gilbert and Barker Manufacturing, *Springfield Gas Machine*, 1874, 58. The company's 1867 catalog lists both galvanized iron and copper machines.

59. Gilbert and Barker Manufacturing, *Springfield Gas Machine*, 1870-71, 17.

60. R. G. Dun and Company, Massachusetts, vol. 41, p. 253, June 9 and 17, 1871, R. G. Dun and Company Collection, Baker Library, Harvard Business School, Cambridge, MA; R. G. Dun and Company, Massachusetts, vol. 42, p. 44, Aug. 19, 1871, R. G. Dun and Company Collection, Baker Library, Harvard Business School, Cambridge, MA.

61. Gilbert and Barker Manufacturing, *Springfield Gas Machine,* 1870–71, 14–15.

62. O. Tirrill, *The Underground Gas Machine,* catalog (New York: Underground Gas Machine, 1871), 43 (in fact, this machine became the subject of a patent infringement lawsuit by Gilbert and Barker), 2–5, and 42; Gilbert and Barker Manufacturing, *Springfield Gas Machine,* 1870–71; Gilbert and Barker Manufacturing, *Springfield Gas Machine,* 1874.

63. Gilbert and Barker Manufacturing, *Springfield Gas Machine,* 1870–71, cover page; "Springfield Gas Machine," *American Gas-Light Journal and Chemical Repertory* 16, no. 10 (May 16, 1872): 1; Gilbert and Barker Manufacturing, *Springfield Gas Machine,* 1870–71, cover page. On the copy of the catalog in the Library of Congress, the address and reference to Archer and Pan-Coast are neatly lined out and an address, 554 Broadway, is written in along with the name Leander A. Bevin, Agent.

64. Gilbert and Barker Manufacturing, *Springfield Gas Machine,* 1870–71, insert; Gilbert and Barker Manufacturing, *Springfield Gas Machine,* 1874, 93.

65. Tirrill, *Underground Gas Machine.*

66. Woodruff (1874) quoted in Gilbert and Barker Manufacturing, *Springfield Gas Machine,* 1874, 93–97.

67. Henry A. Oakley and D. A. Heald, Committee, *Report to the Executive Committee of the National Board of Fire Underwriters on Gas Machines* (New York: National Board of Fire Underwriters, 1876), 10.

68. Ibid.

69. Gilbert and Barker Manufacturing, *Springfield Gas Machine,* 1874; Gilbert and Barker Manufacturing, *Springfield Gas Machine,* 1878; Gilbert and Barker Manufacturing Company, *The Springfield Gas Machine,* catalog (Springfield, MA: Gilbert and Barker Manufacturing, 1879), 18. Heald is listed as vice-president of the Home Insurance Company.

70. "Fire Dangers—Carelessness in Handling Gasoline," *New York Times,* Feb. 3, 1873, 5.

71. R. G. Dun and Company, Massachusetts, vol. 42, p. 44, Apr. 8, 1874, R. G. Dun and Company Collection, Baker Library, Harvard Business School, Cambridge, MA. It seems likely that the firm of J. H. Parsons and Company was tied to previous stockholder W. S. Gilbert of Cohoes, New York.

72. R. G. Dun and Company, Massachusetts, vol. 42, p. 44, Apr. 13, 1875, and Apr. 19, 1876, R. G. Dun and Company Collection, Baker Library, Harvard Business School, Cambridge, MA.

73. Trachtenberg, *Incorporation of America,* 39.

74. Gilbert and Barker Manufacturing, *Springfield Gas Machine,* 1874, title page.

75. Nye and Acheson, advertisement for the Springfield Gas Machine, *Santa Cruz Sentinel* 3 (Oct. 2, 1875): 1.

76. Gilbert and Barker Manufacturing, *Springfield Gas Machine,* 1874, 30, 120, 53, 374.

77. Ibid., 6, 7–8.

78. National Publishing Company, *Commerce, Manufactures and Resources of Springfield, Mass.: A Historical, Statistical & Descriptive Review* (n.p.: National Publishing, 1883), 64.

79. The American Institute exhibition was a trade show held yearly in New York City beginning in 1872. It was held at a fairgrounds and amphitheater located at Third Avenue and Sixty-third Street.

80. Gilbert and Barker Manufacturing, *Springfield Gas Machine,* 1874, back cover.

81. James D. McCabe, *The Illustrated History of the Centennial Exhibition* (1876; rpt. Philadelphia: National Publishing Co., 1975), 281; "The Centennial Awards," *New York Times,* Oct. 7, 1876, 3.

82. McCabe, *Illustrated History of the Centennial Exhibition,* 291; Trachtenberg, *Incorporation of America,* 38.

83. "Springfield Gas Machine," *American Gas-Light Journal and Chemical Repertory,* 1.

84. Gilbert and Barker Manufacturing, *Springfield Gas Machine,* 1874, 16–18.

85. Ibid., 56–57.

86. Ibid.

87. "Supreme Court Decisions," *Washington Post,* Nov. 11, 1879, 3.

88. Gilbert and Barker Manufacturing, *Springfield Gas Machine,* 1878, 28.

89. "Supreme Court Decisions," 3.

90. Loring W. Puffer and Sons Insurance Agency to John S. Ames, Sept. 18, 1919, John S. Ames Collection, Stonehill Industrial History Center, Stonehill College, North Easton, MA (hereafter cited as John S. Ames Collection).

91. Walworth Manufacturing Company, *The Walworth Gas Machine* (Boston: Gunn, Bliss, 1880), 3.

92. Gilbert and Barker were themselves sued for patent infringement in an 1870 case that made it to the Supreme Court in 1882. This case involved an 1870 patent infringement suit related to a gas burner. Theodore Clough argued that Gilbert and Barker were manufacturing a burner that he had originally patented, and Gilbert and Barker replied that they had "improved" on Clough's earlier design. The court found that "none of the prior forms of the burner introduced in evidence anticipate the arrangement covered by the claim of the Barker patent," and the decree of the lower court was affirmed in favor of Gilbert and Barker. *Clough v. Barker* (1882), in *Cases Argued and Adjudged in the Supreme Court of the United States. October Term, 1882,* by William T. Otto, vol. 16 (New York: Banks Law, 1904), 166–78; *Clough v. Manufacturing Co.* (1882), in *Cases Argued and Adjudged in the Supreme Court of the United States. October Term, 1882,* by William T. Otto, vol. 16 (New York: Banks Law, 1904), 178–81.

93. Walworth Manufacturing, *Walworth Gas Machine.*

94. Gilbert and Barker Manufacturing, *Springfield Gas Machine,* 1874, 24.

95. Ibid., 36; Gilbert and Barker Manufacturing, *Springfield Gas Machine,* 1878; "The Springfield Gas Machine: New Fields for a Favorite Local Industry," *Springfield Sunday Republican,* Nov. 11, 1888, 5.

96. Gilbert and Barker Manufacturing, *Springfield Gas Machine,* 1878, 14.

97. National Publishing, *Commerce, Manufactures and Resources,* 64.

98. Gilbert and Barker Manufacturing, *Gilbert & Barker M'f'g Co's Price List.*

99. Walworth Manufacturing, *Walworth Gas Machine,* 7, 16; Gilbert and Barker Manufacturing, *Gilbert & Barker M'f'g Co's Price List.*

100. Gilbert and Barker Manufacturing Company, *Directions for the Care of the Springfield Gas Machine,* pamphlet, May 1, 1885 (Springfield, MA: Gilbert and Barker Manufacturing, 1885), Box 6, Folder 648b, Catalogue Collection, Farm Managers' Papers Collection, Shelburne Farms Archives, Shelburne, VT (hereafter cited as Catalogue Collection, Shelburne); Gilbert and Barker Manufacturing Company, *Directions for the Care of the Springfield Gas Machine,* pamphlet, May 15, 1897 (Springfield, MA: Gilbert and Barker Manufacturing, 1897).

101. Tirrill, *Underground Gas Machine.*

102. National Publishing, *Commerce, Manufactures and Resources,* 64.

103. R. G. Dun and Company, Massachusetts, vol. 42, p. 436, Apr. 18, 1883, R. G. Dun and Company Collection, Baker Library, Harvard Business School, Cambridge, MA; R. G. Dun and Company, Massachusetts, vol. 42, p. 436, Oct. 20, 1883, R. G. Dun and Company Collection, Baker Library, Harvard Business School, Cambridge, MA.

104. National Publishing, *Commerce, Manufactures and Resources,* 63.

105. Sanborn Fire Insurance Company, "Springfield, MA, 1886," in *Digital Sanborn Maps, 1867–1970* (Ann Arbor, MI: ProQuest). This engine is likely the steam engine shown on the 1896 Sanborn map (Sanborn Fire Insurance Company, "Springfield, MA, 1896," in *Digital Sanborn Maps, 1867–1970* [Ann Arbor, MI: ProQuest, online subscription service]).

106. National Publishing, *Commerce, Manufactures and Resources,* 63; Sanborn Fire Insurance, "Springfield, MA, 1886."

107. Gilbert and Barker Manufacturing, *Directions for the Care of the Springfield Gas Machine,* 1885; National Publishing, *Commerce, Manufactures and Resources,* 63; Walter Licht, *Industrializing America: The Nineteenth Century* (Baltimore: Johns Hopkins Univ. Press, 1995), 103.

108. Gilbert and Barker Manufacturing, *Springfield Gas Machine,* 1870–71.

109. Tirrill, *Underground Gas Machine;* Excelsior Gas Machine, *Excelsior Gas Machine,* 1872.

110. Williamson and Daum, *American Petroleum Industry,* 232, 208.

111. Ibid., 234.

112. Ibid., 50.

113. Gilbert and Barker Manufacturing, *How Best to Light Our Country Homes and Resorts,* 1906, 6.

114. Williamson and Daum, *American Petroleum Industry,* 235–36.

115. National Publishing, *Commerce, Manufactures and Resources,* 64; "Springfield Gas Machine: New Fields for a Favorite Local Industry," 5; Ralph W. Hidy and Muriel E. Hidy, *History of Standard Oil Company (New Jersey): Pioneering in Big Business, 1882–1911* (New York: Harper & Brothers, 1955), 46–47.

116. Licht, *Industrializing America,* 141.

117. Hidy and Hidy, *History of Standard Oil Company,* 46–47, 49, 107.

118. Ibid., 108–16.

119. Biographical Review, *Leading Citizens of Hampden County,* 407; "Charles Gilbert Dies at Home at Great Barrington."

120. "Charles Gilbert Dies at Home at Great Barrington." Gilbert moved to Great Barrington, Massachusetts, where he pursued his agricultural interests, owning several large cattle and horse farms in the Great Barrington area after his retirement. He reportedly specialized in Jersey cattle and had several race horses; his Jersey "Pride's Olga" took first place in the Pan-American Exposition in Buffalo, New York, in 1901.

121. R. G. Dun and Company, Massachusetts, vol. 42, p. 436, June 6, 1885, and Aug. 21, 1886, R. G. Dun and Company Collection, Baker Library, Harvard Business School, Cambridge, MA.

122. New York, New Haven, and Hartford Railroad Company to Cyrus Lothrop, invoice, Jan. 7, 1911, Stonehill Industrial History Center, Stonehill College, North Easton, MA.

123. "Light Your Paper Mill with a Springfield Gas Machine," advertisement, *Paper World,* Oct. 1882, 39; "Springfield Gas Machine: New Fields for a Favorite Local Industry," 5.

124. International Publishing, *Leading Manufacturers and Merchants of Central and Western Massachusetts: Historical and Descriptive Review of the Industrial Enterprises of Worcester, Hampden, Hampshire, Berkshire, and Franklin Counties* (New York: International, 1887), 202.

125. Charles F. Warner, ed., *Picturesque Hampden* (Northampton, MA: Picturesque, 1892), 150.

126. The company's work force in 1886 was listed as "50 hands" on the Sanborn map. Sanborn Fire Insurance, "Springfield, MA, 1886."

127. "Springfield Gas Machine: New Fields for a Favorite Local Industry," 5; Warner, *Picturesque Hampden*, 150–51.

128. Gilbert and Barker Manufacturing, *Directions for the Care of the Springfield Gas Machine*, 1897, 1; Gilbert and Barker Manufacturing Company to O. Ames and Son Corporation, North Easton, Massachusetts, invoice, Aug. 1, 1904, Ames Family Collection, Stonehill Industrial History Center, Stonehill College, North Easton, MA (hereafter cited as Ames Family Collection).

129. Hidy and Hidy, *History of Standard Oil Company*, 205–9, 217–19, 225, 301.

130. Gilbert and Barker Manufacturing, *Directions for the Care of the Springfield Gas Machine*, 1897, 13; Gilbert and Barker Manufacturing, *How Best to Light Our Country Homes and Resorts*, 1906, 44.

131. Gilbert and Barker Manufacturing, *Directions for the Care of the Springfield Gas Machine*, 1897; Gilbert and Barker Manufacturing, *How Best to Light Our Country Homes and Resorts*, 1906, 17–19.

132. Gilbert and Barker Manufacturing, *How Best to Light Our Country Homes and Resorts*, 1888; Gilbert and Barker Manufacturing, *How Best to Light Our Country Homes and Resorts*, 1906.

133. Gilbert and Barker Manufacturing, *How Best to Light Our Country Homes and Resorts*, 1906, 5, 16–17, 21, 22–23, 25.

134. Castaneda, *Invisible Fuel*, 60.

135. Messick, "Gas Machines Started Local Concern," 3E.

136. Gilbarco, "Old Indestructible," *Gilbarco News* 12, no. 2 (1953): 12–13.

137. Gilbert and Barker Manufacturing Company to E. F. Gebhardt, Oct. 16, 1916, Box 7, Folder 81, Shelburne Farms Bill Collection, Farm Managers' Papers Collection, Shelburne Farms Archives, Shelburne, VT (hereafter cited as Shelburne Farms Bill Collection); Cowan, *More Work for Mother*, 95.

138. Gilbert and Barker Manufacturing Company, *The Gilbarco Story: The Story of a Name Famous Throughout the World* (West Springfield, MA: Gilbert and Barker Manufacturing, 1947), n.p.; Hidy and Hidy, *History of Standard Oil Company*, 162.

Chapter 3

1. Gilbert and Barker Manufacturing Company to E. F. Gebhardt; Springfield Gas Machine, *Circular of the Springfield Gas Machine Company;* Gilbert and Barker Manufacturing, *Springfield Gas Machine*, 1870–71; Gilbert and Barker Manufacturing, *Springfield Gas Machine*, 1874; Gilbert and Barker Manufacturing, *Springfield Gas Machine*, 1878; Gilbert and Barker Manufacturing, *How Best to Light Our Country Homes and Resorts*, 1888; Gilbert and Barker Manufacturing, *How Best to Light Our Country Homes and Resorts*, 1906.

2. Gilbert and Barker Manufacturing, *Springfield Gas Machine*, 1874, 32–34.

3. Trachtenberg, *Incorporation of America*, 68.

4. Gilbert and Barker Manufacturing, *Springfield Gas Machine*, 1874, 46.

5. George E. Woodward and Edward G. Thompson, *Woodward's National Architect* (New York: G. E. Woodward, New York, 1869; rpt., New York: Da Capo Press, 1975).

6. "Artificial Light: Petroleum or Air-Gas Machines," *New York Times*, Sept, 21, 1867, 2.

7. Gilbert and Barker Manufacturing, *Springfield Gas Machine*, 1874, 46, 44–46, 58, 46.

8. Gilbert, Barker, and Ives, "Improvement in Apparatus for Carbureting Air," 1.

9. John F. Barker and Charles N. Gilbert, "Improved Apparatus for Carbureting Air," Patent No. 66,777, July 16, 1867 (Washington, DC: U.S. Patent Office, 1867), 1.

10. Gilbert and Barker Manufacturing, *Springfield Gas Machine*, 1870–71, 11.

11. Barker and Gilbert, "Improved Apparatus for Carbureting Air," Patent No. 66,777, p. 1.

12. Gilbert and Barker Manufacturing, *How Best to Light Our Country Homes and Resorts*, 1906, 32.

13. Barker and Gilbert, "Improved Apparatus for Carbureting Air," Patent No. 66,777, p. 4.

14. Ibid.; John F. Barker and Charles N. Gilbert, "Improved Apparatus for Carbureting Air," Patent No. 93,267, Aug. 3, 1869 (Washington, DC: U.S. Patent Office, 1869).

15. Gilbert and Barker Manufacturing, *Springfield Gas Machine*, 1870–71, 11–12; John F. Barker and Charles N. Gilbert, "Improved Apparatus for Carbureting Air," Patent No. 93,795, Aug. 17, 1869 (Washington, DC: U.S. Patent Office, 1869).

16. Gilbert and Barker Manufacturing, *Springfield Gas Machine*, 1874, 42.

17. Barker and Gilbert, "Improved Apparatus for Carbureting Air," Patent No. 93,267, pp. 2–3.

18. Barker and Gilbert, "Improved Apparatus for Carbureting Air," Patent No. 93,795, p. 1.

19. Ibid., p. 1; Gilbert and Barker Manufacturing, *Springfield Gas Machine*, 1874, 22.

20. Barker and Gilbert, "Improved Apparatus for Carbureting Air," Patent No. 93,795, p. 2.

21. Gilbert and Barker Manufacturing, *Springfield Gas Machine*, 1874, 22.

22. Ibid., 22–23. Barker patented another version of this mechanism in 1886 (John F. Barker, "Rotary Air-Pump," Patent No. 341,871, Mar. 18, 1886. [Washington, DC: U.S. Patent Office, 1886]).

23. Gilbert and Barker Manufacturing, *Springfield Gas Machine*, 1870–71, 12.

24. John F. Barker, Assignor to the Gilbert and Barker Manufacturing Company, "Improvement in Apparatus for Carbureting Air," Patent No. 115,562, June 6, 1871 (Washington, DC: U.S. Patent Office, 1871). The system required sufficient water power, that is, a head of water at least four feet deep.

25. Gilbert and Barker Manufacturing, *Springfield Gas Machine*, 1874, 28.

26. Barker, "Improvement in Apparatus for Carbureting Air," 2.

27. Gilbert and Barker Manufacturing, *Springfield Gas Machine*, 1878, 14; Gilbert and Barker Manufacturing, *How Best to Light Our Country Homes and Resorts*, 1888, 13. The illustration of the new water-driven air pump was added to the 1878 catalog without being integrated into the text, suggesting that it had been very recently introduced.

28. Gilbert and Barker Manufacturing, *Springfield Gas Machine*, 1878, 14; John F. Barker, "Improvement in Air-Forcing Apparatus," Patent No. 221,385, Nov. 11, 1879 (Washington, DC: U.S. Patent Office, 1879).

29. Robert Hewson, "Improvement in Automatic Air-Blasts for Gas Machines," Patent No. 162,922, May 4, 1875 (Washington, DC: U.S. Patent Office, 1875).

30. Gilbert and Barker Manufacturing Company, "Air Pump of the Springfield Gas Machine," flyer, May 1, 1890, Box 6, Folder 648b, Catalogue Collection, Shelburne.

31. Barker, "Improvement in Air-Forcing Apparatus," Patent No. 221,385; John F. Barker, "Air-Pump," Patent No. 433,064, July 29, 1890 (Washington, DC: U.S. Patent Office, 1890). Barker's 1879 patent describes the machine: "The letter *a* denotes the cylinder containing the rotary meter-wheel [essentially the same as the component for the weight-driven pump], and *b*, denotes the small water-wheel on the same shaft with the meter-wheel. The water for driving the water-wheel finds access to the wheel through the tube *c*, in which is a cock, *d*, to control or shut off the flow of water."

32. Gilbert and Barker Manufacturing Company, "Plan and Elevation Showing Two Hundred Light Springfield Gas Machine with a Ten Barrel Storage Tank, June 29, 1900, Order #590," Hill-Stead Museum Archives, Farmington, CT. A red oil, probably this paraffin oil, was found on the ground surface below both water-driven air pumps at Sagamore (having leaked from the machine).

33. Gilbert and Barker Manufacturing, "Air Pump of the Springfield Gas Machine."

34. Gilbert and Barker Manufacturing, *Springfield Gas Machine*, 1874, 120.

35. Ibid.

36. Ibid., 42.

37. Gilbert and Barker Manufacturing, 1870–71 catalog, 12; 1874 catalog, 58–60.

38. Gilbert and Barker Manufacturing, *How Best to Light Our Country Homes and Resorts*, 1888, 1; Gilbert and Barker Manufacturing, *How Best to Light Our Country Homes and Resorts*, 1906, 1.

39. Gilbert and Barker Manufacturing, *Springfield Gas Machine*, 1874, 51.

40. National Publishing, *Commerce, Manufactures and Resources*, 64. The Springfield Mixing Regulator is illustrated in the company's 1885 publication *Directions for the Care of the Springfield Gas Machine*, 1885, 12.

41. Gilbert and Barker Manufacturing, *Directions for the Care of the Springfield Gas Machine*, 1885, 12; Gilbert and Barker Manufacturing, *How Best to Light Our Country Homes and Resorts*, 1906, 15–16.

42. John F. Barker, "Apparatus for Mixing Aeriform Fluids," Patent No. 234,904, Nov. 30, 1880 (Washington, DC: U.S. Patent Office, 1880).

43. Ibid.

44. John F. Barker, "Apparatus for Mixing Aeriform Fluids," Patent No. 252,009, Jan. 10, 1882 (Washington, DC: U.S. Patent Office, 1882); John F. Barker. "Apparatus for Mixing Aeriform Fluids," Patent No. 254,589, Mar. 7, 1882 (Washington, DC: U.S. Patent Office, 1882); Edwin A. Burdick, Assignor to the Gilbert and Barker Manufacturing Company, "Apparatus for Reducing the Gravity of Hydrocarbon Gas," Patent No. 244,850, July 26, 1881 (Washington, DC: U.S. Patent Office, 1881); Charles F. Hovey, Assignor to the Gilbert and Barker Manufacturing Company, "Apparatus for Reducing the Density of Illuminating-Gases," Patent No. 273, 843, Mar. 13, 1883 (Washington, DC: U.S. Patent Office, 1883).

45. Gilbert and Barker Manufacturing, *Directions for the Care of the Springfield Gas Machine*, 1885. This catalog lists the machine's patent dates as November 30, 1880, July 26, 1881, January 10, 1882, and March 13, 1883.

46. Gilbert and Barker Manufacturing, *Directions for the Care of the Springfield Gas Machine,* 1897, 13.

47. Gilbert and Barker Manufacturing Company, "The Combination Mixing Regulator," flyer, May 1, 1890, Box 6, Folder 648b, Catalogue Collection, Shelburne; Gilbert and Barker Manufacturing Company, *The Combination Mixing Regulator,* catalog, n.d., Box 6, Folder 648b, Catalogue Collection, Shelburne; Gilbert and Barker Manufacturing, *Directions for the Care of the Springfield Gas Machine,* 1897, 13.

48. Gilbert and Barker Manufacturing, *Combination Mixing Regulator,* catalog; Gilbert and Barker Manufacturing, *Directions for the Care of the Springfield Gas Machine,* 1897, 13; Gilbert and Barker Manufacturing, *How Best to Light Our Country Homes and Resorts,* 1906, 14.

49. Edward J. Frost, "Device for Regulating the Quality of Carbureted Vapor or Gas," Patent No. 381,619, Apr. 24, 1888 (Washington, DC: U.S. Patent Office, 1888); Edward J. Frost and Benjamin Middleditch, Assignors to Joseph H. Berry, "Device for Regulating Quality of Gas," Patent No. 577,739, Feb. 23, 1897 (Washington, DC: U.S. Patent Office, 1897).

50. Gilbert and Barker Manufacturing, *How Best to Light Our Country Homes and Resorts,* 1906, 16.

51. Gilbert and Barker Manufacturing, *Combination Mixing Regulator,* catalog; Gilbert and Barker Manufacturing, *Directions for the Care of the Springfield Gas Machine,* 1897, 13. Interestingly, the company's *Directions for the Care of the Springfield Gas Machine* advertises both the Springfield Mixing Regulator and the Combination Mixing Regulator on facing pages. By the publication of their 1906 catalog, they were only marketing the Combination Mixing Regulator.

52. John F. Barker, "Improvement in Gas-Burners," Patent No. 105,768, July 26, 1870 (Washington, DC: U.S. Patent Office, 1870), 1.

53. John F. Barker, "Improvement in Gas-Burners," Patent No. 186,920, Feb. 6, 1877 (Washington, DC: U.S. Patent Office, 1877); John F. Barker, "Gas-Burner," Patent No. 231,256, Aug. 17, 1880 (Washington, DC: U.S. Patent Office, 1880).

54. John F. Barker, "Gas-Burner," Patent No. 238,205, Mar. 1, 1881 (Washington, DC: U.S. Patent Office, 1881).

55. Gilbert and Barker Manufacturing, *Springfield Gas Machine,* 1874, 70.

56. "The Welsbach Incandescent Gas Burner," display ad, *New York Times,* June 20, 1888, 2; Gilbert and Barker Manufacturing Company, *Springfield Incandescent Gas Burner,* catalog, n.d., Box 6, Folder 648a, Catalogue Collection, Shelburne.

57. John F. Barker, "Gas-Burner," Patent No. 575,318, Jan. 19, 1897 (Washington, DC: U.S. Patent Office, 1897); John F. Barker, "Gas-Burner," Patent No. 628,063, July 4, 1899 (Washington, DC: U.S. Patent Office, 1899).

58. Barker, "Gas-Burner," Patent No. 575,318, p. 1.

59. Gilbert and Barker Manufacturing, *Springfield Incandescent Gas Burner,* 4.

60. Ibid., 2.

61. Gilbert and Barker Manufacturing Company, *Are You Aware? Some Suggestions Relating to Gas for Light and Fuel,* pamphlet, n.d. (post-1895), Box 6, Folder 648b, Catalogue Collection, Shelburne.

62. Gilbert and Barker Manufacturing, *Springfield Incandescent Gas Burner,* 8.

63. Ibid.

64. Ibid., 10.

65. Gilbert and Barker Manufacturing, *Gilbert & Barker M'f'g Co's Price List.*

66. Gilbert and Barker Manufacturing, *Springfield Gas Machine,* 1874, 89, 41, 87.

67. John F. Barker and Charles N. Gilbert, "Improved Apparatus for Carbureting Air," Patent No. 66,777; Gilbert and Barker Manufacturing, *Springfield Gas Machine,* 1870-71, Plate No. 2.

68. Gilbert and Barker Manufacturing, *Springfield Gas Machine,* 1874, 75, 77.

69. Gilbert and Barker Manufacturing, *Springfield Gas Machine,* 1878, 22, 24–25, 22.

70. See chapter 4.

71. Gilbert and Barker Manufacturing, *Springfield Gas Machine,* 1878, 24–25; Gilbert and Barker Manufacturing, *Directions for the Care of the Springfield Gas Machine,* 1885, 10, 11.

72. Gilbert and Barker Manufacturing, *Springfield Gas Machine,* 1878, 24–25; Gilbert and Barker Manufacturing, *Directions for the Care of the Springfield Gas Machine,* 1897, 9.

73. Gilbert and Barker Manufacturing, *Springfield Gas Machine,* 1878, 24–25.

74. Gilbert and Barker Manufacturing Company, *The New Feed Springfield Gas Machine,* flyer, Mar. 1, 1883 (Springfield, MA: Gilbert and Barker Manufacturing, 1883); Gilbert and Barker Manufacturing, *Directions for the Care of the Springfield Gas Machine,* 1897, 10. The firm patented the machine itself in 1894.

75. Gilbert and Barker Manufacturing, *Directions for the Care of the Springfield Gas Machine,* 1897, 10; Gilbert and Barker Manufacturing, *New Feed Springfield Gas Machine.*

76. Gilbert and Barker Manufacturing, *Directions for the Care of the Springfield Gas Machine,* 1897, 11.

77. Gilbert and Barker Manufacturing, *Are You Aware?;* Gilbert and Barker Manufacturing, *New Feed Springfield Gas Machine.*

78. Gilbert and Barker Manufacturing, *Springfield Gas Machine,* 1874, 8; Gilbert and Barker Manufacturing, *Springfield Gas Machine,* 1870–71, 6–7.

79. Gilbert and Barker Manufacturing, *Springfield Gas Machine,* 1870–71, 6; Gilbert and Barker Manufacturing, *Springfield Gas Machine,* 1874, 8.

80. Gilbert and Barker Manufacturing, *Springfield Gas Machine,* 1874, 9; Gilbert and Barker Manufacturing, *Springfield Gas Machine,* 1870–71, 32.

81. Lamson (1868) quoted in Gilbert and Barker Manufacturing, *Springfield Gas Machine,* 1874, 32–33; Gilbert and Barker Manufacturing, *Springfield Gas Machine,* 1870–71, 22.

82. Gilbert and Barker Manufacturing, *Springfield Gas Machine,* 1874, 43, 5; Gilbert and Barker Manufacturing, *Springfield Gas Machine,* 1878, 33; Gilbert and Barker Manufacturing, *Springfield Gas Machine,* 1874, 10.

83. Gilbert and Barker Manufacturing, *Are You Aware?;* Gilbert and Barker Manufacturing, *How Best to Light Our Country Homes and Resorts,* 1906, 48, 41, 54.

84. Gilbert and Barker Manufacturing, *How Best to Light Our Country Homes and Resorts,* 1906, 40.

85. S. John Lea (1903) quoted in Gilbert and Barker Manufacturing, *How Best to Light Our Country Homes and Resorts,* 1906, 39. In that all of these comments speak of both light quality and cost, it is difficult to separate the two.

86. Gilbert and Barker Manufacturing, *Are You Aware?;* Gilbert and Barker Manufacturing, *Springfield Gas Machine,* 1870–71, 29.

87. Trachtenberg, *Incorporation of America,* 133.

88. Gilbert and Barker Manufacturing, *How Best to Light Our Country Homes and Resorts,* 1906, 41.

89. Pauline Ames Plimpton, ed., *Oakes Ames: Jottings of a Harvard Botanist, 1874–1950* (Cambridge: Botanical Museum of Harvard Univ., 1979), 43, 44.

90. William B. McCash and June H. McCash, *The Jekyll Island Club: Southern Haven for American's Millionaires* (Athens: Univ. of Georgia Press, 1989), 72.

91. Gilbert and Barker Manufacturing, *Springfield Gas Machine,* 1874, 26, 43, 16.

92. Ibid., 51; Gilbert and Barker Manufacturing, *Springfield Gas Machine,* 1878, 48.

93. Plimpton, *Oakes Ames,* 44.

94. Gilbert and Barker Manufacturing, *Springfield Gas Machine,* 1878, 41, 46.

95. Gilbert and Barker Manufacturing, *Springfield Gas Machine,* 1870–71, 13.

96. Worthington, *Beyond the City Lights,* 3, 7.

97. C.A. Robbe to Alexander Stephens, "Statement for Gas Fixtures," Nov. 1, 1875, photocopy on file at Alexander Stephens State Park, Crawfordville, GA.

98. Gilbert and Barker Manufacturing, *Springfield Gas Machine,* 1874, 51; Gilbert and Barker Manufacturing, *Springfield Gas Machine,* 1878, 46, 51; Gilbert and Barker Manufacturing, *How Best to Light Our Country Homes and Resorts,* 1906, 45.

99. U.S. Bureau of the Census, *The Statistical History of the United States, from Colonial Times to the Present* (New York: Basic Books, 1976), 322.

100. Gilbert and Barker Manufacturing, *Springfield Gas Machine,* 1870–71, 7; Gilbert and Barker Manufacturing, *Springfield Gas Machine,* 1874, 10.

101. Gilbert and Barker Manufacturing, *Springfield Gas Machine,* 1874, 9, 11, 15.

102. Gilbert and Barker Manufacturing, *Springfield Gas Machine,* 1878, 23, 50.

103. Gilbert and Barker Manufacturing, *Springfield Gas Machine,* 1874, 17.

104. Gilbert and Barker Manufacturing, *How Best to Light Our Country Homes and Resorts,* 1906, 19, 51.

105. Gilbert and Barker Manufacturing, *Springfield Gas Machine,* 1874, 17.

106. Gilbert and Barker Manufacturing, *Springfield Gas Machine,* 1870–71, 4; Gilbert and Barker Manufacturing, *Springfield Gas Machine,* 1878, 23; Gilbert and Barker Manufacturing, *Springfield Gas Machine,* 1874, 45.

107. Gilbert and Barker Manufacturing, *Springfield Gas Machine,* 1874, 7.

108. Oakley and Heald, *Report to the Executive Committee of the National Board of Fire Underwriters.*

109. John C. Paige and Company, Boston, "Insurance Policy on Buildings, Contents and Other Property Situated in North Easton , Mass.—John S. Ames. September 21st, 1919," John S. Ames Collection.

110. National Board of Fire Underwriters, *Rules and Regulations of the National Board of Fire Underwriters for the Construction, Installation and Use of Gasolene Vapor Gas Lighting Machines, Lamps and Systems* (New York: National Board of Fire Underwriters, 1910); National Board of Fire Underwriters, *Rules and Regulations of the National Board of Fire Underwriters for the Construction, Installation and Use of Gasolene Vapor Gas Lighting Machines, Lamps and Systems* (New York: National Board of Fire Underwriters, 1926).

111. National Board of Fire Underwriters, *Rules and Regulations,* 1910, 1, 2.

112. Ibid., 2–3.

113. Gilbert and Barker Manufacturing, *Springfield Gas Machine,* 1870–71, 27, 41, 43.

114. Gilbert and Barker Manufacturing, *How Best to Light Our Country Homes and Resorts*, 1906, 21.

Chapter 4

1. W. Barksdale Maynard, *Architecture in the United States, 1800–1850* (New Haven: Yale Univ. Press, 2002).

2. Gilbert and Barker Manufacturing, *How Best to Light Our Country Homes and Resorts*, 1888 and 1906.

3. Gary Cross and Rick Szostak, *Technology and American Society: A History* (Upper Saddle River, NJ: Pearson, Prentice Hall, 2005), 149.

4. Ibid.; Cowan, *Social History of American Technology*, 211; John F. Kasson, *Civilizing the Machine: Technology and Republican Values, 1776–1900* (New York: Penguin Books, 1976), 189.

5. Gilbert and Barker Manufacturing, *Springfield Gas Machine*, 1874, 71; Gilbert and Barker Manufacturing, *Springfield Gas Machine*, 1878, 44.

6. *New York Times*, Apr. 29, 1858, 8

7. Woodward and Thompson, *Woodward's National Architect*. The Empire State Gas Machine was an early design that combined the gas generator and air pump in one system, requiring that it be placed outside the building to be lighted.

8. *New York Times*, Apr. 25, 1875, 5; *New York Times*, Apr. 24, 1881, 10.

9. *New York Times*, Mar. 29, 1903, 21; *New York Times*, May 31, 1890, 7; *New York Times*, Jan. 30, 1887, 11.

10. *Journal of Outdoor Life, Travel, Nature Study, Shooting, Fishing, Yachting* 38, no. 22 (June 2, 1892): 536; Gilbert and Barker Manufacturing, *Springfield Gas Machine*, 1874, 92.

11. Bowers, *Lengthening the Day*, 39; Schivelbusch, *Disenchanted Night*, 153.

12. Bowers, *Lengthening the Day*, 41; Alice Taylor, *Quench the Lamp* (Dingle, Ireland: Brandon, 1990), 109.

13. Ierley, *Comforts of Home*, 139; Trachtenberg, *Incorporation of America*, 123.

14. Robert Louis Stevenson (1881) quoted in Bowers, *Lengthening the Day*, 3; Richards, *Gas-Consumer's Guide*, 26; Castaneda, *Invisible Fuel*, 29.

15. Robert Louis Stevenson, *A Child's Garden of Verses and Underwoods* (1906; rpt., Whitefish, MT: Kessinger, 2009).

16. Bowers, *Lengthening the Day*, 41; William T. O'Dea, *The Social History of Lighting* (London: Wyman and Sons, 1958), 67.

17. Bowers, *Lengthening the Day*, 53; Richards, *Gas-Consumer's Guide*, 26.

18. Bowers, *Lengthening the Day*, 51–54; Elizabeth Barrett Browning quoted in O'Dea, *Social History of Lighting*, 9; Taylor, *Quench the Lamp*, 109.

19. O'Dea, *Social History of Lighting*, 9; Schivelbusch, *Disenchanted Night*, 44.

20. Edgar Allen Poe quoted in Myers, *Gaslighting in America*, 35; Frederika Bremer quoted in Moss, *Lighting for Historic Buildings*, 100.

21. Richards, *Gas-Consumer's Guide*, 113–24, 139–40.

22. Gilbert and Barker Manufacturing, *Springfield Gas Machine*, 1870–71, 26; Strasser, *Never Done*, 61.

23. Schivelbusch, *Disenchanted Night*, 185–86; Ierley, *Comforts of Home*, 140.

24. Clark, "Domestic Architecture as an Index to Social History," 536–37; Schivelbusch, *Disenchanted Night*, 185–86.

25. Cowan, *Social History of American Technology*, 196.

26. Plimpton, *Oakes Ames*, 44.

27. The Civilian Conservation Corps (CCC) may have done this work during the mid-1930s reconstruction project to begin creating Alexander H. Stephens State Park and continued with subsequent restoration work on the house.

28. Thomas E. Schott, *Alexander H. Stephens of Georgia* (Baton Rouge: Louisiana State Univ. Press, 1988), 16, 24, 28, 40, 38.

29. Martha F. Norwood, *Liberty Hall: Taliaferro County, Georgia, A History of the Structures Known as Liberty Hall and Their Owners from 1827 to the Present* (Atlanta: Georgia Dept. of Natural Resources, Historic Preservation Section, 1977), 108; Schott, *Alexander H. Stephens*, 9, 108.

30. Schott, *Alexander H. Stephens*, 48, 61. Stephens was intimately involved in the debates over the war with Mexico, the Compromise of 1850, the Kansas-Nebraska Act, and the ongoing debates over slavery and the southern prerogatives (Norwood, *Liberty Hall*, 150–55).

31. Schott, *Alexander H. Stephens*, 273–75. One of the county's largest and wealthiest planters by 1860, he clearly could have afforded to retire to a life of leisure.

32. Ibid., 322–23. After voting against the ordinance of secession, Stephens signed the document when it overwhelmingly passed in January 1861.

33. Ibid., 323–29, 451–52, 458.

34. Ibid., 471–72. The first volume (completed in 1868) sold over sixty-four thousand copies, and the second (completed in 1870) sold over twenty thousand copies.

35. Ibid., 516–20.

36. Norwood, *Liberty Hall*, 12, 18; *Augusta Constitutionalist* (1858) quoted in Norwood, *Liberty Hall*, 12. The article noted that Stephens had recently added "two rooms, intended for library and bed-chamber and a small and airy passage."

37. Norwood, *Liberty Hall*, 236–39. Norwood incorrectly identifies the Canster Gas Works pamphlet as the "instruction book for the Springfield Gas Machine at Liberty Hall." The description of this machine, while a cold process gasoline system, is clearly different from the Springfield and it would make little sense for Gilbert and Barker to provide instructions for this machine.

38. Norwood, *Liberty Hall*, 236–39, 21.

39. Schott, *Alexander H. Stephens*, 58.

40. Norwood, *Liberty Hall*, 169; Schott, *Alexander H. Stephens*, 483.

41. Norwood, *Liberty Hall*, 236–39.

42. Ibid., 236.

43. Herbert (1875) quoted in Norwood, *Liberty Hall*, 23; Schott, *Alexander H. Stephens;* Herbert (1875) quoted in Norwood, *Liberty Hall*, 22.

44. Gilbert and Barker Manufacturing, *Springfield Gas Machine*, 1878, 67; Gilbert and Barker Manufacturing Company, *The Springfield Gas Machine* (New York: Wheat and Cornett, 1875), 119–20; Gilbert and Barker Manufacturing, *Springfield Gas Machine*, 1878, 67; Gilbert and Barker Manufacturing, *How Best to Light Our Country Homes and Resorts*, 1888.

45. C.A. Robbe to Alexander Stephens, "Statement for Gas Fixtures," Nov. 1, 1875, photocopy on file at Alexander Stephens State Park, Crawfordville, GA. Gilbert and Barker listed the No. 2 machine at $400.00; thus the fixtures, pipes, and labor cost approximately $191.64.

46. C. A. Robbe to Alexander Stephens, Nov. 1, 1875, photocopy on file at Alexander Stephens State Park, Crawfordville, GA; Alexander H. Stephens to C. A. Robbe, Sept. 21, 1875, on file at East Central Georgia Public Library, Augusta; Herbert (1875) quoted in Norwood, *Liberty Hall*, 22.

47. The main house at Liberty Hall did not have a basement but just a crawl space. The wing added to the back of the house has a small English basement.

48. Gilbert and Barker Manufacturing, *Springfield Gas Machine*, 1874, Plate No. 9. The machine's name plate reads "SPRINGFIELD GAS MACHINE, Gilbert & Barker Mfg. Co., Manufacturers and Proprietors, Springfield, Mass., Patented, Jan. 8, 1867, July 16, 1867, April 21, 1868, and August 3, 1869."

49. Gilbert and Barker Manufacturing, *Springfield Gas Machine*, 1874, 120. The machine's four evaporator pans measure forty-three inches in diameter and would have yielded an evaporating surface of approximately ninety-eight square feet.

50. Gilbert and Barker Manufacturing, *Springfield Gas Machine*, 1874, 19–20, 23. A plate on the air pump reads "SPRINGFIELD GAS MACHINE, Gilbert and Barker Mfg. Co., Springfield, Mass., Patented Aug 3, 1869 and Aug 17, 1869."

51. Gilbert and Barker Manufacturing, *Springfield Gas Machine*, 1874, 119. The weight measures approximately sixteen inches high and twenty-six inches in diameter. In its original configuration, a pipe would have connected the air pump directly to the air intake at the base of the gas generator.

52. Gilbert and Barker Manufacturing, *Directions for the Care of the Springfield Gas Machine*, 1897, 12–13; Gilbert and Barker Manufacturing, *How Best to Light Our Country Homes and Resorts*, 1906, 14–16.

53. Martha Neal, personal communication, 2002. Neal pointed out the ceiling fixture locations that existed prior to the 1988–89 restoration of the house.

54. C. A. Robbe to Alexander Stephens, "Statement for Gas Fixtures." This fitting was marked "Pat. Dec. 18, 1868."

55. This burner was marked "Pat. 1867 REISS 1875"; Neal, personal communication.

56. Herbert (1875) quoted in Norwood, *Liberty Hall*, 23; Schott, *Alexander H. Stephens*, 65.

57. Norwood, *Liberty Hall*, 26.

58. Gilbert and Barker Manufacturing, *Springfield Gas Machine*, 1874; Gilbert and Barker Manufacturing, *Springfield Gas Machine*, 1875; Gilbert and Barker Manufacturing, *Springfield Gas Machine*, 1878, 67.

59. William Y. Thompson, *Robert Toombs of Georgia* (Baton Rouge: Louisiana State Univ. Press, 1966), 12, 13, 14.

60. Ibid., 32–33, 90–91, 93–94. The Texas landholding investments would eventually earn Toombs over $200,000. Mrs. Clement C. Clay commented that Mrs. Toombs spent $2,000 a month on entertainment.

61. Ibid., 56–57, 170–74, 219–24.

62. Ibid., 226–27, 227, 249, 254–57.

63. Pleasant A. Stovall, *Robert Toombs: Statesman, Speaker, Soldier, Sage* (New York: Cassell, 1892), 360; "A Brief History," Robert Toombs House State Historic Site, Washington, GA. The state of Georgia purchased the Toombs House in 1973 and opened the site to the public in 1982.

64. Gilbert and Barker Manufacturing, *Springfield Gas Machine*, 1874; Gilbert and Barker Manufacturing, *Springfield Gas Machine*, 1875, 120; Gilbert and Barker Manufacturing,

Springfield Gas Machine, 1878, 67; Gilbert and Barker Manufacturing, *Springfield Gas Machine*, 1879, 19; Gilbert and Barker Manufacturing, *Springfield Gas Machine*, 1878, 49.

65. Marcia Campbell, Robert Toombs House State Historic Site, to author, June 15, 2005 (the name plates on the generators changed somewhat in the 1880s); Gilbert and Barker Manufacturing, *Springfield Gas Machine*, 1874, 119.

66. Gilbert and Barker Manufacturing, *Springfield Gas Machine*, 1874, 119. The weight measures approximately 20 ½ inches high and 33 inches in diameter; photograph (black and white) labeled "South Wall, Room A-7," 1977–78, Robert Toombs House State Historic Site, Washington, GA; Gilbert and Barker Manufacturing, *Springfield Gas Machine*, 1874, 70; Campbell to author. The 1 ½-inch diameter of these pipes corresponds to a No. 4 Springfield system that lit seventy-plus burners.

67. Campbell to author.

68. Based on the 1874 catalog, a No. 4 Springfield generator could power a maximum of seventy-five burners (Gilbert and Barker Manufacturing, *Springfield Gas Machine*, 1874, 119–20). At least one of the wall fixtures was a single burner, which may suggest a lower total number of burners, although some rooms with no current evidence likely contained at least one fixture.

69. These dates would suggest a post-1875 installation of the system or at least the purchase of these burners.

70. Marked "Patented July 26, 1870, Barker, Springfield, Mass."

71. In that Washington had a municipal gas plant by 1854, these fixtures were likely powered from the municipal system.

72. Stovall, *Robert Toombs*, 360.

73. The property is now within the city limits of Lexington.

74. Donald W. Linebaugh, Nancy O'Malley, and Jeanie Duwan, "'A Brilliant and Pleasant Light': Investigating the Springfield Gas Machine System at Ashland, the Henry Clay Estate, Lexington, KY," *Historical Archaeology* 34, no. 4 (2000): 82–100. This article is drawn from the complete technical report on the project: Donald W. Linebaugh, Nancy O'Malley, and Jeanie Duwan, "'A Brilliant and Pleasant Light': Nineteenth-Century Gas Lighting at Ashland, Lexington, Fayette County, KY," Technical Report No. 413, University of Kentucky Program for Archaeological Research, Lexington, 1999. The house and grounds next served as part of the Agricultural and Mechanical College campus until the dissolution of the college in 1882.

75. Major Henry Clay McDowell, born in Fincastle, Virginia, in 1832, married Anne Clay, the daughter of Henry Clay Jr., the son of the statesman. The family moved to Franklin County, Kentucky, in 1879 and to Lexington in 1882.

76. Thomas D. Clark, Robert Magrish, and Terry L. Green, *Ashland: The Henry Clay Estate* (Greendale, IN: Creative, 1996). The major change to the front stairway also involved adding a decorative gas fixture to the newel post. Sophonisba Preston Breckinridge, *Madeline McDowell Breckinridge: A Leader in the New South* (Chicago: Univ. of Chicago Press, 1921), 12.

77. Breckinridge, *Madeline McDowell Breckinridge*, 16.

78. Linebaugh, O'Malley, and Duwan, "'Brilliant and Pleasant Light': Investigating the Springfield Gas Machine," 42–46. Based on the results of a geophysical survey and preliminary documentary research, archaeologists placed excavation units around the brick pit or chamber and over the pipes connecting the machine to the house.

79. Ibid., 49. The soil layers and modern artifacts excavated from the shaft containing the generator indicated that it had remained open for many years after the McDowells stopped using the machine, probably as late as the 1980s.

80. The generator had an estimated diameter of approximately 69 ½ inches and height of 25 ½ inches. The brass name plate lists patent dates of January 8, 1867, July 16, 1867, April 21, 1868, and August 3, 1869.

81. Notations on plans found at the Hill-Stead site (see chapter 4) indicate that the top pans each held 4 ⅜ inches of gasoline, while the bottom pan held 7 ⅜ inches. We know that the Ashland generator had 4 pans and each inch of fluid in a machine measuring 69 ½ inches in diameter was equal to 17 gallons of gasoline (1 gallon per 231 per cubic inches). Thus the calculation results in a capacity of 348.5 gallons for the Ashland system.

82. Gilbert and Barker Manufacturing, *Directions for the Care of the Springfield Gas Machine*, 1885, 6–7.

83. Linebaugh, O'Malley, and Duwan, "'Brilliant and Pleasant Light': Investigating the Springfield Gas Machine," 42–46, 49, 52. Nancy O'Malley, personal communication, June 30, 2008. The two pipes, measuring two and one and a half inches in diameter (OD), were laid at a depth of approximately three feet, four inches.

84. Kim McBride, personal communication, 1996, 1997.

85. The 1 ½-inch pipe had a maximum run of 150 feet and could supply 70 burners; the 2-inch pipe had a maximum run of 200 feet and could supply 140 burners. Gilbert and Barker Manufacturing, *How Best to Light Our Country Homes and Resorts*, 1906, p. 32. In that the larger of the two pipes was the gas supply to the house and the distance from unit to house was approximately 100 feet, these pipes seem designed for a system of approximately 100 burners.

86. Gilbert and Barker Manufacturing, *Springfield Gas Machine*, 1874, 70.

87. One of the Argand-type burners bears the stamp "PAT. 1867 REISS 1875," identical to a burner found at the Toombs House in Georgia. Another type of burner has a fixed, expanding globe clip with vents. Stamped "PAT. FEB. 22. 87," it perfectly matches a burner illustrated in *Appleton's 'Cyclopedia of Applied Mechanics*. Park Benjamin, *Appleton's 'Cyclopaedia of Applied Mechanics: A Dictionary of Mechanical Engineering and Mechanical Arts* (New York: D. Appleton, 1880), 927.

88. Webster Gasolene Company to H. C. McDowell, Esq., Mar. 27 and Apr. 10, 1889, Box 60, H. C. McDowell Correspondence, 1889, McDowell Family Papers, Special Collections, M. I. King Library, University of Kentucky, Lexington (hereafter cited as McDowell Family Papers).

89. Detroit Heating and Lighting Company to H. C. McDowell, Esq., Nov. 5, 1889, McDowell Family Papers.

90. *Kentucky News Leader*, Apr. 19, 1892.

91. Henry C. McDowell Jr. to Anne Clay McDowell, Dec. 23, 1906, McDowell Family Papers.

92. Henry C. McDowell Jr. to Anne Clay McDowell, May 29, 1907, McDowell Family Papers. In an agreement dated May 28, 1907, between Anne Clay McDowell and the Kentucky Railway Company, the company agreed to "construct a line of poles and to string wires thereon sufficient to light Ashland as herein agreed to be done, from the end of Main Street to Ashland." The "contract for Electric Service" between the two parties specified that "16 C.P. [candle power] Incandescent Lamps" would be provided and maintained by Kentucky Railway Company. Kentucky Railway Company to Anne Clay McDowell, "Contract for Electric Service," May 28, 1907, Archives, Henry Clay Memorial Foundation, Lexington, KY.

93. Beverly Bridger, *Mrs. Vanderbilt's Sagamore* (Raquette Lake, NY: Sagamore Institute of the Adirondacks, n.d.), 15. Durant oversaw the construction of the Transcontinental Railroad from the East Coast to Promontory Point in the mid- to late 1860s.

94. Bridger, *Mrs. Vanderbilt's Sagamore,* 15–16.

95. It is also possible that the next owner was responsible for installing both machines. Ibid., 16–17.

96. Ibid., 17–18.

97. Gilbert and Barker Manufacturing, *How Best to Light Our Country Homes and Resorts,* 1906, 34.

98. Bridger, *Mrs. Vanderbilt's Sagamore,* 1; Margaret's first marriage to physician Smith Hollis McKim ended in divorce. Bridger, *Mrs. Vanderbilt's Sagamore,* 21–22; the Vanderbilts always sought to preserve the rustic ambiance of the camp and so had all utility lines buried underground.

99. Gilbert and Barker Manufacturing, *Directions for the Care of the Springfield Gas Machine,* 1897, 10; Gilbert and Barker Manufacturing Company to the Oliver Ames and Son Corporation, North Easton, Massachusetts, Jan. 6, 1904, Ames Family Collection.

100. Gilbert and Barker Manufacturing Company to the Oliver Ames and Son Corporation.

101. Gilbert and Barker Manufacturing, *How Best to Light Our Country Homes and Resorts,* 1906, 32. At 34 gallons per inch, the Main Lodge generator would have held approximately 548 gallons. The bulk tank would have held approximately 420 gallons.

102. Gilbert and Barker Manufacturing, *Springfield Gas Machine,* 1874, 10.

103. Gilbert and Barker Manufacturing, *Directions for the Care of the Springfield Gas Machine,* 1897, 4; Gilbert and Barker Manufacturing, *How Best to Light Our Country Homes and Resorts,* 1906, 9.

104. A plate on the air pump is marked "Springfield Gas Machine, Gilbert and Barker Mfg. Co., Springfield, Mass., Pat. Aug 3rd and Aug 17th, 1869," and serial number 10287. A smaller plate reads "Patented May 4, 1875, July 29, 1890."

105. Gilbert and Barker Manufacturing, *Directions for the Care of the Springfield Gas Machine,* 1897; Gilbert and Barker Manufacturing, *How Best to Light Our Country Homes and Resorts,* 1906, 14–15. The mixing regulator is very similar to the model shown in the 1897 operating instructions and the 1906 Springfield catalog.

106. The illustrated plates read "Combination Mixing Regulator, Pat'd April 24, '88, Other Pats. Pending, Manufactured for Gilbert & Barker Mfg. Co." The name plate on the Main Lodge mixer is missing, but the outline can be seen and is identical in shape to the plate shown in the 1897 instructions and the 1906 catalog. Gilbert and Barker Manufacturing, *Directions for the Care of the Springfield Gas Machine,* 1897; Gilbert and Barker Manufacturing, *How Best to Light Our Country Homes and Resorts,* 1906, 14–15.

107. Another flat-flame burner found in the basement is marked "Gilbert & Barker Mfg. Co., Springfield, Mass., PatD Mch. 1, 1881." This is an adjustable burner used to further regulate the air-gas mixture at the fixture itself (see chapter 3). The first floor contained twenty wall fixtures and one ceiling fixture with six burners. The second floor shows evidence of eleven wall and ceiling fixtures, but as several of these rooms have undergone remodeling, this number is likely low. The third floor contained eight wall fixtures. The basement has seven fixture locations and the exterior porches have evidence of eleven fixtures.

108. The first floor of the Conference or New Laundry building has fifteen wall and ceiling fixtures with a total of twenty-six burners, most of which are intact. The second floor has eight fixtures in the hallways and bathrooms, all with single burners, and the porch has three fixtures.

109. Several of the burners are marked "Gilbert & Barker Mfg. Co., Springfield Gas Machine, Pat. Mar. 1, 1881."

110. Gilbert and Barker Manufacturing, *How Best to Light Our Country Homes and Resorts*, 1906, 32, 15.

111. The Play House contains few gas fixtures or burners. Capped pipes indicate the locations of sixteen wall fixtures; one of these retains the original single-burner fixture. The porch that wraps around the structure bears evidence of five fixtures, four retaining their original single-burner fixtures.

112. The first and second floors had forty-four wall fixtures, as indicated by capped pipes in the walls, and the porch that wraps around the Wigwam had six fixtures.

113. Gilbert and Barker Manufacturing, *Springfield Incandescent Gas Burner*. These fixtures are marked "Gilbert and Barker Mfg. Co. Makers, Pat. Jan. 19, 1897, Nov. 21, 1899."

114. Evidence of ten wall fixtures in the form of capped pipes exists, as do two actual fixtures located on the front porch, both with their original single burners. These fixtures are identical to those on the outside of the Play House.

115. William Cronon, "The Trouble with Wilderness; or, Getting Back to the Wrong Nature," in *Uncommon Ground: Toward Reinventing Nature*, ed. William Cronon, 69–90 (New York: W. W. Norton, 1995), 69–74.

116. Along with a physical inspection of the house, this drawing provided the basis for interpreting the Springfield system at Hill-Stead.

117. Mark Allan Hewitt, *The Architect and the American Country House: 1890–1940* (New Haven: Yale Univ. Press, 1990), 157.

118. Helen Hall, *Hill-Stead Museum* (Farmington, CT: Hill-Stead Museum, 1988), 36–41. Madeline McDowell, daughter of Major Henry Clay McDowell of Ashland, also attended Miss Porter's School in Farmington about this same time (1889–90) (Breckinridge, *Madeline McDowell Breckinridge*, 165).

119. Hewitt, *Architect and the American Country House*, 157; Hall, *Hill-Stead Museum*, 41–42.

120. Hall, *Hill-Stead Museum*, 41–42.

121. Gilbert and Barker Manufacturing, *Are You Aware?*; Gilbert and Barker Manufacturing, *How Best to Light Our Country Homes and Resorts*, 1906. Listed as Whittemore, Harris (J. H.).

122. Pope (1898) quoted in Hewitt, *Architect and the American Country House*, 158; Hall, *Hill-Stead Museum*, 45; Hewitt, *Architect and the American Country House*, 160.

123. Hall, *Hill-Stead Museum*, 46; Henry James (1913) quoted in Hall, *Hill-Stead Museum*, 47.

124. Theodate Pope to Alfred Pope, June 28, 1899, or 1900, No. 772, Hill-Stead Museum Archives, Farmington, CT.

125. The company provided plans for "Order #590" as of June 29, 1900. Hewitt, *Architect and the American Country House*, 160; Gilbert and Barker Manufacturing, *How Best to Light Our Country Homes and Resorts*, 1906, 71.

126. Theodate became a registered architect in the state of New York in 1916.

127. Hall, *Hill-Stead Museum*, 47, 49–50. Between 1919 and 1930 she designed and oversaw the construction of Avon Old Farms School, worked on commissions for private homes, and designed the Westover School in Middlebury, Connecticut.

128. Sanborn Fire Insurance Company, "Farmington, Connecticut, 1924," in *Digital Sanborn Maps, 1867–1970* (Ann Arbor, MI: ProQuest).

129. Gilbert and Barker Manufacturing, *How Best to Light Our Country Homes and Resorts,* 1906, 32. Preliminary remote sensing revealed the routes of the pipes, suggesting that the gas generator lies buried in the lawn on the south side of the house, adjacent to the driveway running between the house and the Sunken Garden.

130. Gilbert and Barker Manufacturing, "Plan and Elevation Showing Two Hundred Light Springfield Gas Machine." The top three pans each held 4 3/8 inches of gasoline, while the bottom pan held 7 3/8 inches. Each inch of fluid equaled approximately 34 gallons of gasoline.

131. At 1,100 cubic feet of air-gas from every six gallons of gasoline, the Pope's Springfield system could generate about 183,000 cubic feet of air-gas per filling.

132. Gilbert and Barker Manufacturing, *How Best to Light Our Country Homes and Resorts,* 1906, 9–10, 15; Gilbert and Barker Manufacturing, "Plan and Elevation Showing Two Hundred Light Springfield Gas Machine."

133. "Inventory, 'Hillstead,' Alfred A. Pope, Farmington, Conn., July 31, 1909," Alfred Atmore Pope Collection, Hill-Stead Museum, Farmington, CT. Hazen and Hazen, *Keepers of the Flame,* 180.

134. Theodate Pope to Alfred Pope.

135. Gilbert and Barker Manufacturing, *Are You Aware?;* Gilbert and Barker Manufacturing, *How Best to Light Our Country Homes and Resorts,* 1906. Listed as Whittemore, Harris (J. H.).

136. Joe Sherman, *The House at Shelburne Farms: The Story of One of American's Great Country Estates* (Forest Dale, VT: Paul S. Eriksson, 1992), 13, 15.

137. Ibid., 18. Olmsted provided a sketch in 1887 that showed separate areas for a farm, a park, and a forest linked by a network of roads.

138. Ibid., 19–20.

139. Ibid., 20, 21–22.

140. Ibid., 24–25, 26, 27–28.

141. William Seward Webb, *Shelburne Farms Stud, Shelburne, Chittenden County, Vermont* (New York: G. P. Putnam's Sons, 1893), 21, 22.

142. Ibid., 21.

143. Sherman, *House at Shelburne Farms,* 72–73. Lila Webb took over more of the decision making and management, along with Webb's lawyer, E. F. Gebhardt.

144. Sanborn Fire Insurance Company, "Burlington, Vermont, 1894," in *Digital Sanborn Maps, 1867–1970* (Ann Arbor, MI: ProQuest).

145. Gilbert and Barker Manufacturing Company, "Plan Drawing," n.d., Shelburne Farms Archives, Shelburne, VT. The Hill-Stead plans indicate that the top pans each held 4 3/8 inches of gasoline, while the bottom pan held 7 3/8 inches. A 30-inch-high generator would have had five or six evaporator pans, and each inch of fluid in a machine measuring 108 inches in diameter was equal to 42 gallons of gasoline (1 gallon per 231 cubic inches). Thus the calculation results in a figure of 1,066 gallons for the generator.

146. Gilbert and Barker Manufacturing Company to Shelburne Farms, invoices, Apr. 15, 1913, June 12, 1914, and Apr. 22, 1915, Box 7, Folder 81, Shelburne Farms Bill Collection. Prices fluctuated during this period from .27 to .30 per gallon.

147. Gilbert and Barker Manufacturing, *Directions for the Care of the Springfield Gas Machine,* 1897. Unlike the "New Feed" machine, the Shelburne system required a brick vault or shaft for access to the generator.

148. Gilbert and Barker Manufacturing, "Plan Drawing"; Sanborn Fire Insurance, "Burlington, Vermont, 1894"; Gilbert and Barker Manufacturing, "Air Pump of the Springfield Gas Machine"; Gilbert and Barker Manufacturing, *Directions for the Care of the Springfield Gas Machine*, 1885.

149. Gilbert and Barker Manufacturing Company, *The Combination Mixing Regulator*, pamphlet (Springfield, MA: Gilbert and Barker Manufacturing, ca. 1900), Box 6, Folder 648a, Catalogue Collection, Shelburne. Although neither of the brochures is dated, they refer to patent dates that indicate they postdate 1899.

150. Gilbert and Barker Manufacturing Company to Shelburne Farms, invoices, Sept. 9, 1913, Sept. 15, 1914, and Nov. 8, 1915, Box 7, Folder 81, Shelburne Farms Bill Collection.

151. Gilbert and Barker Manufacturing, *Springfield Incandescent Gas Burner*, 7, 11. The archive collection includes four brochures for the Springfield Incandescent Gas Burner. One of these illustrates both the No. 95 mantles and the 6-piece mica chimneys that the Webbs repeatedly ordered.

152. Gilbert and Barker Manufacturing Company, *Our Automatic Instantaneous Water Heater*, pamphlet, n.d., Box 7, Folder 849a, Catalogue Collection, Shelburne.

153. Sanborn Fire Insurance Company, "Burlington, Vermont, 1900," in *Digital Sanborn Maps, 1867–1970* (Ann Arbor, MI: ProQuest); Dale Company, *Combination Fixtures (Gas and Electric)*, catalog, n.d., Catalogue Collection, Shelburne; E. P. Gleason Manufacturing Company, *Combination and Electric Only Lighting Fixtures*, catalog, n.d., Catalogue Collection, Shelburne.

154. Sanborn Fire Insurance Company, "Burlington, Vermont, 1906," in *Digital Sanborn Maps, 1867–1970* (Ann Arbor, MI: ProQuest); Sanborn Fire Insurance Company, "Burlington, Vermont, 1912," in *Digital Sanborn Maps, 1867–1970* (Ann Arbor, MI: ProQuest); Sanborn Fire Insurance Company, "Burlington, Vermont, 1919," in *Digital Sanborn Maps, 1867–1970* (Ann Arbor, MI: ProQuest).

155. C. C. Ramsdell to E. F. Gebhardt, Oct. 16, 1916, Box 7, Folder 81, Shelburne Farms Bill Collection.

156. Webb, *Shelburne Farms Stud*, 21.

157. Robert F. Brown, ed., *History of Easton, Massachusetts*, vol. 2, *1886–1974* (Easton, MA: Easton Historical Society, 1974), 265.

158. Shepley, Rutan and Coolidge, Architects, "F. L. Ames Esq. Residence, N. Easton," architectural drawings and plans, n.d., microfilm, Widener Library, Harvard University; Shepley, Rutan and Coolidge, Architects, http://www.shepleybulfinch.com/history/.

159. William H. Ames apparently did just this, building a dynamo at his Spring Hill property (Greg Galer, personal communication, 2009).

160. George R. Burris to J. McKion, May 2, 1919, Letters, G[31/1], 1911–19, John S. Ames Collection.

161. Loring W. Puffer and Sons to John S. Ames, insurance policy, Sept. 21, 1919, Letters, G[31/1], 1911–19, John S. Ames Collection. Spring Hill may have had two gas machines, as John's 1919 insurance policy on the house permitted the use of either a Springfield or Walworth gas machine.

162. Hazel L. Varella, *Growing Up at "Sheep Pasture"* (Easton, MA: Easton Historical Society, 1976); Brown, *History of Easton*, 265.

163. Although the vault or buried generator may have survived below ground.

164. Gilbert and Barker Manufacturing, *How Best to Light Our Country Homes and Resorts*, 1906, 35, 26, and title page.

165. Gilbert and Barker Manufacturing, *Are You Aware?* Oliver Ames's comment that he put a system in his country home six years ago would date the brochure at 1910, as the summer house known as High Wall was constructed in 1904.

166. These statistics are based on a comprehensive list of installations derived from company publications, maps, and site visits—as such, they represent only a sample of the total installations.

167. Gilbert and Barker Manufacturing, *Springfield Gas Machine*, 1878, 45.

168. Maynard, *Architecture in the United States*, 125, 127 (Henry Bradshaw Fearon quote, 1818), 125 (William H. Ranlett quote, 1847), 130.

169. *New York Times*, Apr. 1, 1892, 7; *New York Times*, Apr. 11, 1903, 14.

170. Robert B. Mackay, Anthony K. Baker, and Carol A. Traynor, *Long Island Country Houses and Their Architects, 1860–1940* (New York: W. W. Norton, 1997), 206; Gilbert and Barker Manufacturing, *Springfield Gas Machine*, 1874, 36.

171. "Bradish Johnson," obituary, *New York Times*, Nov. 5, 1892.

172. Gilbert and Barker Manufacturing, *Springfield Gas Machine*, 1870–71, 43.

173. Maynard, *Architecture in the United States*, 130.

174. John M. Groff, "Green Country Towns: The Development of Philadelphia's Main Line, 1870–1915," MA thesis, Winterthur Program in Early American Culture, University of Delaware, Newark; Maynard, *Architecture in the United States*, 134–35.

175. Gilbert and Barker Manufacturing, *Springfield Gas Machine*, 1874, 71; Gilbert and Barker Manufacturing, *Springfield Gas Machine*, 1878, 39.

176. Clark W. Bryan, *The Book of Berkshire: Describing and Illustrating Its Hills and Homes* (Great Barrington, MA: Clark W. Bryan, 1886), 27; "H. H. Cook Bequeaths Millions in Cash," *New York Times*, Oct. 25, 1905.

177. "A Fine Display at Sloanes," *New York Times*, Oct. 21, 1889, 5; "Death of Mr. William Sloane," *New York Times*, May 24, 1879, 12; "Windsor Arcade, Famous Fifth Avenue Landmark, Giving Way to Great Commercial Structure," *New York Times*, May 7, 1911.

178. Gilbert and Barker Manufacturing, *How Best to Light Our Country Homes and Resorts*, 1906, 37. These more modest "cottages" still contained many more rooms than the average residence of the period.

179. Kathleen Maher, *P. T. Barnum (1810–1891)—The Man, the Myth, the Legend*. Barnum Museum, Bridgeport, CT, http://www.barnum-museum.org/manmythlegend.htm/.

180. Kathleen Maher, personal communication, July 21, 2008. Barnum had Waldermere built about 1867 in the Frederick Law Olmsted–designed Seaside Park.

181. Gilbert and Barker Manufacturing, *Springfield Gas Machine*, 1870–71, 22.

182. Gilbert and Barker Manufacturing, *How Best to Light Our Country Homes and Resorts*, 1888.

183. Russell Sturgis, John W. Root, Bruce Price, Donald G. Mitchell, Samuel Parsons Jr., and W. A. Linn, *Homes in City and Country* (New York: Charles Scribner's Sons, 1893), 75.

184. Frank T. Lent, *Sound Sense in Suburban Architecture* (New York: W. T. Comstock, 1895), iii, 83.

185. Lent, *Sound Sense in Suburban Architecture*.

186. Gilbert and Barker Manufacturing, *Springfield Gas Machine*, 1870–71, 25–26.

187. Cowan, *More Work for Mother*, 90; Lent, *Sound Sense in Suburban Architecture*.

188. Gilbert and Barker Manufacturing, *How Best to Light Our Country Homes and Resorts*, 1888.

189. Strasser, *Never Done*, 46.

190. Gilbert and Barker Manufacturing, *How Best to Light Our Country Homes and Resorts*, 1906, 56, 34.

191. Ibid., 50.

192. Gilbert and Barker Manufacturing, *Springfield Gas Machine*, 1874, 25.

193. Cowan, *More Work for Mother*, 100.

194. Catharine E. Beecher and Harriet Beecher Stowe, *The American Woman's Home* (New York: J. B. Ford, 1869; rpt., Watkins Glen, NY: American Life Foundation, 1975).

195. Ibid., 362–63.

196. Gilbert and Barker Manufacturing, *Springfield Gas Machine*, 1874, 8–10.

197. Beecher and Stowe, *American Woman's Home*, 432.

198. General Electric advertisement, *McClure's*, 1917, illustrated in Strasser, *Never Done*, 77.

Chapter 5

1. Melville (1813) quoted in Worthington, *Beyond the City Lights*, 3.

2. Trachtenberg, *Incorporation of America*, 56.

3. Schivelbusch, *Disenchanted Night*, 8.

4. James Marston Fitch, *American Building, 2: The Environmental Forces that Shape It* (New York: Schocken Books, 1975), 89.

5. Bowers, *Lengthening the Day*, 41; Trachtenberg, *Incorporation of America*, 56.

6. Cowan, *Social History of American Technology*, 191.

7. Gilbert and Barker Manufacturing, *Springfield Gas Machine*, 1874, 16; Gilbert and Barker Manufacturing, *Springfield Gas Machine*, 1878, 8.

8. Gilbert and Barker Manufacturing, *How Best to Light Our Country Homes and Resorts*, 1906, 24.

9. Springfield Gas Machine, *Circular of the Springfield Gas Machine Company*.

10. Warner, *Picturesque Hampden*, 150.

11. J. W. Wheelock to J. G. Benton, Sept. 22, 1866, Springfield Armory National Historic Site, Springfield, MA.

12. William Cooke to Col. T. T. Saidley, Feb. 16, 1866; William Cooke to Col. T. T. Saidley, Apr. 4, 1866; and William Cooke to Col. J. G. Benton, Mar. 11, 1867, all in Springfield Armory National Historic Site, Springfield, MA.

13. Gilbert and Barker Manufacturing, *Springfield Gas Machine*, 1870–71, 36.

14. Sanborn Fire Insurance Company, "Meriden, Connecticut, 1884, 1891," in *Digital Sanborn Maps, 1867–1970* (Ann Arbor, MI: ProQuest).

15. Arthur Gilman, ed., *The Cambridge of Eighteen Hundred and Ninety-Six: A Picture of the City and Its Industries Fifty Years After Its Incorporation* (Cambridge: Riverside Press, 1896), 389. An 1869 company catalog indicates that it produced tinware, tin toys, tinned iron wares, and many other metal products that would have required soldering. Dover Stamping Company, *Tinware, Tin Toys, Tinned Iron Wares, Tinners' Material, Enameled Stove Hollow Ware, Tinners' Tools and Machines* (Boston: Dover Stamping; rpt., Princeton, NJ: Pyne Press, 1971).

16. Gilbert and Barker Manufacturing, *Springfield Gas Machine*, 1878, 13.

17. "Scientific Methods Applied to Bicycling-Making," *Harper's Magazine Advertiser*, 1895, 29–32.

18. Warner, *Picturesque Hampden*, 150. The Victor, a chain-drive "safety" bicycle, had quickly caught on and replaced older high-wheel or "boneshaker" bicycles.

19. Sanborn Fire Insurance Company, "Chicopee Falls, Massachusetts, 1891," in *Digital Sanborn Maps, 1867–1970* (Ann Arbor, MI: ProQuest).

20. Richard Edwards, *New York's Great Industries* (New York: Historical, 1884; rpt., Ayer, 1973), 306.

21. Letter, Dec. 7, 1873, in Gilbert and Barker Manufacturing, *Springfield Gas Machine*, 1878, 13.

22. Sanborn Fire Insurance Company, "Wallingford, Connecticut, 1885," in *Digital Sanborn Maps, 1867–1970* (Ann Arbor, MI: ProQuest).

23. Gilbert and Barker Manufacturing, *Springfield Gas Machine*, 1878, 13.

24. "June 18, 1983: American Optical Celebrates 150th Anniversary," Mass Moments, http://massmoments.org/.

25. "American Optical Manufacturing Company of Southbridge, Massachusetts," Antique Spectacles, http://www.antiquespectacles.com/american_optical/american_optical.htm/.

26. Letter, Feb. 11, 1903, in Gilbert and Barker Manufacturing, *How Best to Light Our Country Homes and Resorts*, 1906, 42.

27. Sanborn Fire Insurance Company, "Southbridge, Massachusetts, 1888, 1894," in *Digital Sanborn Maps, 1867–1970* (Ann Arbor, MI: ProQuest).

28. George W. Howard, *The Monumental City: Its Past History and Present Resources* (Baltimore: J. D. Ehlers, 1873), 723, 724.

29. Gilbert and Barker Manufacturing, *Springfield Gas Machine*, 1874, xx.

30. Gilbert and Barker Manufacturing, *How Best to Light Our Country Homes and Resorts*, 1906, 25.

31. Edward F. Heite, *Archaeological Data Recovery on the Collins, Geddes Cannery Site, Road 356A, Lebanon, North Murderkill Hundred, Kent County, Delaware*, Delaware Dept. of Transportation Archaeology Series No. 83, Delaware Dept. of Transportation, 1990, 19, 20–22, 25.

32. Dean A. Doerrfeld, *The Canning Industry in Delaware, 1860–1940: A Historic Context* (Newark: Center for Historic Architecture and Engineering, Univ. of Delaware, 1994), 75.

33. Gilbert and Barker Manufacturing, *Springfield Gas Machine*, 1878, 8.

34. Heite, *Archaeological Data Recovery*, 28; redrawn from E. F. Kirwan Manufacturing Company, *Catalogue of Cans, Machinery, and General Canning House Supplies Manufactured and Sold by E. F. Kirwan Mfg. Co.* (Baltimore: Kirwan, 1890).

35. Sanborn Fire Insurance Company, "Milford, Kent Co., Delaware, 1885," in *Digital Sanborn Maps, 1867–1970* (Ann Arbor, MI: ProQuest).

36. Sanborn Fire Insurance Company, "Camden, Delaware, 1897," in *Digital Sanborn Maps, 1867–1970* (Ann Arbor, MI: ProQuest).

37. Mary B. Sim, *Commercial Canning in New Jersey: History and Early Development* (Trenton: New Jersey Agricultural Society, 1951), 102.

38. Sanborn Fire Insurance Company, "Williamstown, New Jersey, 1895," in *Digital Sanborn Maps, 1867–1970* (Ann Arbor, MI: ProQuest); Sanborn Fire Insurance Company, "Williamstown, New Jersey, 1900," in *Digital Sanborn Maps, 1867–1970* (Ann Arbor, MI: ProQuest);

Sanborn Fire Insurance Company, "Williamstown, New Jersey, 1905," in *Digital Sanborn Maps, 1867–1970* (Ann Arbor, MI: ProQuest).

39. Doerrfeld, *Canning Industry*, 75.

40. Franklin Ellis, *History of Columbia County, New York* (Philadelphia: Everts & Ensign, 1878), 239.

41. Gilbert and Barker Manufacturing, *Springfield Gas Machine*, 1878, 15.

42. Sanborn Fire Insurance Company, "Philmont, New York, 1887," in *Digital Sanborn Maps, 1867–1970* (Ann Arbor, MI: ProQuest).

43. Ibid.

44. Gilbert and Barker Manufacturing, *How Best to Light Our Country Homes and Resorts*, 1906, 55.

45. Sanborn Fire Insurance, "Philmont, New York, 1887."

46. Elizabeth M. Sharpe, *In the Shadow of the Dam: The Aftermath of the Mill River Flood of 1874* (New York: Simon and Schuster, 2004), 28.

47. Moses King, *King's Handbook of Boston* (Cambridge, MA: Moses King, 1881), 304.

48. Jacqueline Field, Marjorie Senechal, and Madelyn Shaw, *American Silk, 1830–1930: Entrepreneurs and Artifacts* (Lubbock: Texas Tech Univ. Press, 2007), 6. The firm received medals at the Centennial Exhibition for their new thread and for silk-making machinery (6).

49. King, *King's Handbook of Boston*, 304.

50. Gilbert and Barker Manufacturing, *Springfield Gas Machine*, 1874, 31; Gilbert and Barker Manufacturing, *Springfield Gas Machine*, 1878, 15.

51. Sanborn Fire Insurance Company, "Northampton, Massachusetts, 1902," in *Digital Sanborn Maps, 1867–1970* (Ann Arbor, MI: ProQuest).

52. Howard, *Monumental City*, 865.

53. Gilbert and Barker Manufacturing, *Springfield Gas Machine*, 1870–71, 43.

54. Gilbert and Barker Manufacturing, *Springfield Gas Machine*, 1874, 15.

55. Ibid., xx; Gilbert and Barker Manufacturing, *Springfield Gas Machine*, 1878, 45.

56. Wallace E. Jarrell, *The Randolph Hornets in the Civil War* (Jefferson, NC: McFarland, 2004), 69, 107.

57. Ibid., 69.

58. Sanborn Fire Insurance Company, "Concord, North Carolina, 1885," in *Digital Sanborn Maps, 1867–1970* (Ann Arbor, MI: ProQuest).

59. *Paper World*, Oct. 1882, 28, 39.

60. Lyman Horace Weeks, *A History of Paper-Manufacturing in the U.S., 1690–1916* (New York: Lockwood Trade Journal, 1916), 241.

61. 2004 Berkshire Towns Guide, "Lee," *Berkshire Eagle*, http://www.berkshireeagle.com/.

62. Weeks, *History of Paper-Manufacturing in the U.S.*, 242–43.

63. 2004 Berkshire Towns Guide, "Lee"; Dard Hunter, *Papermaking: The History and Technique of an Ancient Craft* (New York: Dover, 1978).

64. Gilbert and Barker Manufacturing, *Springfield Gas Machine*, 1878, 41.

65. Dunn County Historical Society, "The Knapp, Stout & Co., Company," http://www.dunnhistory.org/history/exkscc.html/.

66. Gilbert and Barker Manufacturing, *Springfield Gas Machine*, 1874, 62.

67. Sanborn Fire Insurance Company, "Menomonie, Wisconsin, 1885," in *Digital Sanborn Maps, 1867–1970* (Ann Arbor, MI: ProQuest). This use of electric light in a rather remote

location is interesting given the early date; the map indicates that the firm generated its own electricity, illustrating a "turbine wheel for electric power."

68. Susan E. Winter, "Mill Settlement Patterns Along the Antietam Creek Drainage, Washington County, Maryland," in *Spatial Patterning in Historical Archaeology: Selected Studies of Settlement*, ed. Donald W. Linebaugh and Gary G. Robinson, 65–90, William and Mary Center for Archaeological Research, Occasional Papers in Archaeology, No. 2 (Williamsburg, VA: King and Queen Press, 1994).

69. Sanborn Fire Insurance Company, "Ellicott City, Maryland, 1887," in *Digital Sanborn Maps, 1867–1970* (Ann Arbor, MI: ProQuest); Sanborn Fire Insurance Company, "Ellicott City, Maryland, 1894," in *Digital Sanborn Maps, 1867–1970* (Ann Arbor, MI: ProQuest).

70. Endangered Durham, "Blackwell's Durham Tobacco/American Tobacco Co.," http://endangereddurham.blogspot.com/2008/03/blackwells-durham-tobacco-american.html/.

71. National Historic Landmark Program, "Blackwell, W. T., and Company Tobacco Factory," http://tps.cr.nps.gov/nhl/.

72. Endangered Durham, "Blackwell's Durham Tobacco/American Tobacco Co."

73. National Historic Landmark Program, "Blackwell, W. T."

74. Gilbert and Barker Manufacturing, *Springfield Gas Machine*, 1874, 1875.

75. Sanborn Fire Insurance Company, "Durham, North Carolina, 1884," in *Digital Sanborn Maps, 1867–1970* (Ann Arbor, MI: ProQuest).

76. Herbert Packard DeForest and Edward Craig Bates, *The History of Westborough, Massachusetts* (Westborough, MA: Town of Westborough, 1891), 451.

77. Ellery Bicknell Crane, *Historic Homes and Institutions and Genealogical and Personal Memoirs of Worcester County, Massachusetts* (New York: Lewis Historical, 1907), 177.

78. Sanborn Fire Insurance Company, "Westboro, Massachusetts, 1888," in *Digital Sanborn Maps, 1867–1970* (Ann Arbor, MI: ProQuest).

79. Porter Sargent, *A Handbook of New England* (Boston: Porter E. Sargent, 1916), 283.

80. Sanborn Fire Insurance Company, "Winsted, Connecticut, 1887," in *Digital Sanborn Maps, 1867–1970* (Ann Arbor, MI: ProQuest); *American Electrical Directory of 1886: Being a Complete Book Reference of the Electric Lighting Interests of North America* (Ft. Wayne, IN: Star Iron Tower, 1886), 175.

81. Sanborn Fire Insurance Company, "Winsted, Connecticut, 1904," in *Digital Sanborn Maps, 1867–1970* (Ann Arbor, MI: ProQuest).

82. American Clock and Watch Museum, "William L. Gilbert Clock Co.," Charity Advantage, http://www.charityadvantage.com/American_Clock_and_Watch_MJHBMHC/gilbert.asp/; Gilbert and Barker Manufacturing, *Springfield Gas Machine*, 1874.

83. Sanborn Fire Insurance, "Winsted, Connecticut, 1887"; Sanborn Fire Insurance, "Winsted, Connecticut, 1904."

84. Springfield Gas Machine, *Circular of the Springfield Gas Machine Company*, 3.

85. Ibid., 4.

86. Gilbert and Barker Manufacturing, *Springfield Gas Machine*, 1870–71, 12.

87. Gilbert and Barker Manufacturing, *Springfield Gas Machine*, 1874, 26, 28.

88. Ibid., 28.

89. Gilbert and Barker Manufacturing, *Springfield Incandescent Gas Burner*, 10.

90. Sanborn Fire Insurance Company, "Rye, New York, 1892," in *Digital Sanborn Maps, 1867–1970* (Ann Arbor, MI: ProQuest).

91. Gilbert and Barker Manufacturing, *Springfield Gas Machine*, 1874.

92. Sanborn Fire Insurance, "Rye, New York, 1892."

93. Richard Hill, "A Tour of Unity Church," Unity Church of North Easton, http://www.unity-church.com/building/tour.htm/.

94. Oliver Ames Jr., "1875 Diary of Oliver Ames, Jr. 1807–1877," transcribed by Sarah L. Ames, 2002, Stonehill Industrial History Center, Stonehill College, North Easton, MA.

95. Sanborn Fire Insurance Company, "North Easton, Massachusetts, 1903," in *Digital Sanborn Maps, 1867–1970* (Ann Arbor, MI: ProQuest).

96. George R. Burris to John S. Ames, May 14, 1914, Letters, G[31/1], 1911–19, John S. Ames Collection.

97. Gilbert and Barker Manufacturing, *Springfield Gas Machine*, 1874, 27.

98. Springfield Gas Machine, *Circular of the Springfield Gas Machine Company*, 6–7.

99. Gilbert and Barker Manufacturing, *How Best to Light Our Country Homes and Resorts*, 1906, 52.

100. Sanborn Fire Insurance Company, "Hanover, New Hampshire, 1884," in *Digital Sanborn Maps, 1867–1970* (Ann Arbor, MI: ProQuest).

101. Gilbert and Barker Manufacturing, *Springfield Gas Machine*, 1874, 73–74.

102. Ibid., 74.

103. "An Historical Map of Mount Holyoke College, 1837–1987," South Hadley, Massachusetts, Alumnae Association of Mount Holyoke College, 1986.

104. Ibid.

105. Gilbert and Barker Manufacturing, *How Best to Light Our Country Homes and Resorts*, 1906, 30.

106. At the time research into Mount Holyoke's system began, no one at the college knew of its existence and use. Staff in many departments, especially the Archives and Buildings and Grounds, supported this research and expressed a great deal of interest in the project.

107. Sanborn Fire Insurance Company, "South Hadley, Massachusetts, 1905," in *Digital Sanborn Maps, 1867–1970* (Ann Arbor, MI: ProQuest); Sanborn Fire Insurance Company, "South Hadley, Massachusetts, 1910," in *Digital Sanborn Maps, 1867–1970* (Ann Arbor, MI: ProQuest).

108. Gilbert and Barker Manufacturing, *How Best to Light Our Country Homes and Resorts*, 1906, 28.

109. "Library Note," *Mount Holyoke* 14, no. 2 (Oct. 1904): 83.

110. Descending into the vault, after doing appropriate air testing for confined spaces, I was able to measure and draw the setup before the air quality detector indicated that the level of toxic fumes required my immediate departure.

111. Chest-high water in the vault prevented a more accurate measurement.

112. Barker, "Gas-Burner," Patent No. 238,205.

113. Ironically, the new oil tank had been planned for the location of the gas machine prior to its discovery. The environmental issues of old gasoline required that the machine and contaminated soil be removed prior to the start of the new project.

114. Gilbert and Barker Manufacturing, *Directions for the Care of the Springfield Gas Machine*, 1897, 10.

115. R. D. Kimball, "Map of Mt. Holyoke College Showing Electric Duct Lines and Gas Pipes" (Boston: R. D. Kimball, Engineer, July 1902).

116. Likely explaining the thin gasoline slick that would reportedly appear on the lake's surface each spring.

117. Kimball, "Map of Mt. Holyoke College."

118. The college constructed a new physics building in 1932 and renamed it Shattuck Hall in 1954.

119. Gilbert and Barker Manufacturing, *How Best to Light Our Country Homes and Resorts*, 1906, 30–31.

120. Rev. J. Wilbur Chapman, D.D., "The Life and Work of Dwight Lyman Moody," chap. 14, Bible Believers, http://www.biblebelievers.com/moody/14.html/.

121. Gilbert and Barker Manufacturing, *How Best to Light Our Country Homes and Resorts*, 1906, 41.

122. Hotchkiss School, http://www.hotchkiss.org/AboutHotchkiss/HistoryTra.asp/.

123. Gilbert and Barker Manufacturing, *How Best to Light Our Country Homes and Resorts*, 1906, 57.

124. Lois Stevenson Greenwood, *Winchendon Years, 1764–1964* (Winchendon, MA: Town of Winchendon, 1970), 121.

125. Abijah Perkins Marvin, *History of Worcester County, Massachusetts, Embracing a Comprehensive History of the County from Its First Settlement to the Present Time, with a History and Description of Its Cities and Towns* (Boston: C. F. Jewett, 1879), 544–47.

126. It bears a plaque reading "Springfield Gas Machine, Patd. Aug. 3rd, Aug. 17, 1869."

127. The Combination Mixing Regulator was originally patented in 1888 and then modified in 1897. It is the later version that is installed at Murdock.

128. Glen C. Wheeler, *Images of America: Winchendon* (Dover, NH: Arcadia, 1997), 69.

129. Brown, *Inventing New England Regional Tourism*, 1–13.

130. Gilbert and Barker Manufacturing, *Springfield Gas Machine*, 1874, 5.

131. Tolles, *Grand Resort Hotels of the White Mountains: A Vanishing Architectural Legacy* (Boston: David R. Godine, 1998), 19, 20.

132. Ibid., 67, 111.

133. Gilbert and Barker Manufacturing, *How Best to Light Our Country Homes and Resorts*, 1888; Gilbert and Barker Manufacturing, *How Best to Light Our Country Homes and Resorts*, 1906, 20.

134. Gilbert and Barker Manufacturing, *Springfield Gas Machine*, 1878, 35.

135. Dana Cudmore, *The Remarkable Howe Caverns Story* (Woodstock, NY: Overlook Press, 1990), x.

136. "Howe's Cave," *New York Times*, Aug. 4, 1874, 3.

137. Gilbert and Barker Manufacturing, *Springfield Gas Machine*, 1878, 35; Gilbert and Barker Manufacturing, *Springfield Gas Machine*, 1874, 56.

138. Reverend Horace Hovey (1896) quoted in Cudmore, *Remarkable Howe Caverns Story*, 59–60.

139. Gilbert and Barker Manufacturing, *How Best to Light Our Country Homes and Resorts*, 1888; Gilbert and Barker Manufacturing, *How Best to Light Our Country Homes and Resorts*, 1906, 24.

140. Henry R. Burt (1874) quoted in Tolles, *Grand Resort Hotels of the White Mountains*, 70.

141. Ibid., 66.

142. Gilbert and Barker Manufacturing, *Springfield Gas Machine*, 1878, 33.

143. Tolles, *Grand Resort Hotels of the White Mountains*, 66.

144. Gilbert and Barker Manufacturing, *How Best to Light Our Country Homes and Resorts*, 1906, 46.

145. Tolles, *Grand Resort Hotels of the White Mountains,* 126.

146. Ibid., 55, 58.

147. Ibid., 160, 161.

148. Gilbert and Barker Manufacturing, *How Best to Light Our Country Homes and Resorts,* 1906, 53.

149. Tolles, *Grand Resort Hotels of the White Mountains,* 213.

150. Gerald Chapman, *A History of the Red Lion Inn in Stockbridge, Massachusetts* (Stockbridge, MA: Red Lion Inn, 1987), 5, 13–14.

151. Charles W. Bryan (1886) quoted in Chapman, *History of the Red Lion Inn,* 14.

152. Ibid., 24.

153. *Pittsfield Sun* (1896) quoted in Chapman, *History of the Red Lion Inn,* 16.

154. *Valley Gleaner* (1897) quoted in Chapman, *History of the Red Lion Inn,* 17.

155. Ibid.; *Courier* (1897) quoted in Chapman, *History of the Red Lion Inn,* 17; *Valley Gleaner,* Feb. 10, 1897, quoted in Chapman, *History of the Red Lion Inn,* 22.

156. Chapman, *History of the Red Lion Inn,* 17 (kerosene lamps), and 14, 21, 32 (gas fixtures).

157. Allen Treadway (1905) quoted in Gilbert and Barker Manufacturing, *How Best to Light Our Country Homes and Resorts,* 1906, 40.

158. Choate (n.d.) quoted in Chapman, *History of the Red Lion Inn,* 23.

159. Ibid.

160. Ibid., 24.

161. Sanborn Fire Insurance Company, "Stockbridge, Massachusetts, 1911," in *Digital Sanborn Maps, 1867–1970* (Ann Arbor, MI: ProQuest).

162. Chapman, *History of the Red Lion Inn,* 26–27.

163. Both the 1885 and 1897 editions of Gilbert and Barker's *Directions for the Care of the Springfield Gas Machine* illustrate partial and fully buried versions of the generator.

164. Allen Treadway (1905) quoted in Gilbert and Barker Manufacturing, *How Best to Light Our Country Homes and Resorts,* 1906, 40.

165. A plate on the air pump is marked "Springfield Gas Machine, Gilbert and Barker Mfg. Co., Springfield, Mass., Pat. Aug 3rd and Aug 17th, 1869," and the serial number is 10557. A smaller plate reads "Patented May 4, 1875, July 29, 1890."

166. Gilbert and Barker Manufacturing, *How Best to Light Our Country Homes and Resorts,* 1906, 15.

167. *Valley Gleaner* (1897) quoted in Chapman, *History of the Red Lion Inn,* 17; *Courier* (1897) quoted in Chapman, *History of the Red Lion Inn,* 17.

168. "Henry Morrison Flagler," Flagler Museum, Palm Beach, FL, http://www.flagler.org/biography.html/.

169. Susan R. Braden, *The Architecture of Leisure: The Florida Resort Hotels of Henry Flagler and Henry Plant* (Gainesville: Univ. Press of Florida, 2002), 23–24.

170. Cruft owned the Maplewood Hotel in Bethlehem, New Hampshire (Tolles, *Grand Resort Hotels of the White Mountains,* 124). Another Florida hotel owned by Cruft was the Magnolia Hotel, built by James McGuire in the early 1880s and located near Jacksonville (Braden, *Architecture of Leisure,* 143).

171. Braden, *Architecture of Leisure,* 145. The Florida House (begun ca. 1848) in St. Augustine, although not a Flagler hotel, also used a Springfield Gas Machine (Braden, *Architecture of Leisure,* 143).

172. Braden, *Architecture of Leisure*, 145.

173. Gilbert and Barker Manufacturing, *How Best to Light Our Country Homes and Resorts*, 1906, 53.

174. Ibid., 33.

175. Braden, *Architecture of Leisure*, 209, 210–12, 215–16.

176. Ibid., 217, 218, 221.

177. Ibid., 221.

178. Gilbert and Barker Manufacturing, *How Best to Light Our Country Homes and Resorts*, 1906, 33.

179. Braden, *Architecture of Leisure*, 215, 157, 217.

180. Ibid., 33–34.

181. Ibid., 299–301, 303, 301–2, 297–99.

182. Leo C. Lillie, *Historic Grand Haven and Ottawa County* (1931; rpt., Grand Haven, MI: Tri-Cities Historical Society, 1980), 330–31.

183. Ibid., 330–31; Franklin Everett, *Memorials of the Grand River Valley* (Chicago: Chicago Legal News, 1878), 422, 335.

184. Lillie, *Historic Grand Haven and Ottawa County*, 335; Wallace K. Ewing and David H. Seibold, *Images of America: The Grand Haven Area, 1860–1960* (Chicago: Arcadia, 2002), 10.

185. Gilbert and Barker Manufacturing, *Springfield Gas Machine*, 1874, 100.

186. Sanborn Fire Insurance Company, "Grand Haven, Michigan, 1883," in *Digital Sanborn Maps, 1867–1970* (Ann Arbor, MI: ProQuest).

187. Lillie, *Historic Grand Haven and Ottawa County*, 368.

188. Ibid., 330–31.

189. *History of Muskegon County, Michigan* (Chicago: H. R. Page, 1882), 96.

190. *Grand Haven Herald*, June 29, 1872.

191. Gilbert and Barker Manufacturing, *Springfield Gas Machine*, 1874, 13.

192. William Barton McCash and June Hall McCash, *Jekyll Island Club Historic District* (Jekyll Island, GA: Jekyll Island Authority, 1986), 4.

193. Ibid., 4, 1–13.

194. The initial offering was oversubscribed and the club eventually capped membership at one hundred.

195. McCash and McCash, *Jekyll Island Club Historic District*, 4.

196. "A New Winter Club: One of the Old Sea Islands Secured by a Party of Wealthy Gentlemen," *New York Times*, Apr. 23, 1886.

197. McCash and McCash, *Jekyll Island Club*, 17.

198. William A. Cleveland quoted in McCash and McCash, *Jekyll Island Club*, 17.

199. Ibid., 18, 21.

200. Ibid., 26.

201. Sanborn Fire Insurance Company, "Brunswick, Georgia, 1893," in *Digital Sanborn Maps, 1867–1970* (Ann Arbor, MI: ProQuest); Sanborn Fire Insurance Company, "Brunswick, Georgia, 1898," in *Digital Sanborn Maps, 1867–1970* (Ann Arbor, MI: ProQuest); Sanborn Fire Insurance Company, "Brunswick, Georgia, 1908," in *Digital Sanborn Maps, 1867–1970* (Ann Arbor, MI: ProQuest).

202. McCash and McCash, *Jekyll Island Club*, 69, 72.

203. Ibid., 94, 97.

204. Springfield Gas Machine, *Circular of the Springfield Gas Machine Company.*

205. Cowan, *Social History of American Technology,* 189.

206. Gilbert and Barker Manufacturing, *Springfield Gas Machine,* 1874, 16; Gilbert and Barker Manufacturing, *Springfield Gas Machine,* 1878, 8.

207. Mount Holyoke College, *Catalogue of Mount Holyoke College in South Hadley, Mass., 1899–1900* (Springfield, MA: Press of Springfield, 1899), 59, Five College Archives Digital Access Project, http://clio.fivecolleges.edu/mhc/catalogs/.

208. Trachtenberg, *Incorporation of America,* 69.

Chapter 6

1. Wright, *Lexington,* 112–13.

2. *Visitor* (Bradford, NH), Jan. 8, 1898.

3. Olin J. Ferguson, *Electric Lighting* (New York: McGraw-Hill, 1920), 63.

4. Strasser, *Never Done,* 81.

5. David E. Nye, *Electrifying America: Social Meanings of a New Technology* (Cambridge, MA: MIT Press, 1990), 23, 24.

6. Bowers, *Lengthening the Day,* 66.

7. Hale, "Diary of an Industry," 32.

8. O'Dea, *Short History of Lighting,* 12.

9. Bowers, *Lengthening the Day,* 99. One of these new firms, the United State Electric Lighting Company, was founded in 1877 by inventor and gas machine developer Hiram S. Maxim. Maxim, who had worked as a designer and engineer of gas machines for Gilbert and Barker in their early years, was chief engineer for the company, which initially produced arc lighting systems.

10. Ibid., 150–61.

11. O'Dea, *Short History of Lighting,* 13; P. A. Redhead, "History of Vacuum Devices," pp. 284–85, National Research Council, Ottawa, Canada, http://www.chem.elte.hu/departments/altkem/vakuumtechnika/CERN23.pdf/.

12. O'Dea, *Short History of Lighting,* 58.

13. Ibid., 14.

14. Cross and Szostak, *Technology and American Society,* 177.

15. Schivelbusch, *Disenchanted Light,* 57.

16. Edison quoted in George S. Bryan, *Edison: The Man and His Work* (Garden City, NY: Garden City Publishing, 1926), 111.

17. Robert Conot, *A Streak of Luck: The Life and Legend of Thomas Alva Edison* (New York: Sea View Books, 1979), appendix.

18. "Springfield Gas Machine: New Fields for a Favorite Local Industry," 5; Warner, *Picturesque Hampden,* 150.

19. Charles Morton (1878) quoted in Phillip McGuire, "Technology and Commerce: The Gas Light Industry's Response to Edison's Electric Bulb," *Potomac Review* 6, no. 2 (1973): 70.

20. Ibid., 71.

21. Schivelbusch, *Disenchanted Night,* 60, 51.

22. Conot, *Streak of Luck,* appendix; Paul Israel, *Edison: A Life of Invention* (New York: John Wiley & Sons, 1998), 199, 218.

23. Nye, *Electrifying America*, 2.

24. J. J. Moore of Springfield, Massachusetts (1899), quoted in the *New York Times*, 1899, 9.

25. Bowers, *Lengthening the Day*, 160.

26. McGuire, "Technology and Commerce," 71.

27. C. O. Bond, "Gas as an Illuminant," *Proceedings of the American Gas Light Institute* 9 (1914): 1350–54.

28. Strasser, *Never Done*, 76.

29. Vivian B. Lewis, "The Use of Gas from a Hygienic Standpoint—Theory and Practice in Favor of Gas," *Illuminating Engineer* 2 (1907): 560.

30. Bond, "Gas as an Illuminant," 1350–54.

31. McGuire, "Technology and Commerce," 71–72.

32. Gerhard, *American Practice*, 123.

33. *Financial Times*, Mar. 21, 1887, quoted in Schivelbusch, *Disenchanted Night*, 48.

34. Schivelbusch, *Disenchanted Night*, 48; Moss, *Lighting for Historic Buildings*, 102–3.

35. Myers, *Gaslighting in America*, 206.

36. Ibid.

37. Gilbert and Barker Manufacturing, *How Best to Light Our Country Homes and Resorts*, 1906, 17–19, 45.

38. Boston Consolidated Gas Company, "Gas Equipment in the Ideal Home," flyer included with letter from Boston Consolidated Gas Company to J. A. Ames, Aug. 3, 1916, Letters, B[123/1] 1916–19, Stonehill Industrial History Center, Stonehill College, North Easton, MA.

39. Boston Consolidated Gas Company to J. A. Ames (letter), Aug. 3, 1916, Letters, B[123/1] 1916–19, Stonehill Industrial History Center, Stonehill College, North Easton, MA.

40. Strasser, *Never Done*, 72.

41. Boston Consolidated Gas Company to J. A. Ames.

42. Gilbert and Barker Manufacturing Company to Owners of Gas Machines, Apr. 21, 1911, F302, Incoming Correspondence, 1906—Gi to Gre, Arnold B. Tofias Industrial Archives, Stonehill Industrial History Center, Stonehill College, North Easton, MA (hereafter cited as Tofias Industrial Archives).

43. Gilbert and Barker Manufacturing, *Are You Aware?*

44. Nye, *Electrifying America*, 145, 176.

45. Tirrill Gas Machine Lighting Company, "Tirrill's Equalizing Gas Machine," advertisement, *Country Life in America*, Apr. 1903; Nye, *Electrifying America*, 145.

46. Sanborn Fire Insurance Company, "Ellicott City, Maryland, 1887 and 1894," in *Digital Sanborn Maps, 1867–1970* (Ann Arbor, MI: ProQuest).

47. Erik Larson, *The Devil in the White City: Murder, Magic, and Madness at the Fair that Changed America* (New York: Crown, 2003), 254; Trachtenberg, *Incorporation of America*, 209.

48. Gilbert and Barker Manufacturing, *How Best to Light Our Country Homes and Resorts*, 1906, 33.

49. Sanborn Fire Insurance Company, "North Hadley, Massachusetts, 1905 and 1929," in *Digital Sanborn Maps, 1867–1970* (Ann Arbor, MI: ProQuest).

50. Linebaugh, O'Malley, and Duwan, "'Brilliant and Pleasant Light': Investigating the Springfield Gas Machine," 49.

51. Nye, *Electrifying America*, 17.

52. Ibid., 13, 193.

53. Gilmore Electric Company to Mr. S. S. Erley (c/o Oliver Ames & Sons Corp.), Sept. 22, 1905, F302, Incoming Correspondence, 1906—Gi to Gre, Tofias Industrial Archives.

54. Joseph J. Schroeder Jr., ed., *Sears, Roebuck & Co. 1908 Catalogue No. 117* (Northfield, IL: Digest Books, 1971), 363–66; Joseph J. Schroeder Jr., ed., *1923 Sears, Roebuck Catalogue* (Northfield, IL: Digest Books, 1973), 731.

55. Gilbert and Barker Manufacturing Company to E. F. Gebhardt.

56. Gilbert and Barker Manufacturing Company, *Vent*, Apr. 1932, 8.

57. Gilbert and Barker Manufacturing Co., *The Gilbarco Story*; Hidy and Hidy, *History of Standard Oil Company*, 162.

58. Gilbert and Barker Manufacturing Company to E. F. Gebhardt.

59. Hidy and Hidy, *History of Standard Oil Company*, 162.

60. Gilbert and Barker Manufacturing, *Gilbarco Story*.

61. Cummings, "From Appomattox to Versailles" (1919), 3.

62. Gilbert and Barker Manufacturing, *Gilbarco Story*.

63. Ibid.

64. Cummings, "From Appomattox to Versailles" (1919), 4, 5.

65. Henrietta M. Larson, Evelyn H. Knowlton, and Charles S. Popple, *History of Standard Oil Company (New Jersey): New Horizons, 1927–1950* (New York: Harper & Brothers, 1971), 276.

66. Standard Oil Company of New Jersey, advertisement, *Washington Post*, June 16, 1932, 10.

67. Andrew P. Baird, ed., "Keep on Keeping on," *Vent* 16, no. 4 (Jan. 1932): 8; Gilbert and Barker Manufacturing, *Gilbarco Story*.

68. Gilbert and Barker Manufacturing, *Gilbarco Story*; Hidy and Hidy, *History of Standard Oil Company*, 435.

69. J. N. Cummings. "From Appomattox to Versailles," *Vent* 4, no. 1 (Jan. 1920): 10.

70. Gilbert and Barker Manufacturing, *Gilbarco Story*.

71. Bob Lee, *Gilbert & Barker Gilbarco, Inc.: A Pictorial History of the Growth of a Great Company* (Detroit: Harlo Press, 1989), 11; Sanborn Fire Insurance Company, "Springfield, MA, 1911," in *Digital Sanborn Maps, 1867–1970* (Ann Arbor, MI: ProQuest); Sanborn Fire Insurance Company, "Springfield, MA, 1931," in *Digital Sanborn Maps, 1867–1970* (Ann Arbor, MI: ProQuest).

72. George Sweet Gibb and Evelyn H. Knowlton, *History of Standard Oil Company (New Jersey): The Resurgent Yeats, 1911–1927* (New York: Harper & Brothers, 1956), 5–7.

73. "R. H. M'Nall Dead: Industrial Figure," *New York Times*, Apr. 14, 1930, 17; Gibb and Knowlton, *History of Standard Oil*, 9.

74. Gibb and Knowlton, *History of Standard Oil*, 133.

75. See John A. Jakle and Keith A. Sculle, *The Gas Station in America* (Baltimore: Johns Hopkins Univ. Press, 1994), 48–84, for a discussion of oil companies expansion into distribution and retailing.

76. Gibb and Knowlton, *History of Standard Oil*, 184.

77. Lee, *Gilbert & Barker Gilbarco*, 12.

78. Messick, "Gas Machines Started Local Concern," 3E.

79. Gilbert and Barker Manufacturing, *Gilbarco Story*; Lee, *Gilbert & Barker Gilbarco*, 12.

80. Gibb and Knowlton, *History of Standard Oil*, 566, 686.

81. Gilbert and Barker Manufacturing, *Gilbarco Story*; "Gilbert & Barker, Special to the Wall Street Journal," *Wall Street Journal*, Dec. 13, 1930, 12.

82. Larson, Knowlton, and Popple, *History of Standard Oil Company (New Jersey)*, 517–18.

83. Lee, *Gilbert & Barker Gilbarco*, 13.

84. Messick, "Gas Machines Started Local Concern," 3E.

85. Lee, *Gilbert & Barker Gilbarco*, 14.

86. "Business Brief: Gilbarco, Inc.," *Wall Street Journal*, July 28, 1987, 1.

87. Lee, *Gilbert & Barker Gilbarco*, 259–63.

88. Messick, "Gas Machines Started Local Concern," 3E.

89. Gilbarco, "Old Indestructible," 12–13.

BIBLIOGRAPHY

Primary Sources

Catalogs and Pamphlets

Dale Company. *Combination Fixtures (Gas and Electric).* Catalog, n.d. Shelburne Farms Catalogue Collection, Farm Managers' Papers Collection. Shelburne Farms Archives, Shelburne, VT.

E. P. Gleason Manufacturing Company. *Combination and Electric Only Lighting Fixtures.* Catalog, n.d. Shelburne Farms Catalogue Collection, Farm Managers' Papers Collection. Shelburne Farms Archives, Shelburne, VT.

Excelsior Gas Machine Company. *The Excelsior Gas Machine, Described and Compared.* Catalog. New York: Less & Skeen, 1872.

General Acetylene Company. *Daylight by Night: The General Acetylene Generator.* New York: Press of Robert L. Stillson Co., ca. 1905.

Gilbert and Barker Manufacturing Company. *The Springfield Gas Machine, Gilbert and Barker Manufacturing Co.* Catalog. Springfield, MA: Gilbert and Barker Manufacturing, 1870–71.

———. *The Springfield Gas Machine.* Catalog. New York: Excelsior Printing, 1874.

———. *The Springfield Gas Machine.* Catalog. New York: Wheat and Cornett, 1875.

———. *The Springfield Gas Machine.* Catalog. New York: Wheat and Cornett, 1878.

———. *The Springfield Gas Machine.* Catalog. Springfield, MA: Gilbert and Barker Manufacturing, 1879.

———. *Gilbert & Barker M'f'g Co's Price List.* Pamphlet, July 1, 1882. New York: Gilbert and Barker Manufacturing, 1882.

———. *The New Feed Springfield Gas Machine.* Pamphlet, Mar. 1, 1883. Springfield, MA: Gilbert and Barker Manufacturing, 1883.

———. *Directions for the Care of the Springfield Gas Machine.* Pamphlet, May 1, 1885. Springfield, MA: Gilbert and Barker Manufacturing, 1885. Box 6, Folder 648a. Catalogue Collection, Farm Managers' Papers Collection. Shelburne Farms Archives, Shelburne, VT.

———. *How Best to Light Our Country Homes and Resorts.* Catalog. New York: Press of Fleming, Brewster & Alley, 1888. Box 6, Folder 648a. Catalogue Collection, Farm Managers' Papers Collection. Shelburne Farms Archives, Shelburne, VT.

———. *Air Pump of the Springfield Gas Machine.* Pamphlet, May 1, 1890. Box 6, Folder 648b. Shelburne Farms Bill Collection, Farm Managers' Papers Collection. Shelburne Farms Archives, Shelburne, VT.

———. *Directions for the Care of the Springfield Gas Machine.* Pamphlet, May 15, 1897. Springfield, MA: Gilbert and Barker Manufacturing, 1897.

———. *The Combination Mixing Regulator.* Pamphlet. Springfield, MA: Gilbert and Barker Manufacturing, ca. 1900. Box 6, Folder 648a. Catalogue Collection, Farm Managers' Papers Collection. Shelburne Farms Archives, Shelburne, VT.

———. *How Best to Light Our Country Homes and Resorts.* Catalog. Springfield, MA: Gilbert and Barker Manufacturing, 1906.

———. *Are You Aware? Some Suggestions Relating to Gas for Light and Fuel.* Pamphlet, n.d. (post-1895). Box 6, Folder 648a. Catalogue Collection, Farm Managers' Papers Collection. Shelburne Farms Archives, Shelburne, VT.

———. *The Combination Mixing Regulator.* Catalog, n.d. Box 6, Folder 648b. Catalogue Collection, Farm Managers' Papers Collection. Shelburne Farms Archives, Shelburne, VT.

———. *Our Automatic Instantaneous Water Heater.* Pamphlet, n.d. Box 6, Folder 648b. Catalogue Collection, Farm Managers' Papers Collection. Shelburne Farms Archives, Shelburne, VT.

———. *Springfield Incandescent Gas Burner.* Catalog, n.d. Box 6, Folder 648a. Catalogue Collection, Farm Managers' Papers Collection. Shelburne Farms Archives, Shelburne, VT.

Maryland Portable Gas Company. *Gas Apparatus of the Maryland Portable Gas Company.* Catalog. Baltimore: Murphy, 1855. Trade Catalog Collection, Baker Library, Harvard Business School, Cambridge, MA.

Maxim Gas Machine Company. *The Maxim Steam Gas Machine for Lighting Buildings, Towns and Cities.* Catalog. New York: Maxim Gas Machine, 1873.

Springfield Gas Machine Company. *Circular of the Springfield Gas Machine Company of Springfield, Mass.: Manufacturers of Portable Gas Machines and Contractors for the Erection of Gas Works . . .* trade catalog. Springfield, MA: Samuel Bowles, 1867. Trade Catalog Collection, Baker Library, Harvard Business School, Cambridge, MA.

Sunlight Gas Machine Company. *Acetylene: The Catalogue of the Sunlight Gas Machine Company.* New York: Sunlight Gas Machine, 1906. Trade Catalog Collection, Baker Library, Harvard Business School, Cambridge, MA.

Tirrill, O. *The Underground Gas Machine.* Catalog. New York: Underground Gas Machine, 1871.

Walworth Manufacturing Company. *The Walworth Gas Machine.* Boston: Gunn, Bliss, 1880.

Letters, Papers, and Newspaper Articles

Ames, Oliver, Jr. "1875 Diary of Oliver Ames, Jr. 1807–1877." Transcribed by Sarah L. Ames, 2002. Stonehill Industrial History Center, Stonehill College, North Easton, MA.

Boston Consolidated Gas Company. "Gas Equipment in the Ideal Home." Flyer included with letter from Boston Consolidated Gas Company to J. A. Ames. Aug. 3, 1916. Letters, B[123/1] 1916–19. Stonehill Industrial History Center, Stonehill College, North Easton, MA.

"Bradish Johnson." Obituary. *New York Times,* Nov. 5, 1892.

Burris, George R., to John S. Ames. May 14, 1914. Letters, G[31/1], 1911–19. John S. Ames Collection. Stonehill Industrial History Center, Stonehill College, North Easton, MA.

Burris, George R., to J. McKion. May 2, 1919. Letters, G[31/1], 1911–19. John S. Ames Collection. Stonehill Industrial History Center, Stonehill College, North Easton, MA.

"Business Brief: Gilbarco, Inc." *Wall Street Journal,* July 28, 1987, 1.

"The Centennial Awards." *New York Times,* Oct. 7, 1876, 3.

"Charles Gilbert Dies at Home at Great Barrington." *Berkshire Courier,* Jan. 18, 1934.

Classified advertisement. *New York Times*, Apr. 25, 1875, 5.
———. *New York Times*, Apr. 24, 1881, 10.
———. *New York Times*, Jan. 30, 1887, 11.
———. *New York Times*, May 31, 1890, 7.
———. *New York Times*, Apr. 1, 1892, 7.
———. *New York Times*, Apr. 11, 1903, 14.
Clough v. Barker (1882). In *Cases Argued and Adjudged in the Supreme Court of the United States. October Term, 1882*, by William T. Otto. Vol. 16. New York: Banks Law, 1904.
Clough v. Manufacturing Co. (1882). In *Cases Argued and Adjudged in the Supreme Court of the United States. October Term, 1882*, by William T. Otto, 178–81. Vol. 16. New York: Banks Law, 1904.
"Coates' Improved Portable as Works for the Country." *New York Times*, Apr. 29, 1858, 8.
"Death of Mr. William Sloane." *New York Times*, May 24, 1879, 12.
Detroit Heating and Lighting Company to H. C. McDowell, Esq. Nov. 5, 1889. Box 60, H. C. McDowell Correspondence, 1889. McDowell Family Papers. Special Collections. M. I. King Library, University of Kentucky, Lexington.
"A Fine Display at Sloanes." *New York Times*, Oct. 21, 1889, 5.
"Fire Dangers—Carelessness in Handling Gasoline." *New York Times*, Feb. 3, 1873, 5.
"A Gasoline Blow-Up. Terrible Explosion and Loss of Life." *Washington Post*, Mar. 11, 1878, 1.
"A Gentleman's Country Place." *New York Times*, Mar. 29, 1903, 21.
Gilbert and Barker Manufacturing Company. *Vent*, Apr. 1932, 8.
Gilbert and Barker Manufacturing Company to E. F. Gebhardt. Oct. 16, 1916. Box 7, Folder 81. Shelburne Farms Bill Collection, Farm Managers' Papers Collection. Shelburne Farms Archives, Shelburne, VT.
Gilbert and Barker Manufacturing Company to O. Ames and Son Corporation, North Easton, Massachusetts. Invoice. Aug. 1, 1904. Ames Family Collection. Stonehill Industrial History Center, Stonehill College, North Easton, MA.
Gilbert and Barker Manufacturing Company to Owners of Gas Machines. Apr. 21, 1911. F302, Incoming Correspondence, 1906—Gi to Gre. Arnold B. Tofias Industrial Archives. Stonehill Industrial History Center, Stonehill College, North Easton, MA.
Gilbert and Barker Manufacturing Company to Shelburne Farms. Invoices, Apr. 15, 1913, Sept. 9, 1913, June 12, 1914, Sept. 15, 1914, Apr. 22, 1915, and Nov. 8, 1915. Box 7, Folder 81. Shelburne Farms Bill Collection, Farm Managers' Papers Collection. Shelburne Farms Archives, Shelburne, VT.
Gilbert and Barker Manufacturing Company to the Oliver Ames and Son Corporation, North Easton, Massachusetts. Jan. 6, 1904. Ames Family Collection. Stonehill Industrial History Center, Stonehill College, North Easton, MA.
"Gilbert & Barker, Special to the Wall Street Journal." *Wall Street Journal*, Dec. 13, 1930, 12.
Gilbert, Charles. Alumni Records, Beloit College. Oct., 19, 1928. Beloit College Archives, Beloit, WI.
Gilmore Electric Company to Mr. S. S. Erley (c/o Oliver Ames & Sons Corp.). Sept. 22, 1905. F302, Incoming Correspondence, 1906—Gi to Gre. Arnold B. Tofias Industrial Archives. Stonehill Industrial History Center, Stonehill College, North Easton, MA.
Grand Haven Herald. June 29, 1872.

Detroit Free Press. "The Combination Gas Machine." *Detroit Free Press* 42, no. 229 (May 13, 1877): 1.

"H. H. Cook Bequeaths Millions in Cash." *New York Times*, Oct. 25, 1905.

"An Historical Map of Mount Holyoke College, 1837–1987." South Hadley, Massachusetts, Alumnae Association of Mount Holyoke College, 1986.

"Howe's Cave." *New York Times*, Aug. 4, 1874, 3.

"Inventory, 'Hillstead,' Alfred A. Pope, Farmington, Conn., July 31, 1909." Alfred Atmore Pope Collection. Hill-Stead Museum, Farmington, CT.

John C. Paige and Company, Boston. "Insurance Policy on Buildings, Contents and Other Property Situated in North Easton , Mass.—John S. Ames. September 21st, 1919." John S. Ames Collection. Stonehill Industrial History Center, Stonehill College, North Easton, MA.

Journal of Outdoor Life, Travel, Nature Study, Shooting, Fishing, Yachting 38, no. 22 (June 2, 1892): 536.

Kentucky Railway Company to Anne Clay McDowell. "Contract for Electric Service." May 28, 1907. Archives, Henry Clay Memorial Foundation, Lexington, KY.

Kentucky University. Minutes of the Board of Curators (1866–77). Special Collections. Transylvania University, Lexington, KY.

"Library Note." *Mount Holyoke* 14, no. 2 (Oct. 1904).

"Light Your Paper Mill with a Springfield Gas Machine." Advertisement. *Paper World*, Oct. 1882, 39.

Loring W. Puffer and Sons to John S. Ames. Insurance policy, Sept. 21, 1919. Letters, G[31/1], 1911–19. John S. Ames Collection. Stonehill Industrial History Center, Stonehill College, North Easton, MA.

Loring W. Puffer and Sons Insurance Agency to John S. Ames. Sept. 18, 1919. John S. Ames Collection. Stonehill Industrial History Center, Stonehill College, North Easton, MA.

McDowell, Henry C., Jr. to Anne Clay McDowell. Dec. 23, 1906. McDowell Family Papers. Special Collections. M. I. King Library, University of Kentucky, Lexington.

———. May 29, 1907. McDowell Family Papers. Special Collections. M. I. King Library, University of Kentucky, Lexington.

Messick, J. Holton. "Gas Machines Started Local Concern on 75 Years of Expanding Production." *Springfield Sunday Union and Republican*, Jan. 28, 1940, 3E.

Moore, J. J., Springfield, Massachusetts (1899). Quoted in the *New York Times*, 1899, 9.

Mount Holyoke College. *Catalogue of Mount Holyoke College in South Hadley, Mass., 1899–1900.* Springfield, MA: Press of Springfield, 1899. Five College Archives Digital Access Project, http://clio.fivecolleges.edu/mhc/catalogs/.

"A New Winter Club: One of the Old Sea Islands Secured by a Party of Wealthy Gentlemen." *New York Times*, Apr. 23, 1886.

New York, New Haven, and Hartford Railroad Company to Cyrus Lothrop. Invoice, Jan. 7, 1911. Stonehill Industrial History Center, Stonehill College, North Easton, MA.

Nye and Acheson. Advertisement for the Springfield Gas Machine. *Santa Cruz Sentinel* 3 (Oct. 2, 1875): 1.

Pope, Theodate, to Alfred Pope. June 28, 1899, or 1900. No. 722. Hill-Stead Museum Archives, Farmington, CT.

Ramsdell, C. C., to E. F. Gebhardt. Oct. 16, 1916. Box 7, Folder 81. Shelburne Farms Bill Collection, Farm Managers' Papers Collection. Shelburne Farms Archives, Shelburne, VT.

R. G. Dun and Company. Massachusetts. Vols. 41 and 42. R. G. Dun and Company Collection, Baker Library, Harvard Business School, Cambridge, MA.

"R. H. M'Nall Dead: Industrial Figure." *New York Times*, Apr. 14, 1930, 17.

Robbe, C.A., to Alexander Stephens. Nov. 1, 1875. Photocopy on file at Alexander Stephens State Park, Crawfordville, GA.

———. "Statement for Gas Fixtures." Nov. 1, 1875. Photocopy on file at Alexander Stephens State Park, Crawfordville, GA.

"Scientific Methods Applied to Bicycling-Making." *Harper's Magazine Advertiser*, 1895, 29–32.

"South Wall, Room A-7." Photograph (black and white), 1977–78. Robert Toombs House State Historic Site, Washington, GA.

"The Springfield Gas Machine." Advertisement. *Appletons' Journal of Literature, Science and Art*, May 29, 1869, 286.

"The Springfield Gas Machine." Advertisement. *Scientific American*, May 29, 1869, 352.

"The Springfield Gas Machine." *Paper World*, Oct. 1882, 28.

"The Springfield Gas Machine: New Fields for a Favorite Local Industry." *Springfield Sunday Republican*, Nov. 11, 1888.

Standard Oil Company of New Jersey. Advertisement. *Washington Post*, June 16,1932, 10.

Stephens, Alexander H., to C. A. Robbe. Sept. 21, 1875. On file at East Central Georgia Public Library, Augusta.

"Supreme Court Decisions." *Washington Post*, Nov. 11, 1879, 3.

Tirrill Gas Machine Lighting Company. "Tirrill's Equalizing Gas Machine." Advertisement. *Country Life in America*, Apr. 1903.

Visitor (Bradford, NH), Jan. 8, 1898.

Webster Gasolene Company to H. C. McDowell, Esq. Mar. 27 and Apr. 10, 1889. Box 60, H. C. McDowell Correspondence, 1889. McDowell Family Papers. Special Collections. M. I. King Library, University of Kentucky, Lexington.

"The Welsbach Incandescent Gas Burner." Advertisement. *New York Times*, June 20, 1888, 2.

Wheelock, J. W., to J. G. Benton. Sept. 22, 1866. Springfield Armory National Historic Site, Springfield, MA.

"Windsor Arcade, Famous Fifth Avenue Landmark, Giving Way to Great Commercial Structure." *New York Times*, May 7, 1911.

Maps and Drawings

G. H. Bailey and Company. "View of Springfield, Mass., 1875." Boston: G. H. Bailey, 1875.

Gilbert and Barker Manufacturing Company. "Plan and Elevation Showing Two Hundred Light Springfield Gas Machine with a Ten Barrel Storage Tank, June 29, 1900, Order #590." Hill-Stead Museum Archives, Farmington, CT.

———. "Plan Drawing." N.d. Shelburne Farms Archives, Shelburne, VT.

Kimball, R. D. "Map of Mt. Holyoke College Showing Electric Duct Lines and Gas Pipes." Boston: R. D. Kimball, Engineer, July 1902.

Sanborn Fire Insurance Company. *Digital Sanborn Maps, 1867–1970*. Ann Arbor, MI: ProQuest.

Shepley, Rutan and Coolidge, Architects. "F. L. Ames Esq. Residence, N. Easton." Architectural drawings and plans, n.d. Microfilm. Widener Library, Harvard University.

Patents

Archer, Ellis S. "Improvement in Apparatus for Carbureting Air." Patent No. 44,060. Sept. 6, 1864. Washington, DC: U.S. Patent Office.

Bancroft, N. W. "Improved Gas-Machine." Patent No. 82,786. Oct. 6, 1868. Washington, DC: U.S. Patent Office.

Barker, John F. "Improvement in Gas-Burners." Patent No. 105,768. July 26, 1870. Washington, DC: U.S. Patent Office.

———. "Improvement in Gas-Burners." Patent No. 186,920. Feb. 6, 1877. Washington, DC: U.S. Patent Office.

———. "Improvement in Air-Forcing Apparatus." Patent No. 221,385. Nov. 11, 1879. Washington, DC: U.S. Patent Office.

———. "Gas-Burner." Patent No. 231,256. Aug. 17, 1880. Washington, DC: U.S. Patent Office.

———. "Apparatus for Mixing Aeriform Fluids." Patent No. 234,904. Nov. 30, 1880. Washington, DC: U.S. Patent Office.

———. "Gas-Burner." Patent No. 238,205. Mar. 1, 1881. Washington, DC: U.S. Patent Office.

———. "Apparatus for Mixing Aeriform Fluids." Patent No. 252,009. Jan. 10, 1882. Washington, DC: U.S. Patent Office.

———. "Apparatus for Mixing Aeriform Fluids." Patent No. 254,589. Mar. 7, 1882. Washington, DC: U.S. Patent Office.

———. "Rotary Air-Pump." Patent No. 341,871. May 18, 1886. Washington, DC: U.S. Patent Office.

———. "Gas-Burning Device." Patent No. 407,323. July 23, 1889. Washington, DC: U.S. Patent Office.

———. "Laboratory Burner, Specifically Useful with Gasoline Gas." Patent No. 423,896. Mar. 25, 1890. Washington, DC: U.S. Patent Office.

———. "Air-Pump." Patent No. 433,064. July 29, 1890. Washington, DC: U.S. Patent Office.

———. "Gas-Burner." Patent No. 434,013. Aug. 12, 1890. Washington, DC: U.S. Patent Office.

———. "Gas-Burner." Patent No. 575,318. Jan. 19, 1897. Washington, DC: U.S. Patent Office.

———. "Gas-Burner." Patent No. 628,063. July 4, 1899. Washington, DC: U.S. Patent Office.

———. "Gas-Burner." Patent No. 637,441. Nov. 21, 1899. Washington, DC: U.S. Patent Office.

———, Assignor to the Gilbert and Barker Manufacturing Company. Patent No. 115,562. June 6, 1871. Washington, DC: U.S. Patent Office.

———. "Gas-Burner." Patent No. 628,064. July 4, 1899. Washington, DC: U.S. Patent Office.

———. "Gas-Burner." Patent No. 628,065. July 4, 1899. Washington, DC: U.S. Patent Office.

Barker, John F., and Charles N. Gilbert. "Improved Apparatus for Cabureting Air and Gas." Patent No. 66,777. July 16, 1867. Washington, DC: U.S. Patent Office.

———. "Improved Apparatus for Cabureting Air." Patent No. 93,267. Aug. 3, 1869. Washington, DC: U.S. Patent Office.

———. "Improved Apparatus for Cabureting Air." Patent No. 93,795. Aug. 17, 1869. Washington, DC: U.S. Patent Office.

Bassett, John A. "Improved Apparatus for Carbureting Gas." Patent No. 39,541. Mar. 18, 1863. Washington, DC: U.S. Patent Office.

Burdick, Edwin A., Assignor to the Gilbert and Barker Manufacturing Company. "Apparatus for Reducing the Gravity of Hydrocarbon Gas." Patent No. 244,850. July 26, 1881. Washington, DC: U.S. Patent Office.

Cunningham, Charles. "Benzole Vapor Apparatus." Patent No. 12,535. Mar. 13, 1855. Washington, DC: U.S. Patent Office.

Fogarty, Thomas B. "Improvement in Gas Machines." Patent No. 179,652. July 11, 1876. Washington, DC: U.S. Patent Office.

Foster, William, Jr., and George P. Ganster. "Improved Portable Gas Apparatus." Patent No. 84,941. Dec. 15, 1868. Washington, DC: U.S. Patent Office.

Frost, Edward J. "Device for Regulating the Quality of Cabureted Vapor or Gas." Patent No. 381,619. Apr. 24, 1888. Washington, DC: U.S. Patent Office.

Frost, Edward J., and Benjamin Middleditch, Assignors to Joseph H. Berry. "Device for Regulating Quality of Gas." Patent No. 577,739. Feb. 23, 1897. Washington, DC: U.S. Patent Office.

Gilbert, Charles N., John F. Barker, and E. N. Ives, Assignors to the New England Portable Gas Works Company. "Improvement in Apparatus for Cabureting Air." Patent No. 61,004. Jan. 8, 1867. Washington, DC: U.S. Patent Office.

Hewson, Robert. "Improvement in Automatic Air-Blasts for Gas Machines." Patent No. 162,922. May 4, 1875. Washington, DC: U.S. Patent Office.

Hovey, Charles F., Assignor to the Gilbert and Barker Manufacturing Company. "Apparatus for Reducing the Density of Illuminating-Gases." Patent No. 273,843. Mar. 13, 1883. Washington, DC: U.S. Patent Office.

Kitchen, George H., and Scotto C. Nash. "Improved Portable Gas Apparatus." Patent No. 84,283. Nov. 24, 1868. Washington, DC: U.S. Patent Office.

Maxim, Hiram. "Improved Gas-Machine." Patent No. 81,922. Sept. 8, 1868. Washington, DC: U.S. Patent Office.

———. "Gas Machine." Patent No. 245,736. Aug. 16, 1881. Washington, DC: U.S. Patent Office.

Maxim, Hiram, and James Radley. "Improved Apparatus for Making Illuminating-Gas from Gasoline." Patent No. 89,588. May 4, 1869. Washington, DC: U.S. Patent Office.

Miscellaneous patents for acetylene gas machines, 1895–1903. Washington, DC: U.S. Patent Office, 1895–1903.

Ofeldt, Frank W. "Improvement in Carburetors." Patent No. 131,369. Sept. 17, 1872. Washington, DC: U.S. Patent Office.

———. "Improvement in Gas Apparatus for Carbureting Air." Patent No. 162,848. May 4, 1875. Washington, DC: U.S. Patent Office.

Prichard, Iva. "Improvement in Gas-Machines." Patent No. 74, 132. Feb. 4, 1868. Washington, DC: U.S. Patent Office.

Sprang, Jacob D. "Improved Gas-Machine." Patent No. 85,185. Dec. 22, 1868. Washington, DC: U.S. Patent Office.

Stebbins, M. C. "Improved Portable Gas-Apparatus." Patent No. 83,419. Oct. 27, 1868. Washington, DC: U.S. Patent Office.

Secondary Sources

Accum, Frederick Christian. *A Practical Treatise on Gas-Light*. London: R. Ackerman, 1815.

American Clock and Watch Museum. "William L. Gilbert Clock Co." Charity Advantage. http://www.charityadvantage.com/American_Clock_and_Watch_MJHBMHC/gilbert.asp/.

American Electrical Directory of 1886: Being a Complete Book Reference of the Electric Lighting Interests of North America. Ft. Wayne, IN: Star Iron Tower, 1886.

"American Optical Manufacturing Company of Southbridge, Massachusetts." Antique Spectacles. http://www.antiquespectacles.com/american_optical/american_optical.htm/.

Anderson, Frederick Paul. *Specifications for Lighting by Electricity or by Electricity and Gas, the Streets, Alleys, Squares and Public Places of the City of Lexington, Kentucky.* Lexington, KY: Jas. M. Byrnes, 1896.

———. *General Specifications for the Installation of Electric Light, Power, Heating and Steam Plants in the Court House of Fayette County, Lexington, KY.* Lexington, KY, n.d.

Archer and Warner. *A Familiar Treatise on Candles, Lamps and Gas Lights: With Incidental Matters, Prepared for the Use of Their Customers, by Archer and Warner, Manufacturers of Gas Fixtures, Chandeliers, Lamps, Girandoles, etc.* Philadelphia: Archer and Warner, 1850.

Baird, Andrew P., ed. "Keep on Keeping on." *Vent* 16, no. 4 (Jan. 1932): 8.

Barraclough, Kenneth C. *Steelmaking Before Bessemer.* London: Metals Society, 1984.

Beecher, Catharine E., and Harriet Beecher Stowe. *The American Woman's Home.* 1869. Rpt. Watkins Glen, NY: American Life Foundation, 1975.

Benjamin, Park. *Appletons' 'Cyclopaedia of Applied Mechanics: A Dictionary of Mechanical Engineering and Mechanical Arts.* New York: D. Appleton, 1880.

Biographical Review. *The Leading Citizens of Hampden County, Massachusetts.* Boston: Biographical Review, 1895.

Bond, C. O. "Gas as an Illuminant." *Proceedings of the American Gas Light Institute* 9 (1914): 1350–54.

Bowers, Brian. *Lengthening the Day: A History of Lighting Technology.* Oxford: Oxford Univ. Press, 1998.

Braden, Susan R. *The Architecture of Leisure: The Florida Resort Hotels of Henry Flagler and Henry Plant.* Gainesville: Univ. Press of Florida, 2002.

Breckinridge, Sophonisba Preston. *Madeline McDowell Breckinridge: A Leader in the New South.* Chicago: Univ. of Chicago Press, 1921.

Brown, Dona. *Inventing New England: Regional Tourism in the Nineteenth Century.* Washington, DC: Smithsonian Institution Press, 1995.

Brown, Robert F., ed. *History of Easton, Massachusetts.* Vol. 2, *1886–1974.* Easton, MA: Easton Historical Society, 1974.

Bridger, Beverly. *Mrs. Vanderbilt's Sagamore.* Raquette Lake, NY: Sagamore Institute of the Adirondacks, n.d.

Bryan, Clark W. *The Book of Berkshire: Describing and Illustrating its Hills and Homes.* Great Barrington, MA: Clark W. Bryan, 1886.

Bryan, George S. *Edison: The Man and His Work.* Garden City, NY: Garden City Publishing, 1926.

Bryant, James W., ed. *Mining and Petroleum Standard and American Gas-Light Journal,* Nov. 2, 1864, 136.

Campbell, Marcia. Personal communication. June 15, 2005.

Carson, Barbara G. *Ambitious Appetites: Dining, Behavior and Patterns of Consumption in Federal Washington.* Washington, DC: American Institute of Architects Press, 1990.

Castaneda, Christopher J. *Invisible Fuel: Manufactured and Natural Gas in America, 1800–2000.* New York: Twayne, 1999.

Chandler, Alfred D., Jr. *Scale and Scope: The Dynamics of Industrial Capitalism*. Cambridge: Harvard Univ. Press, 1990.

Chapman, Gerald. *A History of the Red Lion Inn in Stockbridge, Massachusetts*. Stockbridge, MA: Red Lion Inn, 1987.

Chapman, Rev. J. Wilbur, D.D. "The Life and Work of Dwight Lyman Moody," chap. 14. Bible Believers. http://www.biblebelievers.com/moody/14.html/.

Childs, Marquis. *The Farmer Takes a Hand: The Electric Power Revolution in Rural America*. Garden City, NY: Doubleday, 1961.

Christian, Robert. *On the construction of oil and coal gas burners and the circumstances that influence the light emitted by the gases during their combustion: with some observations on their relative illuminating power, and on the different modes of ascertaining it*. Edinburgh: A. Constable, 1825.

Clark, Clifford E., Jr. "Domestic Architecture as an Index to Social History: The Romantic Revival and the Cult of Domesticity in America, 1840–1870." In *Material Life in America, 1600–1860*, ed. Robert Blair St. George, 535–49. Boston: Northeastern Univ. Press, 1986.

Clark, Thomas D., Robert Magrish, and Terry L. Green. *Ashland: The Henry Clay Estate*. Greendale, IN: Creative, 1996.

Conot, Robert. *A Streak of Luck: The Life and Legend of Thomas Alva Edison*. New York: Seaview Books, 1979.

Cooke, Lawrence S., ed. *Lighting in America from Colonial Rushlights to Victorian Chandeliers*. New York: University Books, 1860.

Cooke, William, to Col. T. T. Saidley. Feb. 16, 1866; William Cooke to Col. T. T. Saidley, Apr. 4, 1866; and William Cooke to Col. J. G. Benton, Mar. 11, 1867. Springfield Armory National Historic Site, Springfield, MA.

Cornelius and Baker. *Description of the Establishment of Cornelius and Baker, Manufacturers of Lamps, Chandeliers and Gas Fixtures*. Philadelphia: J. B. Chandler, 1860.

———. *Examples of Gas Fixtures and Other Metal Works of Ecclesiastical and Domestic Use Designed after the Manner of Mediaeval Art Works by J. M. Beesley*. Philadelphia: Cornelius, 1977.

Cowan, Ruth Schwartz. "The 'Industrial Revolution' in the Home: Household Technology and Social Change in the Twentieth Century." In *Material Culture Studies in America*, ed. Thomas J. Schlereth, 222–36. Nashville: American Association for State and Local History, 1982.

———. *More Work for Mother: The Ironies of Household Technology from the Open Hearth to the Microwave*. New York: Basic Books, 1983.

———. *A Social History of American Technology*. New York: Oxford Univ. Press, 1997.

Crane, Ellery Bicknell. *Historic Homes and Institutions and Genealogical and Personal Memoirs of Worcester County, Massachusetts*. New York: Lewis Historical, 1907.

Cronon, William. "The Trouble with Wilderness; or, Getting Back to the Wrong Nature." In *Uncommon Ground: Toward Reinventing Nature*, ed. William Cronon, 69–90. New York: W. W. Norton, 1995.

Cross, Gary, and Rick Szostak. *Technology and American Society: A History*. Upper Saddle River, NJ: Pearson, Prentice Hall, 2005.

Cudmore, Dana. *The Remarkable Howe Caverns Story*. Woodstock, NY: Overlook Press, 1990.

Cummings, J. N. "From Appomattox to Versailles." *Vent* 3, no. 11 (Nov. 1919): 2–3.

———. "From Appomattox to Versailles." *Vent* 4, no. 1 (Jan. 1920): 10.

Davis, William C. *The Union that Shaped the Confederacy: Robert Toombs and Alexander H. Stephens*. Lawrence: Univ. Press of Kansas, 2001.

DeForest, Herbert Packard, and Edward Craig Bates. *The History of Westborough, Massachusetts*. Westborough, MA: Town of Westborough, 1891.

Dietz, Ulysses G. *Victorian Lighting: The Dietz Catalogue of 1860*. Watkins Glen, NY: American Life Foundation, 1860.

Doerrfeld, Dean A. *The Canning Industry in Delaware, 1860–1940: A Historic Context*. Newark: Center for Historic Architecture and Engineering, Univ. of Delaware, 1994.

Doherty, Henry Latham. *Principles and Ideas for Doherty Men; Papers, Addresses and Letters*. New York: n.p., 1923.

Dover Stamping Company. *Tinware, Tin Toys, Tinned Iron Wares, Tinners' Material, Enameled Stove Hollow Ware, Tinners' Tools and Machines*. Boston: Dover Stamping. Rpt. Princeton, NJ: Pyne Press, 1971.

Dunn County Historical Society. "The Knapp, Stout & Co., Company." http://www.dunnhistory.org/history/exkscc.html/.

Edwards, Richard. *New York's Great Industries*. 1884. Rpt. New York: Arno Press, 1973.

Ellis, Franklin. *History of Columbia County, New York*. Philadelphia: Everts & Ensign, 1878.

Embury, Aymar, II. *One Hundred County Houses*. New York: Century, 1909.

Endangered Durham. "Blackwell's Durham Tobacco/American Tobacco Co." http://endangereddurham.blogspot.com/2008/03/blackwells-durham-tobacco-american.html/.

Everett, Franklin. *Memorials of the Grand River Valley*. Chicago: Chicago Legal News, 1878.

Ewing, Wallace K., and David H. Seibold. *Images of America: The Grand Haven Area 1860–1960*. Chicago: Arcadia, 2002.

Ferguson, Olin Jerome. *Electric Lighting*. New York: McGraw-Hill, 1920.

Field, Jacqueline, Marjorie Senechal, and Madelyn Shaw. *American Silk, 1830–1930: Entrepreneurs and Artifacts*. Lubbock: Texas Tech Univ. Press, 2007.

Fitch, James Marston. *American Building, 2: The Environmental Forces that Shape It*. New York: Schocken Books, 1975.

Garvey, Dick. "Big Name in Business Born in Local Rooming House." *Springfield Union*, Nov. 3, 1994, 19.

Gerhard, William Paul. *The American Practice of Gas Piping and Gas Lighting in Buildings*. New York: McGraw, 1908.

———. *Gas-lighting and Gas-fitting: A pocket book for gas companies, gas engineers and gas fitters, for manufacturers of gas fixtures and dealers in gas appliances, for gas consumers, architects and builders, health officers and sanitary inspectors*. 4th ed. New York: Van Nostrand, 1913.

Gibb, George Sweet, and Evelyn H. Knowlton. *History of Standard Oil Company (New Jersey): The Resurgent Years, 1911–1927*. New York: Harper & Brothers, 1956.

Gilbarco. "Old Indestructible." *Gilbarco News* 12, no. 2 (1953): 12–13.

———. "These Are the Men Who Started It All." *Gilbarco News* 14, no. 1 (1955): 8.

Gilbert and Barker Manufacturing Company. *The Gilbarco Story: The Story of a Name Famous Throughout the World*. West Springfield, MA: Gilbert and Barker Manufacturing, 1947.

Gilman, Arthur, ed. *The Cambridge of Eighteen Hundred and Ninety-Six: A Picture of the City and Its Industries Fifty Years After Its Incorporation*. Cambridge: Riverside Press, 1896.

Gordon, Robert B., and Patrick M. Malone. *The Texture of Industry: An Archaeological View of the Industrialization of North America*. New York: Oxford Univ. Press, 1994.

Gould, George Glen. *Period Lighting Fixtures*. New York: Dodd, Mead, 1928.

Gowans, Alan. *The Comfortable House: North American Suburban Architecture, 1890–1930*. Cambridge, MA: MIT Press, 1986.

Greenwood, Lois Stevenson. *Winchendon Years, 1764–1964*. Winchendon, MA: Town of Winchendon, 1970.

Groff, John M. "Green Country Towns: The Development of Philadelphia's Main Line, 1870–1915." MA thesis, Winterthur Program in Early American Culture, University of Delaware, Newark.

Grisi, Mark P. "The Automatic Gas Machine Era, 1853–1923: Analysis of Improvements in Patented Illuminating Devices and Processes." Master's thesis, Goucher College, Baltimore, 2003.

Hale, D. "Diary of an Industry." *American Gas Journal* 186, no. 11 (1959): 21–52.

Hall, Helen. *Hill-Stead Museum*. Farmington, CT: Hill-Stead Museum, 1988.

Hayward, Arthur H. *Colonial Lighting*. New York: Dover, 1962.

Hazen, Margaret Hindle, and Robert M. Hazen. *Keepers of the Flame: The Role of Fire in American Culture, 1775–1925*. Princeton: Princeton Univ. Press, 1992.

Hebard, Helen Brigham. *Early Lighting in New England*. Rutland, VT: Charles E. Tuttle, 1964.

Heite, Edward F. *Archaeological Data Recovery on the Collins, Geddes Cannery Site, Road 356A, Lebanon, North Murderkill Hundred, Kent County, Delaware*. Delaware Dept. of Transportation Archaeology Series No. 83. Delaware Dept. of Transportation, 1990.

Hennessey, R. A. S. *The Electric Revolution*. Newcastle-upon-Tyne, England: Oreil Press, 1972.

"Henry Morrison Flagler." Flagler Museum, Palm Beach, FL. http://www.flagler.org/biography.html/.

Hewitt, Mark Alan. *The Architect and the American Country House, 1890–1940*. New Haven: Yale Univ. Press, 1990.

Hidy, Ralph W., and Muriel E. Hidy. *History of Standard Oil Company (New Jersey): Pioneering in Big Business, 1882–1911*. New York: Harper & Brothers, 1955.

Hinckley, C. T. "Everyday Actualities. No. VIII, The Manufacture of Gas and Gas-Fixtures." *Godey's Lady's Book* 47 (1853): 197–203.

History of Muskegon County, Michigan. Chicago: H. R. Page, 1882.

Howard, George W. *The Monumental City: Its Past History and Present Resources*. Baltimore: J. D. Ehlers, 1873.

Hughes, Samuel, and Henry O'Connor. *Gas Works: Their Construction and Arrangement and the Manufacture and Distribution of Coal Gas*. London: Crosby Lockwood and Son, 1904.

Humphreys, C. J. Russell. *Gas as a Source of Light, Heat, and Power*. New York: A. M. Callender, 1886.

Hunter, Dard. *Papermaking: The History and Technique of an Ancient Craft*. New York: Dover, 1978.

Ierley, Merritt. *The Comforts of Home: The American Home and the Evolution of Modern Convenience*. New York: Three Rivers Press, 1999.

International Publishing. *Leading Manufacturers and Merchants of Central and Western Massachusetts: Historical and Descriptive Review of the Industrial Enterprises of Worcester, Hampden, Hampshire, Berkshire, and Franklin Counties*. New York: International, 1887.

Israel, Paul. *Edison: A Life of Invention*. New York: John Wiley & Sons, 1998.

Jakle, John A., and Keith A. Sculle. *The Gas Station in America*. Baltimore: Johns Hopkins Univ. Press, 1994.

Jarrell, Wallace E. *The Randolph Hornets in the Civil War*. Jefferson, NC: McFarland, 2004.

"June 18, 1983: American Optical Celebrates 150th Anniversary." Mass Moments. http://massmoments.org/.

Kasson, John F. *Civilizing the Machine: Technology and Republican Values, 1776–1900*. New York: Penguin Books, 1976.

King, Moses. *King's Handbook of Boston*. Cambridge, MA: Moses King, 1881.

Larson, Erik. *The Devil in the White City: Murder, Magic, and Madness at the Fair that Changed America*. New York: Crown, 2003.

Larson, Henrietta M., Evelyn H. Knowlton, and Charles S. Popple. *History of Standard Oil Company (New Jersey): New Horizons, 1927–1950*. New York: Harper & Brothers, 1971.

Lee, Bob. *Gilbert & Barker Gilbarco, Inc.: A Pictorial History of the Growth of a Great Company*. Detroit: Harlo Press, 1989.

Lent, Frank T. *Sound Sense in Suburban Architecture*. New York, W. T. Comstock, 1895.

Lewis, Arnold. *American Country Houses of the Gilded Age*. New York: Dover, 1982.

Lewis, Vivian B. "The Use of Gas From a Hygienic Standpoint—Theory and Practice in Favor of Gas." *Illuminating Engineer* 2 (1907): 560.

Licht, Walter. *Industrializing America: The Nineteenth Century*. Baltimore: Johns Hopkins Univ. Press, 1995.

Lillie, Leo C. *Historic Grand Haven and Ottawa County*. 1931. Rpt. Grand Haven, MI: Tri-Cities Historical Society, 1980.

Linebaugh, Donald W. "Gas Lighting at the President Lincoln and Soldiers' Home National Monument." University of Maryland, School of Architecture, Planning and Preservation, College Park. Technical report submitted to the National Trust for Historic Preservation, Washington, DC, 2006.

Linebaugh, Donald W., Nancy O'Malley, and Jeanie Duwan. "'A Brilliant and Pleasant Light': Nineteenth-Century Gas Lighting at Ashland, Lexington, Fayette County, KY." Technical Report No. 413. University of Kentucky Program for Archaeological Research, Lexington, 1999.

———. "'A Brilliant and Pleasant Light': Investigating the Springfield Gas Machine System at Ashland, The Henry Clay Estate, Lexington, KY." *Historical Archaeology* 34, no. 4 (2000): 82–100.

Littlefield, Douglas, and Tanis C. Thorne. *The Spirit of Enterprise: The History of Pacific Enterprises from 1886 to 1989*. Los Angeles: Pacific Enterprises, 1990.

"Mace's Benzole Gas Apparatus." *Scientific American* 10, no. 1 (Jan. 27, 1855).

MacKay, Robert B, Anthony K. Baker, and Carol A. Traynor, eds. *Long Island Country Houses and Their Architects, 1860–1940*. New York: W. W. Norton, 1997.

Maher, Kathleen. Personal communication, July 21, 2008.

———. *P. T. Barnum (1810–1891)—The Man, the Myth, the Legend*. N.d. Barnum Museum, Bridgeport, CT. http://www.barnum-museum.org/manmythlegend.htm/.

Maril, Nadja. *American Lighting, 1840–1940*. Atgien, PA: Schiffer, 1995.

Marvin, Abijah Perkins. *History of Worcester County, Massachusetts, Embracing a Comprehensive History of the County from Its First Settlement to the Present Time, with a History and Description of Its Cities and Towns.* Boston: C. F. Jewett, 1879.

Marx, Leo. *The Machine in the Garden: Technology and the Pastoral Ideal in America.* New York: Oxford Univ. Press, 1964.

Maynard, W. Barksdale. *Architecture in the United States, 1800–1850.* New Haven: Yale Univ. Press, 2002.

McBride, Kim. Personal communication, 1996, 1997.

McCabe, James D. *The Illustrated History of the Centennial Exhibition.* 1876. Rpt. Philadelphia: National Publishing Co., 1975.

McCash, William Barton, and June Hall McCash. *Jekyll Island Club Historic District.* Jekyll Island, GA: Jekyll Island Authority, 1986.

———. *The Jekyll Island Club: Southern Haven for American's Millionaires.* Athens: Univ. of Georgia Press, 1989.

McGuire, Phillip. "Technology and Commerce: The Gas Light Industry's Response to Edison's Electric Bulb." *Potomac Review* 6, no. 2 (1973): 70–75.

Moss, Roger W. *Lighting for Historic Buildings.* New York: Preservation Press, John Wiley & Sons, 1988.

Motz, Marilyn Ferris, and Pat Browne, eds. *Making the American Home: Middle-Class Women and Domestic Material Culture, 1840–1940.* Madison: Univ. of Wisconsin Press, 2002.

Mouer, Daniel. "RE: Residential Gas Lighting Systems." Sept. 1, 1998. Historical archaeology LISTSERV mailing list, HISTARCH@asuvm.inre.asu.edu.

Myers, Denys Peter. *Gaslighting in America: A Guide for Historic Preservation.* Washington, DC: Denys Peter Myers, 1978.

National Board of Fire Underwriters. *Rules and Regulations of the National Board of Fire Underwriters for the Construction, Installation and Use of Gasolene Vapor Gas Lighting Machines, Lamps and Systems.* New York: National Board of Fire Underwriters, 1910.

———. *Rules and Regulations of the National Board of Fire Underwriters for the Construction, Installation and Use of Gasolene Vapor Gas Lighting Machines, Lamps and Systems.* New York: National Board of Fire Underwriters, 1926.

National Historic Landmark Program. "Blackwell, W. T., and Company Tobacco Factory." http://tps.cr.nps.gov/nhl/.

National Publishing Company. *Commerce, Manufactures and Resources of Springfield, Mass.: A Historical, Statistical & Descriptive Review.* N.p.: National Publishing, 1883.

Neal, Martha. Personal communication, 2002.

New York Times. "Artificial Light: Petroleum or Air-Gas Machines." *New York Times*, Sept. 21, 1867, 2.

Norman, Oscar E. *The Romance of the Gas Industry.* Chicago: A. C. McClurg, 1922.

Norwood, Martha F. *Liberty Hall: Taliaferro County, Georgia, A History of the Structures Known as Liberty Hall and Their Owners from 1827 to the Present.* Atlanta: Georgia Dept. of Natural Resources, Historic Preservation Section, 1977.

Nye, David E. *Electrifying America: Social Meanings of a New Technology.* Cambridge, MA: MIT Press, 1990.

Oakley, Henry A., and D. A. Heald, Committee. *Report to the Executive Committee of the National Board of Fire Underwriters on Gas Machines.* New York: National Board of Fire Underwriters, 1876.

O'Dea, William T. *A Short History of Lighting.* London: Her Majesty's Stationary Office, 1958.

———. *The Social History of Lighting.* London: Wyman and Sons, 1958.

Plimpton, Pauline Ames, ed. *Oakes Ames: Jottings of a Harvard Botanist, 1874–1950.* Cambridge: Botanical Museum of Harvard Univ., 1979.

Redhead, P. A. "History of Vacuum Devices." Pp. 284–85. Ottawa, Canada, National Research Council. http://www.chem.elte.hu/departments/altkem/vakuumtechnika/CERN23.pdf/.

Richards, William. *The Gas-Consumer's Guide: A Hand-book of Instruction.* Boston: Alexander Moore, 1871.

Romaine, Lawrence B. *A Guide to American Trade Catalogs (1744–1900).* New York: R. R. Bowker, 1960.

Russell, Loris S. *A Heritage of Light: Lamps and Lighting in the Early Canadian Home.* Toronto: Univ. of Toronto Press, 1968.

Sargent, Porter. *A Handbook of New England.* Boston: Porter E. Sargent, 1916.

Schivelbusch, Wolfgang. *Disenchanted Night: The Industrialization of Light in the Nineteenth Century.* Berkeley and Los Angeles: Univ. of California Press, 1988.

Schott, Thomas E. *Alexander H. Stephens of Georgia.* Baton Rouge: Louisiana State Univ. Press, 1988.

Schroeder, Joseph J., Jr., ed. *Sears, Roebuck & Co. 1908 Catalogue No. 117.* Northfield, IL: Digest Books, 1971.

———. *1923 Sears, Roebuck Catalogue.* Northfield, IL: Digest Books, 1973.

Sharpe, Elizabeth M. *In the Shadow of the Dam: The Aftermath of the Mill River Flood of 1874.* New York: Simon and Schuster, 2004.

Sheldon, George W. *Artistic Country-Seats.* New York: D. Appleton, 1886. Rpt. New York: Da Capo Press, 1979.

Shepley, Rutan and Coolidge, 1886–1915. http://www.shepleybulfinch.com/history/.

Sherman, Joe. *The House at Shelburne Farms: The Story of One of America's Great Country Estates.* Forest Dale, VT: Paul S. Eriksson, 1986.

Sim, Mary B. *Commercial Canning in New Jersey: History and Early Development.* Trenton: New Jersey Agricultural Society, 1951.

Smith, Judd. Personal communication, 2002.

"Springfield Gas Machine." *American Gas-Light Journal and Chemical Repertory* 16, no. 10 (May 16, 1872).

Stevenson, Robert Louis. *A Child's Garden of Verses and Underwoods.* 1906. Rpt. Whitefish, MT: Kessinger, 2009.

Stilgoe, John R. *Borderland: Origins of the American Suburb, 1820–1939.* New Haven: Yale Univ. Press, 1988.

Stovall, Pleasant A. *Robert Toombs: Statesman, Speaker, Soldier, Sage.* New York: Cassell, 1892.

Strasser, Susan. *Never Done: A History of American Housework.* New York: Henry Holt, 1982.

Sturgis, Russell, John W. Root, Bruce Price, Donald G. Mitchell, Samuel Parsons Jr., and W. A. Linn. *Homes in City and Country.* New York: Charles Scribner's Sons, 1893.

Taylor, Alice. *Quench the Lamp.* Dingle, Ireland: Brandon, 1990.

Temin, Peter, ed. *Engines of Enterprise: An Economic History of New England.* Cambridge: Harvard Univ. Press, 2000.

Thompson, William Y. *Robert Toombs of Georgia.* Baton Rouge: Louisiana State Univ. Press, 1966.

Thwing, Leroy L. *Flickering Flames: A History of Domestic Lighting through the Ages.* Rutland, VT: Charles E. Tuttle, 1958.

Tolles, Bryant F., Jr. *The Grand Resort Hotels of the White Mountains: A Vanishing Architectural Legacy.* Boston: David R. Godine, 1998.

"A Tour of Unity Church." Unity Church of North Easton. http://www.unity-church.com/building/tour.htm/.

Trachtenberg, Alan. *The Incorporation of America: Culture and Society in the Gilded Age.* New York: Hill and Wang, 1982.

U.S. Bureau of the Census. *The Statistical History of the United States, from Colonial Times to the Present.* New York: Basic Books, 1976.

Varella, Hazel L. *Growing Up at "Sheep Pasture."* Easton, MA: Easton Historical Society, 1976.

Veit, Richard. *Historical Documentation of the Loch Aerie Gasworks, Home Depot—East Whiteland, East Whiteland Township, Chester County, Pennsylvania.* Highland Park, NJ: Cultural Resource Consulting Group, 1997.

Von Abele, Rudolph. *Alexander H. Stephens: A Biography.* New York: Alfred A. Knopf, 1946.

Wahl, Dr. William H. "Gas Machines." *American Gas Light Journal and Chemical Repertory* 17 (1872): 130.

Warner, Charles F., ed. *Picturesque Hampden.* Northampton, MA: Picturesque, 1892.

Watkins, C. Malcolm. "Artificial Lighting in America: 1830–1860." In *Annual Report of the Board of Regents of the Smithsonian Institution, 1951,* by Smithsonian Regents, 385–407. Washington, DC: Government Printing Office, 1952.

Webb, William Seward. *Shelburne Farms Stud, Shelburne, Chittenden County, Vermont.* New York: G. P. Putnam's Sons, 1893.

Webster, Thomas, W. R. Parkes, and D. Meredith. *An Encyclopaedia of Domestic Economy: Comprising Such Subjects as Are Most Immediately Associated with Housekeeping: As, the Construction of Domestic Edifices, with the Modes of Warming, Ventilating, and Lighting Them: A Description of the Various Articles of Furniture: A General Account of the Animal and Vegetable Substances Used as Food.* New York: Harper, 1845.

Weeks, Lyman Horace. *A History of Paper-Manufacturing in the U.S., 1690–1916.* New York: Lockwood Trade Journal, 1916.

Wheeler, Gervase. *Homes for the People in Suburb and Country: The Villa, the Mansion, and the Cottage.* 1855. Rpt. New York: Arno Press, 1972.

Wheeler, Glen C. *Images of America: Winchendon.* Dover, NH: Arcadia, 1997.

Williamson, Harold F., and Arnold R. Daum. *The American Petroleum Industry: The Age of Illumination, 1859–1899.* Evanston, IL: Northwestern Univ. Press, 1959.

Winter, Susan E. "Mill Settlement Patterns Along the Antietam Creek Drainage, Washington County, Maryland." In *Spatial Patterning in Historical Archaeology: Selected Studies of Settlement,* ed. Donald W. Linebaugh and Gary G. Robinson, 65–90. William and Mary Center for Archaeological Research, Occasional Papers in Archaeology, No. 2. Williamsburg, VA: King and Queen Press, 1994.

Woodhead, E., I. Sullivan, and G. Gusset. *Lighting Devices in the National Reference Collection, Parks Canada.* Ottawa: Parks Canada, 1984.

Woodward, George E., and Edward G. Thompson. *Woodward's National Architect.* New York: G. E. Woodward, 1869. Rpt. New York: Da Capo Press, 1975.

Worthington, William E., Jr. *Beyond the City Lights: American Domestic Gas Lighting Systems.* Exhibit at the National Museum of American History, Oct. 17, 1985–Apr. 20, 1986. Washington, DC: Smithsonian Institution Press, 1985.

Wright, John D. *Lexington: Heart of the Bluegrass.* Lexington, KY: Lexington–Fayette County Historic Commission, 1982.

Yergin, Daniel. *The Prize: The Epic Quest for Oil, Money, and Power.* New York: Simon & Schuster, 1990.

INDEX

SGM is an abbreviation for the Springfield Gas Machine, and page numbers in **boldface** refer to illustrations.

A

C

D

E

I

J

K

N

O

T

Y